SAFETY SYSTEMS RELIABILITY

Safety Systems Reliability

A. E. GREEN

National Centre of Systems Reliability,
United Kingdom Atomic Energy Authority,
Culcheth, Warrington, UK

A Wiley–Interscience Publication

JOHN WILEY & SONS

Chichester · New York · Brisbane · Toronto · Singapore

Library of Congress Cataloging in Publication Data:

Green, A. E. (Arthur Eric)
Safety systems reliability.

'A Wiley–Interscience publication.'
Bibliography: p.
Includes index.
1. Atomic power-plants—Reliability. 2. Atomic power-plants—Safety measures. 3. System safety.
I. Title.
TK1078.G74 1983 621.48′35 82–24863
ISBN 0 471 90144 X

British Library Cataloguing in Publication Data:

Green, A. E.
Safety systems reliability.
1. Atomic power-plants—Great Britain—Safety measures
I. Title
363.1′79 TK1362.G7

ISBN 0 471 90144 X

Typeset by Preface Ltd, Salisbury, Wilts
Printed by The Pitman Press, Bath

Contents

	Page
Preface	ix
CHAPTER 1 GENERAL INTRODUCTION TO SAFETY AND RELIABILITY	1
1.1 Introduction	1
1.2 Reliability Policy	2
1.3 Factors Affecting Reliability and Risk	12
1.4 Generalized Protective System	19
1.5 Main Objectives of the Study	21
CHAPTER 2 THE INITIAL APPROACH TO PERFORMANCE ASSESSMENT	24
2.1 Design and Assessment	24
2.2 Theoretical Prediction of System Performance	39
2.3 Performance Characteristics	40
2.4 Interface Problems	44
2.5 Information and Communication	46
CHAPTER 3 VARIABILITY IN PERFORMANCE AND THE ASSESSMENT OF RELIABILITY	47
3.1 Variability of Performance	47
3.2 Performance Spectrum Method	48
3.3 Variability Region	50
3.4 Reliability Model	52
3.5 Hazard Probability	56
3.6 Application of Reliability to Example Protective System	56
3.7 The Human Factor	60
3.8 Operator Response	61
3.9 Human Operator Reliability	69
3.10 Human Factors in Maintenance and Testing	71
3.11 Implementation of Design and its Assessment	79
3.12 The Safety Assessment Report	80

3.13 The Need for a Posterior 83
3.14 Implementing the Reliability Policy 85

CHAPTER 4 RELIABILITY ESTIMATING, MONITORING AND THE INVESTIGATION OF OPERATING PLANT SYSTEMS 93
4.1 Monitoring and Estimating Reliability Characteristics . . 93
4.2 Subjective Data 94
4.3 Reliability Data Banks 101
4.4 The Correlation between Predicted and Practical Results . . 107
4.5 Limitations of Conventional Estimating Techniques . . . 113
4.6 The Bayesian Approach 114
4.7 Analysis of Operating Plant Systems 120
4.8 Gas Cooled-Type Reactor 121
4.9 The Automatic Protective System for the Winfrith Steam Generating Heavy-Water Reactor (SGHWR) 153
4.10 SGHWR Operational Experience over 5 years and some Aspects of Unavailability 167

CHAPTER 5 A TOTAL APPROACH TO RELIABILITY GROWTH AND CONTROL 174
5.1 Reliability Growth and Control 174
5.2 Reliability Growth 179
5.3 Early-Life Plant Defects and Breakdowns 183
5.4 The Overall Reliability Programme 184
5.5 Common-Mode Failure 187
5.6 Some Aspects of an Integrated Structure 191
5.7 Reliability Control System Analogy of Integrated Structures . 193

CHAPTER 6 CONCLUSIONS AND RECOMMENDATIONS FOR FURTHER STUDY 201
6.1 Conclusions 201
6.2 Recommendations for Practitioners 205
6.3 Recommendations for Academic and Similar Institutions . . 206

Appendix 1 Extract from 'Vapour Clouds' 207

Appendix 2 Event and Fault Trees 214

Appendix 3 Failure Modes and Effects Analysis 217
A.3.1 Differential Pressure Transmitter Example . 217
A.3.2 Gamma Radiation Monitor Example . . 218
A.3.3 Four-Character Coding System 221
A.3.4 Extract from a Standard and Code for Information Required on the Reactor Protection System

on the Licensing Procedures in the Federal Republic of Germany 222

Appendix 4 Comparative Reliability Case Study for Manual versus Automatic Control 225

A.4.1 Introduction 225
A.4.2 The Protective System 225
A.4.3 Assessment Process 226
A.4.4 Assessment Results 226

Appendix 5 Mean Failure Rates for Components and Instruments . 229

Appendix 6 Derivation of Bayes Theorem 230

Appendix 7 Proof of Theorem on the Asymptotic Behaviour of Gamma Percentiles 232

Appendix 8 Failure Probability Evaluation of an Automatic Protective System on a PWR by Mme A. Carnino . . . 234

A.8.1 Introduction 234
A.8.2 System Description 234
A.8.3 System Actuation 235
A.8.4 Study of the Event Sequence 235
A.8.5 Reliability Assessment of the System . . 239
A.8.6 Analysis of Control-Rod Drive Mechanisms . 247
A.8.7 Test of the System and Human Factors . . 248
A.8.8 Conclusion 250

Appendix 9 HTGR Core Auxiliary Cooling System (CACS)—Reliability Prediction by V. Joksimovich . . . 252

A.9.1 Introduction 252
A.9.2 Brief Description of CACS 252
A.9.3 Start-up Reliability Prediction 253
A.9.4 Operation until Main Loops are Restored—Reliability Prediction 259
A.9.5 Comparison of Reliability Data with Some Other Sources 268
A.9.6 Acknowledgement 269

References for Main Text 270
References for Appendices 280

INDEX 282

Preface

One of the products of modern technology is a tendency to produce plant in various industries which is of greater capacity, complexity and capital cost. The physical operating conditions have tended to become more stringent and the consequences of unreliable behaviour more severe. Hence, the desire for reliable systems becomes more acute. In reviewing reliability factors relating to the safety of plant it may be deduced that the breaking of the 'domino effect' of a potentially dangerous sequence of events has given rise to the introduction of a high-reliability automatic protective system. In this book the reliability assessment of such a protective system has been investigated from the conceptual design phase to ultimate operation of the system on a plant. In particular, the general typè of automatic protective system installed on nuclear reactors has been studied. The strategies and techniques discussed have a much wider application and to give an integrated approach some of the historical reliability factors relating to safety have been considered.

The importance of the independent safety and reliability assessment process for such a type of protective system is paramount and particular attention has been given to the process of evaluation in this assessment as opposed to deriving the design solution for the system. Basically, this involves validating the adequacy of the reliability of the protective system, and that it should have the capability of performing as intended on demand should an abnormal condition arise on the plant.

An important factor is that over the past two decades there has been a trend to extend from deterministic to probabilistic reliability assessment. The initial approach of a deterministic assessment may assume no variability or failure, but it is necessary to take into account the real conditions of variability and failure using probabilistic techniques to evaluate reliability. From the 1950s onwards the nuclear industry has grown up in an atmosphere where reliability has been of salient importance from a safety point of view. Equally, in the introduction of high-integrity protective systems for reducing the risk to an acceptable level for potentially hazardous chemical processes, the same type of emphasis has been applied. In this book, examples have been given showing the application of the techniques and their limitations in order to derive a prior assessment at the conceptual design phase.

Interface problems involving human factors require to be taken into account in the identification of various assumptions in order to obtain a boundary or envelope approach to the reliability characteristics of system performance. The prior assessment of the achieved performance of the protective system then requires to be compared with a reliability criterion to demonstrate adequacy, which is illustrated in the text. A limiting factor in achieving high reliability leads to the investigation of a common-mode type of failure and the human influence factors which arise, for example in testing and maintenance, can be very important.

The limitations of the assessment techniques and their development are considered as the assessor is 'information limited' due to a lack of data at the complete-system level, hence the need for methods of synthesis for reliability at a level in the system where data can be found available. The correlation between reliability prediction for characteristics such as the failure rate of equipment has been compared with actual field operating data and the accuracy commented upon. This involves data-bank techniques and in the text reference has been made to some aspects of 'soft' or subjective data and an experimental approach with the results has been discussed.

Reliability is essentially a growth process and the complete cycle involving the initial concept of the design through the operation of the plant to its ultimate shutting down and decommissioning, may well be 20, 30 or more years. Hence, from an engineering point of view, the reliability evaluation of plant automatic protective systems and equipment which have been operating for periods up to and over 20 years, have been considered.

For effective management and control of the reliability and protective systems, it is necessary to cater for this growth process. Therefore, a total and overall life cycle approach has been developed and considered in the text for the control of reliability giving an integrated structure through all phases of the life cycle. This is an essential element in forming the basis of a management and control data base for decision-making by the up-dating of the prior estimates to give posteriors at each phase of the life cycle on the information available.

The general conclusions are intended to indicate some areas for further research and development arising in the approach described in the book with applications which will be found useful to practitioners and academic and similar institutions. Clearly, the design engineer, the user and the safety assessor will have specific interests in their own applications of protective systems where risks are involved, and the general principles illustrated in the text should be beneficial in giving generic ideas in such applications. Similarly, students engaged in various studies of system or equipment engineering will need to have an understanding of how to substantiate the reliability of protective systems and plant where safety is paramount. These concepts are further illustrated in the appendices contributed by Mme A. Carnino and Dr V. Joksimovich for which the author is much indebted.

Appreciation is expressed by the author to the United Kingdom Atomic Energy Authority, especially Mr. G. H. Kinchin, the Director of the Safety and Reliability Directorate, for the support and facilities given to enable this work to be undertaken. In addition, the author would like to thank his colleagues, particularly those in the National Centre of Systems Reliability, for the useful discussions and the information given. Acknowledgement is made of the kind co-operation given by Messrs A. J. Bourne, B. K. Daniels, J. C. Moore, I. A. Watson and others, which is much appreciated. The author would like to express his thanks for the advice given by Dr. A. Z. Keller of the University of Bradford. Finally, the author, in undertaking this work, would like to thank his wife Margaret for her understanding and encouragement.

A. E. Green

The National Centre of Systems Reliability,
Safety and Reliability Directorate,
United Kingdom Atomic Energy Authority,
Culcheth

February 1982

CHAPTER 1

General Introduction to Safety and Reliability

1.1 Introduction

In a plant process there is an input of some type which may consist of materials, energy or other items such as information, and within the plant some conversion process takes place. This gives rise to an output of some form such as materials or energy.

A nuclear reactor converts energy from uranium fuel into heat which can then be used for steam raising and the generation of electricity. Similarly a conversion process takes place in a chemical plant which can often be working at high temperatures and pressures. Under abnormal conditions of operation serious financial loss may occur due to plant damage or loss of output which may be accompanied by injury or even death to one or more persons. Technological developments have led to plants becoming more complex, more costly, and often making use of materials that are more reactive than hitherto.

Such developments have been relatively rapid resulting for example in very high mass transfer rates and energy densities also increasing potential energies leading to the creation of 'high-risk' plants for which there exists little or no previous experience. For these situations it is pertinent to consider potential accidents which may be defined as unplanned and uncontrolled events which alter the operation of the plants towards hazardous conditions thus increasing the probability of financial or human loss. This definition, which is based on one by Heinrick[1], involves the concept of a 'near accident'.

Classical methods of investigating accident behaviour which may result in physical damage or injury to the person or persons involved have given rise to industrial safety procedures. These have also been given legal standing in the UK by the Health and Safety at Work Act[2], and in the USA by The Occupational Safety and Health Act of 1970[3]. Not only is safety covered by Statute Law but there has always been, and still is, a 'Duty of Care' under Common Law. Everybody has a 'Duty of Care' to third parties in a variety of circumstances which have been defined by the Common Law and in the past numerous legal actions have been based on this principle.

Where damage to property may occur, then legislation can be introduced and a programme at the national level can evolve — as for example the UK Nuclear Power Programme[4].

Where the safety of people is not involved and the loss is purely financial, accounting methods can then be used to effect a policy of control. In the ultimate, the 'market place' itself may resolve the financial issue. However, safety of plant is often related to questions of reliability which are becoming increasingly important in the market place. It is evident that techniques for 'tailoring' plant proposals and their design must not only meet the financial constraints but they must also satisfy safety requirements in connection with people, the cost of insurance could be a major factor.

The demands for operating at higher temperature and pressure in the commercial field, for instance in the search for rare materials, have not led to the stopping of the development but to moulding it and adapting it to meet the various requirements which may be socially or otherwise imposed.

Risk, as caused by accidents, can be considered as the combination of the probability of an event together with its consequences. In many plants the probability of the accident condition arising will be a combination of some abnormal event occurring and the failure of the particular system designed to prevent such consequences, as stated by Haddon *et al.*[5], and is to be adopted in the present text.

The approach to this general problem will be dictated by many factors, not the least of which is that analysis is 'information limited' and in applying a series of 'what if' questions dealing with low probabilities, there is little if any data but decisions have to be based on prior data. Such prior data may have of necessity been derived from high probability (non-accident) events. This then involves a process of prediction using the information available which may come through some sample testing, leading to some posterior information during the life of the plant, when the confirmed results of the reliability and safety involved will be known.

The acceptance of a design of a plant, to be adequately safe in the face of potential accidents, will be largely determined by the reliability policy being adopted for the plant systems and the means to contain effects of failures of those accidents having the potential to give rise to the most severe consequences.

1.2 Reliability Policy

1.2.1 Historical

Problems of unreliability are not new but date back to antiquity. Over long periods of time designs have been developed by trial and error methods, but nevertheless they may be considered to be very successful; for example the wheel and the axle. However, it will be readily recognized that human beings tend to resent failure in other human beings and this resentment extends to items around them such as equipment or plant on which they depend. Historically such activities have led to the view that whilst it is impossible to prevent every potential accident from materializing, it is essential to limit the frequency of failure and its consequences.

Thus 'reliability policy' may involve many factors including psychological attitudes and reactions. The words used have to be carefully chosen since philosophically it is generally more acceptable to have a reliability policy than unreliability policy as the first could suggest success and the second failure. On the other hand there are indications that the 'spares' market has traditionally been attractive to British industry and this is a further factor for consideration.

A brief survey has been carried out of various industrial activities with a view to obtaining some indication of the reliability policy that emerges, although in some cases the word 'reliability' may not have been used directly it is obvious that once an activity had been made viable then reliability was sought after.

1.2.1.1 Catastrophe Affecting Reliability Philosophy

In the survey undertaken it appears that the thinking at the particular historical period required to be altered or motivated by some form of catastrophe. In the case of loss of human life there appears to be a psychological difference in approach between say ten people being killed individually spread over 1 year as opposed to ten people being killed together at the one time and place. Furthermore the psychological reaction between one, ten or one thousand people being killed simultaneously is highly non-linear. There is a parallel in the case of financial loss where a single large financial loss at a single point in times appears to have a greater motivating influence on the organizations covering such a loss than a series of separate small financial losses.

The author in the study undertaken in this book has been concerned more specifically with the safety aspects which in the ultimate may hazard human life. Hence a brief survey of some such events resulting in catastrophe has been undertaken over the past 100 years or so and some of these are summarized both chronologically and by activity in Fig. 1.1.

These catastrophic events have had the effect of hastening technological development and this is clearly seen in the activity of defence where war has resulted in rapid developments. The First World War gave rise to the development of aircraft which had the first powered flights of only a few seconds' duration by the Wright brothers only about a decade before. Similarly, the Second World War gave rise to the rapid development of radar. In addition such developments gave rise to solving technological problems with a spin-off to other activities. The present subject of reliability owes much of its development to such pioneers as Robert Lüsser who in Germany in the early 1940s in the missile field, found increasing demands arising for reliable electronic equipment and systems.

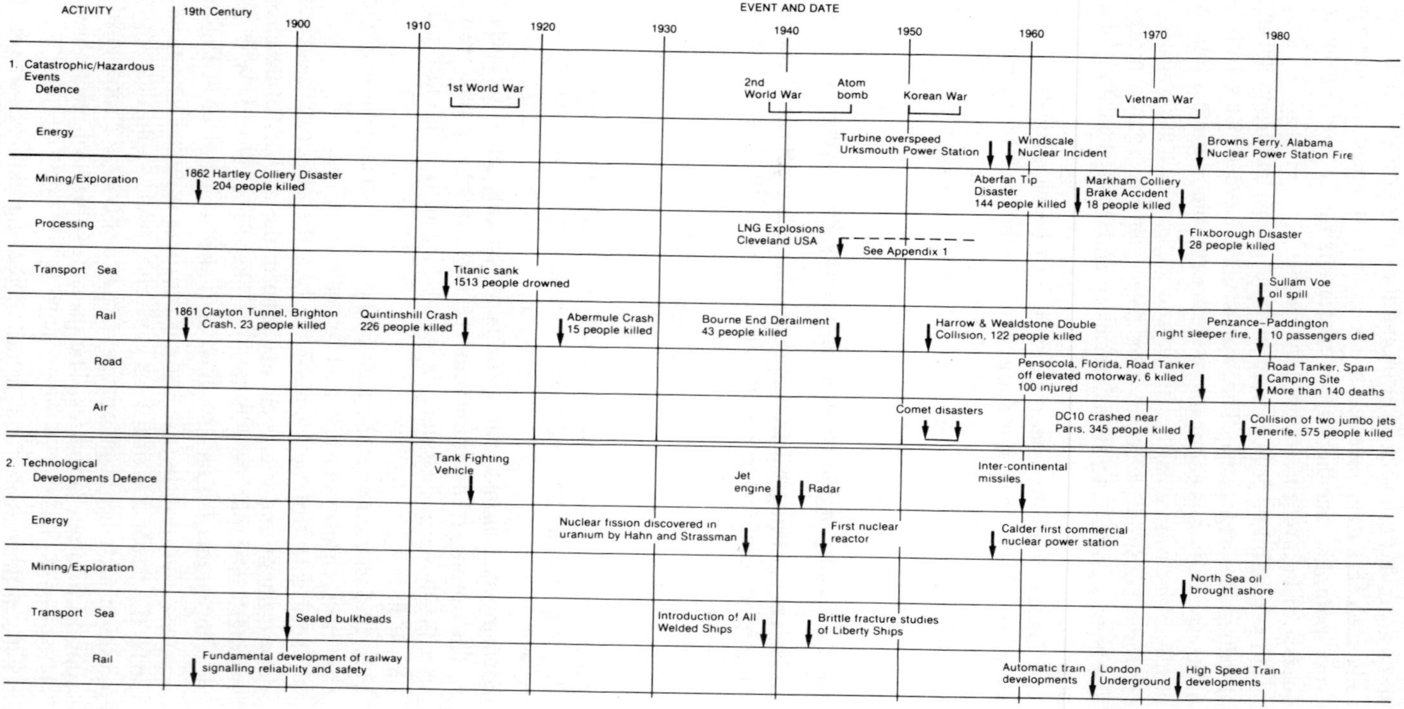

EVENT AND DATE

ACTIVITY	19th Century	1900	1910	1920	1930	1940	1950	1960	1970	1980
1. Catastrophic/Hazardous Events — Defence			1st World War			2nd World War; Atom bomb	Korean War	Vietnam War		
Energy							Turbine overspeed Urksmouth Power Station; Windscale Nuclear Incident		Browns Ferry, Alabama Nuclear Power Station Fire	
Mining/Exploration	1862 Hartley Colliery Disaster 204 people killed							Aberfan Tip Disaster 144 people killed	Markham Colliery Brake Accident 18 people killed	
Processing						LNG Explosions Cleveland USA — See Appendix 1			Flixborough Disaster 28 people killed	
Transport — Sea			Titanic sank 1513 people drowned						Sullam Voe oil spill	
Rail	1861 Clayton Tunnel, Brighton Crash, 23 people killed		Quintinshill Crash 226 people killed	Abermule Crash 15 people killed		Bourne End Derailment 43 people killed	Harrow & Wealdstone Double Collision, 122 people killed		Penzance–Paddington night sleeper fire, 10 passengers died	
Road									Pensocola, Florida, Road Tanker off elevated motorway, 6 killed 100 injured; Road Tanker, Spain Camping Site More than 140 deaths	
Air							Comet disasters		DC10 crashed near Paris, 345 people killed; Collision of two jumbo jets Tenerife, 575 people killed	
2. Technological Developments — Defence			Tank Fighting Vehicle		Jet engine	Radar	Inter-continental missiles			
Energy					Nuclear fission discovered in uranium by Hahn and Strassman	First nuclear reactor	Calder first commercial nuclear power station			
Mining/Exploration									North Sea oil brought ashore	
Transport — Sea		Sealed bulkheads			Introduction of All Welded Ships	Brittle fracture studies of Liberty Ships				
Rail	Fundamental development of railway signalling reliability and safety							Automatic train developments; London Underground	High Speed Train developments	

Category	Events and activities
Road	FIRST MOTOR CAR; Mass production developments of cars; Motorway developments
Air	First powered flights by Wright brothers; First jet-propelled flew in Germany; Autoland; Apollo lunar landing; Use of Tri-jets and Concorde
Measurement and Control	Discovery of electron by J. J. Thompson; Fleming's diode; Lee de Forest's triode; Servo system early example giropilot applied to the automatic steering of ships; Transistor Bell telephone; Automatic control; Developments of integrated circuits; Chips and microelectronics
Computation	1822 1833 Difference Engine by Babbage; First electronic ENIAC (Electronic, Numerical Integrator and Calculator) computer; Minicomputers; Microcomputers
Communications	Braun's CRT with fluorescent screen; Magnetron by A. W. Hull; EMI complete television system; Development of magnetron and radar; Laser proposals by A. L. Schamlow
3. Reliability and Safety Methodology Assessment Techniques	Deterministic methods; Quantitative Analysis; Diakoptic and topology; Fault trees; Farmer Probablistic Criterion; WASH 1400 NUCLEAR; COMMON MODE FAILURE ANALYSIS
Data	Farada; Syrel
Applications	MISSILES; AIRCRAFT; NUCLEAR; CHEMICAL PROCESSING
Organizations	UKAEA, HEALTH and SAFETY BRANCH; SYSTEMS RELIABILITY SERVICE; National Centre of Systems Reliability
Collaborations	OECD TASK FORCE ON RARE EVENTS
Publications and Meetings	IEEE (USA) Reliability; ANNUAL R & M SYMPOSIA (USA); OECD SPECIALIST RELIABILITY MEETINGS; UK NATIONAL RELIABILITY CONFERENCES
4. Certification and Legislation Organization	1860 Lloyds Shipping Register; International Convention for Safety of Life at Sea (SOLAS); International Convention for Safety of Life at Sea; Air Registration Board; Inspectorate of Nuclear Power Ministry of Power; Health and Safety Commission Health and Safety Executive
Act	1872 Coal Mines Regulation Act; Ratified by Department of Board of Trade; Air Navigation Act; Nuclear Installation Licensing and Insurance Act; Occupational Safety and Health Act (USA); Health and Safety at Work Act (UK)

Fig. 1.1 Summary chart of catastrophic events and related activities

In the marine field the development of sealed bulkheads in ship design was a basis for the 46 000-tons White Star liner 'Titanic' to be considered 'unsinkable'. In 1912 it struck an iceberg on its maiden voyage and sank with the loss of 1513 lives. As a result of the disaster the first International Convention for Safety at Sea (SOLAS) was called in London in 1913. This drew up rules that every ship should carry lifeboat space for every person embarked. Also the International Ice Patrol was established to warn shipping of ice in the North Atlantic Ocean shipping lanes. Here is an example of catastrophe interacting with technological developments, safety and protection and legislation.

From Fig. 1.1 the trend of legislation following on catastrophic events may be observed. An important aspect is that for some catastrophic events very few people have lost their lives and yet the current thinking has been to introduce the most stringent of safety legislation. This may be seen in the nuclear power field where the thinking has included the maximum consideration of ecology and genetic effects which may be of significance to future generations unless careful consideration is given today.

The processing industry is an activity which is following on from the lessons and technological developments of the nuclear industry and applying the techniques and developing them for their own specialized problems. This is illustrated in Fig. 1.1 which has resulted in the Canvey Island report[6].

Appendix 1 gives the results of a survey taken from Slater[7] of incidents involving vapour clouds and is included to give just one part of the overall spectrum of events to illustrate the consequences which contribute to changing attitudes and thinking.

The review undertaken shows similar trends in different activities such as bridge and structure failure etc., leading to catastrophe. This leads the author to believe there is a requirement for more research on the general subject of catastrophe. It is of interest to note the Catastrophe Theory formulated in a qualitative way by Thom[8] and the quantitative general branching theory of Thompson and Hunt[9,10]. These theories relate to any system governed by a potential function $V\ (Q_i, \Lambda^j)$ where Q_i represents a set of n state variables or generalized coordinates and Λ^j is a set of K externally controlled parameters. For equilibrium stationary values of V with respect to the Q_i are considered necessary and sufficient whilst for stability minimum values are supposed to be necessary and sufficient. Critical equilibrium states can be identified at which a loss of stability can be expected. It is then possible to project a catastrophe boundary. Changes in Λ^j can be studied searching for the steady evolution of an equilibrium solution which may be initially stable but which can be destroyed by a sudden change from one set of the critical equilibrium states.

In the context of protective systems it is the aim of the system design to operate on demand in the required time and that there is no 'instability hole' during this time of operation where the operating time is increased to infinity in which case there is a 'catastrophe'. Where a system design has been highly optimized then small changes internal and external to it by definition can often lead to this 'optimal' being highly disturbed. This point is commented upon by Chilver[11] with reference to system analysis. However, the optimization of a critical protective system for a high-risk situation can be expected to follow a different course than one for a low-risk situation as is discussed later in this book.

Some of the activities which have been considered historically are discussed in slightly more detail in the following sections.

1.2.2 Coal mining

Some aspects of coal mining have been studied in the UK and it has been noted that for the purpose of hauling coal, one pit shaft was considered to be adequate. In 1862 there took place the Hartley Colliery disaster, as described by McCutcheon[12], when 204 people were killed. There was only a single shaft with men working 600 ft below the ground. Safety depended on this one shaft, 12 ft 6 in in diameter which served as an inlet and outlet, fresh air being sent down one half and foul air being returned up the other. An engine beam fell down the pit and caused a total blockage which sealed the fate of the men below by preventing rescue attempts, and in fact paralysed the pit's system of ventilation, as shown in Fig. 1.2.

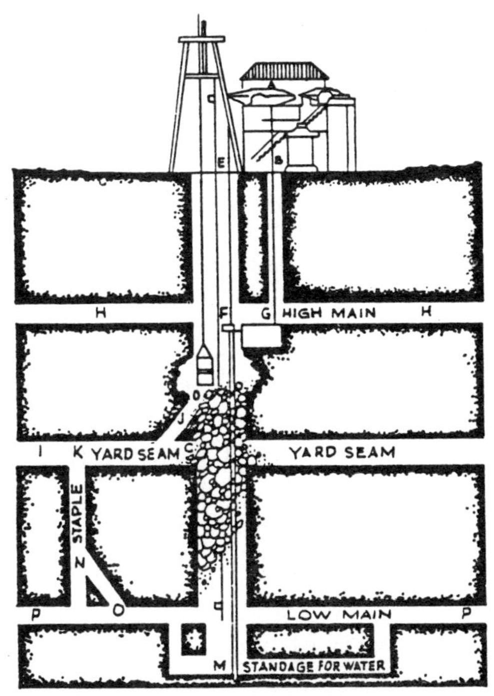

Fig. 1.2 Diagram showing the position of the blockage to the shaft. (J. E. McCutcheon (1963) *The Hartley Colliery Disaster, 1862*)

The foregoing situation caused a catastrophe at Hartley Colliery leading to an immediate review of the policy of sinking shafts, and the passing of an Act of Parliament prohibiting collieries from being worked unless provided with at least two means of exit separated by not less than 10 ft of natural strata. Although the word 'reliability' was not being used, this may be interpreted in the modern reliability terms, that redundancy was then found to be lacking, and experience showed a need for a redundancy of at least two.

It may be noted that apart from the Hartley Colliery disaster, there were other accidents due to two shafts being insufficiently separated, so that under explosion conditions, the force of the blast breached both shafts at a weak point. Once again in modern reliability terminology this common cause may be considered as leading to a 'common-mode failure'. The implications of redundancy and diversity against common-mode failures are stated in more detail later in this book.

More recently in the coal-mining industry, a keen interest has been maintained in developing safety measures in order to protect human life. At the same time it must be borne in mind that because the coal must be taken out of the pit then pit availability must be carefully considered. This is a typical dilemma which will be found to occur in the various industries studied, where there emerges a clear problem of ensuring production while at the same time maintaining safety. An example of some of these factors can be found at Markham Colliery in the UK where in August 1973 a fatal accident occurred in which eighteen men were killed and eleven seriously injured as a result of a cage crashing to the bottom of the pit shaft. This accident was the outcome of a combination of events in so far as there was a failure of a major component in the mechanical braking for the cage winder and, in addition, due to human operation, the electrical braking was lost when the emergency stop was pressed. This combination of events resulting in the accident is considered by Clanzy and Luxmore[13].

As a result of investigations into this accident, various recommendations arose in connection with winding systems such as that operation should not rely on 'single line' components, i.e. there should be redundancy. It was concluded that design analyses should be made of all winding engine-brake components essential for safety. This development is one which is noted to be similar to that in other industries where there have been needs for techniques to be applied in order to justify the adequacy of the design and is a point which will be further studied later in this book.

1.2.3 Power generation

From the earliest days of the conventional power stations where the aim was to generate electricity, e.g. by Edison, power plants have increased in their complexity, output and the economic incentives which can vary according to changes in cost of fuel, equipment etc. However, problems have arisen which are not only related to the availability requirement of the power station, but also the need to ensure the safety of plant and site personnel and possibly the general public. In the case of nuclear power stations the design and choice of a particular type of reactor may be very much dominated by safety considerations.

In the last decade or so the quantification of reliability enabled it to be more usefully employed as a parameter in system planning, design and operation.

In the brief survey undertaken it appears that virtually all power-generating organization collect reliability data in some form or other, particularly for their generating units. Various references to this may be found in the literature, specifically in the USA. The Edison Electric Institute has provided an annual summary of generating units and major equipment availability data. In the UK there have been various nationwide programmes for example in the late 1940s there was a nationwide programme on 'Boiler Availability' of power stations with periodic publication of the information derived.

It is observed that the word 'reliability' has many different meanings, although it is generally used in the context to convey information about the adequacy of a proposed system to perform its intended function. Availability appears to have been the prime concern with reference to various aspects of generating unit adequacy. Broadly this can be defined as the percentage of time a unit is available for service, whether operated or not, and generally it is required to take account of both forced and scheduled outages.

In the case of nuclear-power generating stations it is apparent that although availability is obviously very important, reliability issues connected with safety appear to be even more dominant. This is apparent in the UK from the start of the nuclear-power programme which was outlined in 1955 and is evident in paragraph 19 of *A Programme for Nuclear Power*[4]. This was about the time that Calder Hall was built to produce electricity, it being recognized as the world's first nuclear power station. At the outset of this programme the need for considering the safety of people was met by the building of various means of protection into the plant concerned.

Typically the protective systems built into the type of reactor operated at Calder, which is a gas-cooled reactor, were essentially systems which would shut down the nuclear reactor if any fault condition were to arise. In addition, there are emergency electrical supplies and other important systems. These systems had to meet certain standards of reliability in their operation. Here it would appear that a type of system design dealing with high-risk situations was evolved which led to the high-integrity type of system. At the same time the literature shows the use of independent safety systems to complement the normal engineering systems so as to reduce the consequences of failure and to ensure adequate reliability of the systems. This approach to safety was carried out on subsequent nuclear-power reactors built as part of the British Nuclear Power Programme where the generating stations were operated by the Central Electricity Generating Board and the South of Scotland Electricity Board.

It is interesting to note the interplay between the different reactor types. The objectives set for the early UK fast reactors, as outlined by Moore[14, 15] and by Marsham[16], were similar to those for other types of reactor. These entailed whether the reactor should be free from siting restrictions and components should have high reliability in service. More specifically, the designer should provide ease of access to components so that any necessary alterations or repairs or defects arising in service could be carried out without affecting station outage, in a similar fashion to a conventional power station.

In the UK Second Power Programme 1964[17], it is brought out that nuclear-power stations have a heavy capital cost, but their running costs are low, whereas coal- or oil-fired stations have a lower capital cost but higher running costs. It was also stated that 'if nuclear power stations after 1975 became so cheap in capital cost and running costs that the purchase of an extremely large proportion were justified, pre-1975 stations of all types will be displaced to less intensive service because of the lower running costs of nuclear power stations'. This could be indicative of costs being dependent in part on the standards of reliability achieved and may be part of the optimization process.

It is clear that benefits accrued from the stage control in the designing, building and operation of these nuclear-power plants. The general discipline of this approach put into perspective certain types of systems such as control systems, on which safety depends. In some cases, compared with conventional power stations, higher degrees of control and greater reliability were required. Formulation of a safety policy was taking place, as for example from the experience of the commercial gas-cooled reactors as stated by Silverleaf and Weeks[18].

Questions can be raised as to whether or not the whole or any part of it can be applied to, for example future commercial reactors. Silver and Weeks[18] noted that in 1966 the major contributions to unplanned losses of availability were from large items of mechanical plant, particularly turbines.

In the brief survey similar considerations have been given to various types of reactor such as water and heavy-water moderated reactors in the USA and other countries. However, an important development in the general design and building of this new power plant was the demand for some criteria of their safety assessment, and it is interesting to note the proposed requirement which has been expressed by Farmer[19] and given in Fig. 1.3. Here along one axis is shown increasing consequences and the other axis is related to the probability of occurrence. The consequence is measured by the release of a radioactive fission product expressed in Curies of ^{131}I, and the other axis is measured in reactor years, or the events occurring in a complete reactor programme. It may be noted that the starting point here is taken as a programme of thirty reactors operating for some 30 years, which represents 900 reactors years. This criterion logically permits a reliability assessment to be undertaken and for it to be demonstrated that all consequences and their occurrence produce points which lie below the criterion line.

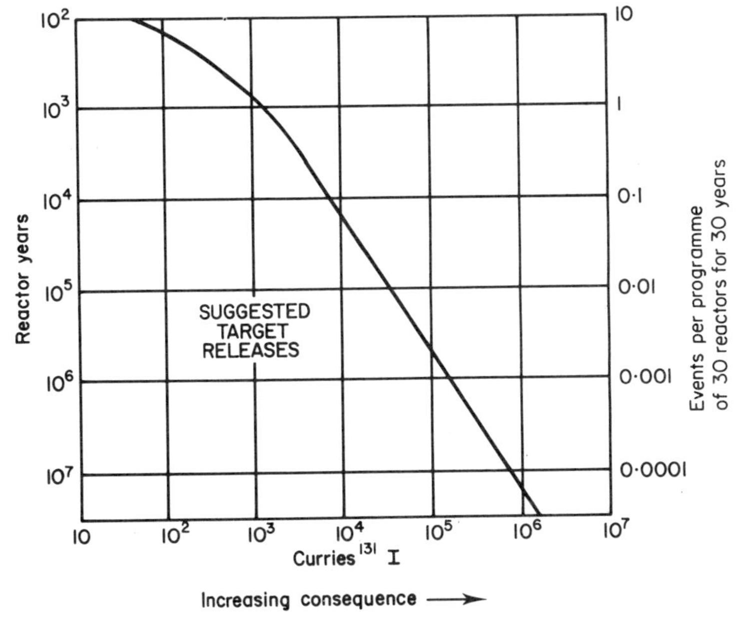

Fig. 1.3 Proposed criterion for nuclear reactors

Earlier attempts were made to derive similar types of criteria, for example as stated by Siddall[20] and more recent thoughts in this type of analysis using a probabilistic approach in connection with the nuclear risk accidents involved in the USA were reviewed as outlined by The Nuclear Regulatory Commission[21]. The important point of this work is that it develops methods for assessing the probability of failure or the probability of a particular sequence, but does not actually give any specific requirements to be met. Consideration has been given to this by the Risk Assessment Review Group in the USA[22].

1.2.4 Chemical plants

In chemical processing plants there is a strong economic incentive to maintain throughput in accordance with the design flow sheet. In the search to produce more of the less easily obtained chemicals to improve efficiency of plant operation, the chemical plants have tended to become increasingly complex and to have increased capital cost. In some cases this has led to higher concentrations of reactive chemicals than previously used, being involved in the processes. Many processes in themselves are exothermic, sometimes only at plant operating temperatures. The consequences of these situations have tended to result in greater damage under fault conditions. Not only can damage to the plant give rise to economic loss of future markets but human life can also be put at stake. This can extend to persons working outside the confines of the factory who may, under certain conditions suffer harm as a result of the emission of poisonous material in some processes.

In June 1974 an explosion took place at the Nypro UK Chemical Plant at Flixborough which killed twenty-eight people. The insurance market is estimated to have paid out £36 million for fire and accident damage following this explosion. This insurance claim was the largest of its kind to be faced by the insurance industry. The literature, as described by Daffer[23], shows that this type of accident suggests that the insurers may have to improve their technical resources to enable adequate assessments to be made.

The implications of The Health and Safety at Work Act as reviewed by the Health and Safety Commission, and applied by the Health and Safety Executive, could be significant, with parallel implications from the Royal Commission on Environmental Pollution[24] of 1976. In this publication a discussion of reactor safety emerges which could raise important issues, not only from the nuclear point of view but which could be considered to be common with chemical plant and other types of installations.

The emergencies which may arise on chemical plants may have initiating causes which are more frequent than others. These may stem from the loss of such services as cooling water, steam, air, inert gas or such events as loss of power supplies. Runaway reaction in the plant may result and in some processes certain conditions may be very difficult and uneconomic to contain, for example detonation. In such cases high reliance may be placed upon high-integrity shutdown systems. It is interesting to note that the current approach made in the design and evaluation of these systems is very similar to that in nuclear-power plants. This approach is illustrated by Stewart and Hensley[25] for high integrity protective systems on hazardous chemical plants.

1.3 Factors Affecting Reliability and Risk

In the accidents which may affect plant various factors lead to limiting the risk and consequently to showing that the reliability is adequate. There is a balance to be struck between the economic and humanitarian aspects. On the purely economic front a safe plant may be required but the standards to be met may be different from those set purely by humanitarian and ecological considerations. Hence there emerges a technological reliability to be met by the plant which is not just economically imposed. This is illustrated in Fig. 1.4 where financial requirements are shown separate from technological requirements. The general question of 'Acceptable Risk' is obviously salient and the application of science in the determination of safety has been widely discussed in the open literature as for example by Lowrance[26].

Figure 1.5 demonstrates both the formal and technical factors considered in quantified reliability assessment and the major factors which could enter into societal and individual judgements and decision-making about what the reliability/safety/risk should be.

It is considered very worthwhile and probably necessary in many cases to gain an understanding of the latter (on the right-hand side) as well as the former (on the left-hand side) which are more familiar to engineers, in order to appreciate the subject properly.

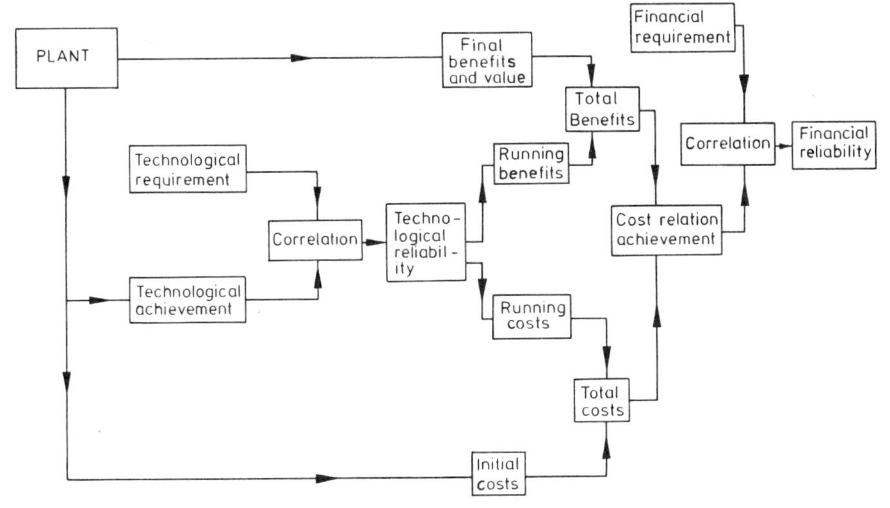

Fig. 1.4 Financial and technological aspects of reliability

The individual factors are shown at two levels in Fig. 1.5, (a) indicated by upper-case letters, (b) indicated by lower-case letters. There is a high degree of interrelatedness between many of them which is indicated by overlapping rectangles, circles etc. The factors considered are at a *very high* hierarchical level.

The relationship to 'benefits' is indicated because this is crucial to an understanding of the judgemental process, although details of the 'benefit' relationship structure have not been developed like those for reliability and risk, as this is not considered necessary for the subject of this particular book.

The critical and psychological factors are shown as having an hierarchical structure and fundamental flow process — moving from the top circle to the bottom circle on the right-hand side. In the case of the formal and technical factors of quantified reliability and risk assessment, factors 1, 2, 3, 4 and 5 precede 6, but not necessarily in any particular order. All these factors are shown to be affecting the availability and safety assessments through the final synergy of analysis, availability/risk criteria and actual experience of plant availability and risk. These assessments are processed by various societal institutions before receiving qualified acceptance and approval in some form. This itself has a feedback effect on all the major factors and many of the sub-factors, as indicated by the dashed lines feeding back around the bottom and left-hand edges of Fig. 1.5.

There is probably a greater understanding of the structures in which the formal and technical factors are operative than the structures in which the social and psychological factors come into play. In some technologies there is a strong body of professional appreciation of the formal and technical factors and their relationships. This is in strong contrast to the understanding of the relationships of social and psychological factors which affect any particular

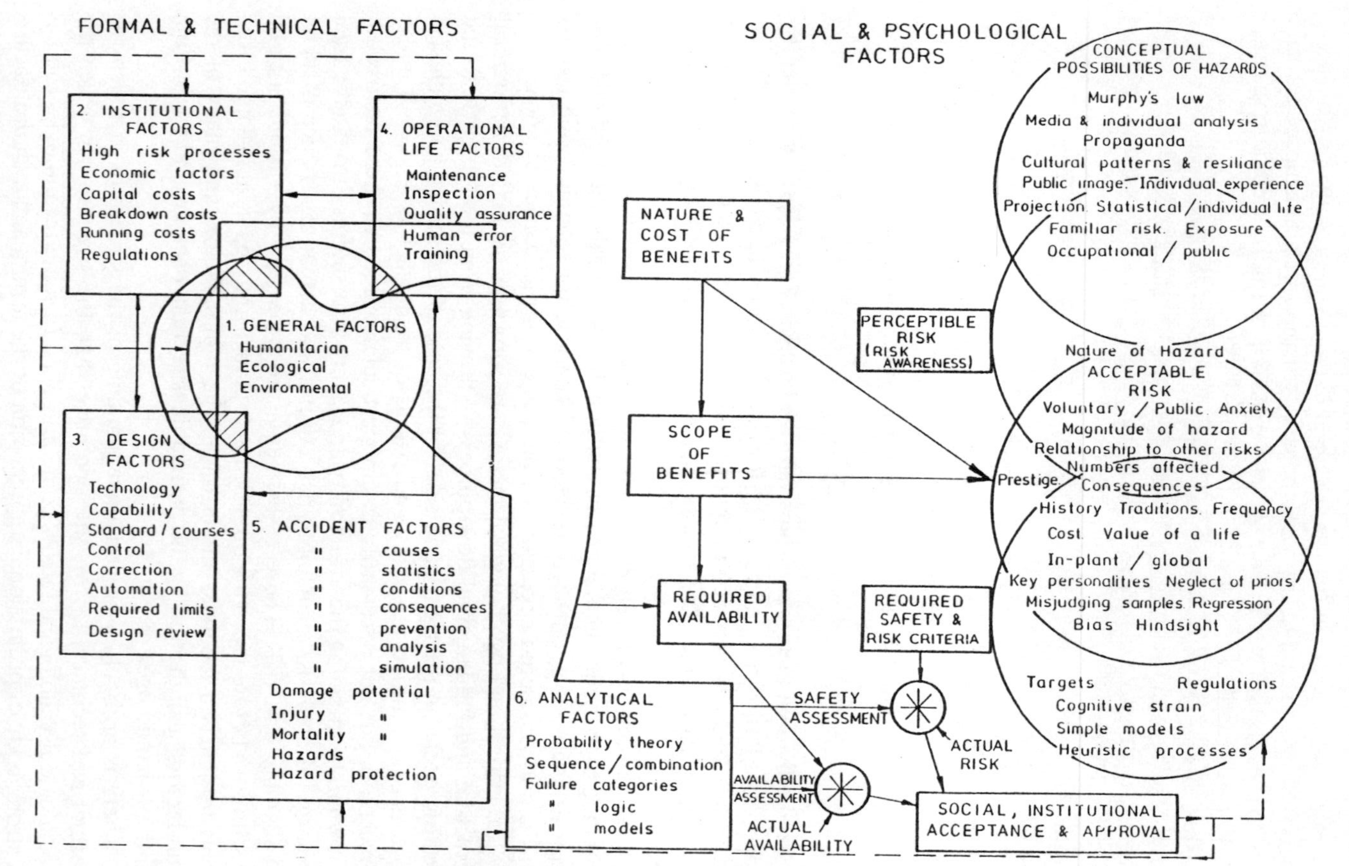

Fig. 1.5 Relationship of factors affecting reliability and risk

plant application or technology in general. The evidence for this is clear from the current controversies which surround some technologies far more than others. Although some of the factors are individually appreciated, a thorough synergistic understanding of the complex structural relationship of all the significant factors is yet to emerge and needs much more specific research. The simple structure indicated in Fig. 1.5 has been derived after consideration of Refs 27 – 34. Although analysis of some factors has been reported the simple structure shown in Fig. 1.5 is the first attempt to indicate a relationship between all the most significant factors. However, logical development of this is required before it can be useful, e.g. through being able to insert the known parameters affecting a particular technology and thereby predict society's reaction to the risks involved.

The various factors which have been discussed obviously lead to reliability technology which may be considered as a total technology. This means that it

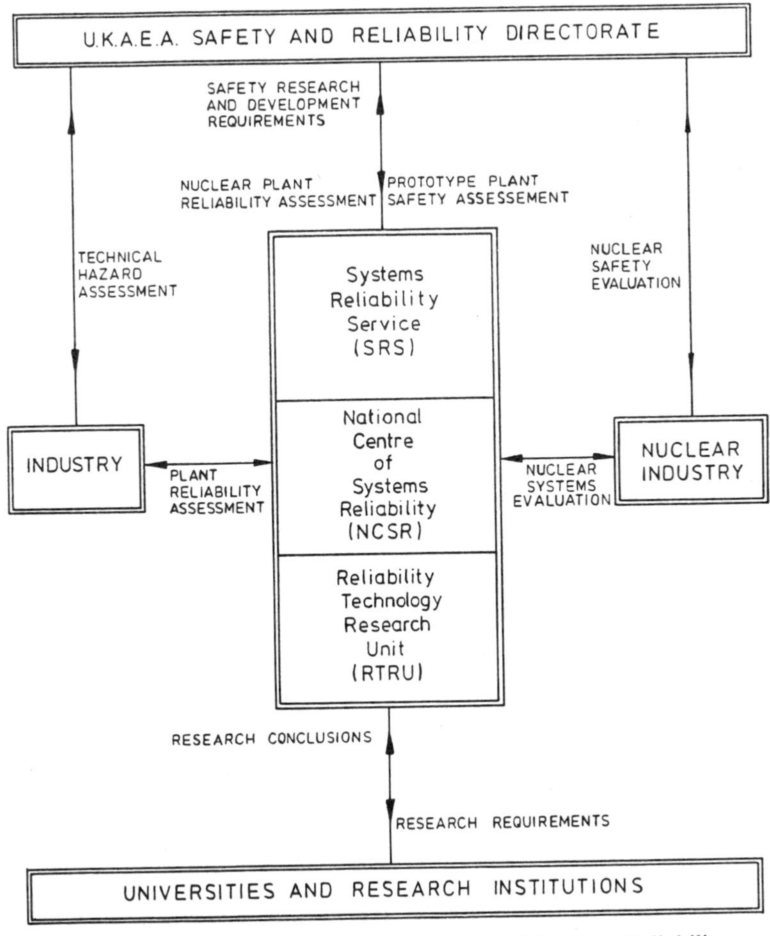

Fig. 1.6 Organization of National Centre of Systems Reliability

is multi-disciplinary applied on a systems engineering basis. Referring to the historical developments briefly reviewed in Section 1.2.1, it may be noted from Fig. 1.6 that in the UK the National Centre of Systems Reliability was formed with the interrelated functions of providing a Systems Reliability Service (SRS) with a Reliability Technology Research Unit (RTRU) giving an integrated approach to searching for solutions to reliability problems occurring in industry. More recently a technology transfer unit was incorporated. A further important feature is the provision of a wide-ranging Reliability Data Bank known as 'SYREL' which is also operating. Figure 1.6 shows the organization of this National Centre and the general connections with the various organizations for the development of reliability technology. The Associate Membership of the SRS has been studied and Fig. 1.7 has been prepared showing the geographical spread of the Associate Members and their principal activities. It is of some interest to note the comparison between the historical events discussed earlier and the principal activities which are undertaken by the Associate Members. Furthermore, there is a common ground arising from the factors affecting reliability and risk which makes it beneficial to have a cooperative approach to the development of the appropriate technology. In other countries similar approaches have been made by operating at a national level, for example in France there is the Centre National de Fiabilité du Centre National d'Etudes des Télécommunications (CNET) which operates a reliability data bank concentrating particularly on electronics and associated telecommunication equipment. In addition to undertaking functions such as testing, it cooperates with the Electricité de France (EDF) and the Commissariat à l'Energie Atomique (CEA).

The method by which the technological reliability of the plant is going to be maintained is obviously one which may permit of different design solutions. These may extend from concentrating on the cause of an accident condition and attempting to minimize the probability of it arising, to mitigating against the effects should the accident arise. In general terms a 'control' is applied which implies 'prevention' and also 'correction'. This introduces the idea of a 'domino consequence' where the last thing to happen is the actual consequence which may be plant damage or injury to human beings but will not take place if the sequence of initiating events does not follow on in a domino fashion. This sequential action can be stopped by preventing the sequence continuing and to control the resulting effect by applying some form of correction. Therefore, the 'domino effect' is broken and the accident condition is not allowed to develop. This is shown in Fig. 1.8(a) and (b), where one of the events, C, in the causal chain of A to E is withdrawn making the preceding events ineffective.

The general experience and knowledge gained on the various factors in the accident sequence have led to selecting methods of approach to prevent the complete sequence taking place. It is clear that in the broad category of accidents the combination of human failure with other factors such as environment can be readily tackled with programmes of safety training. Where

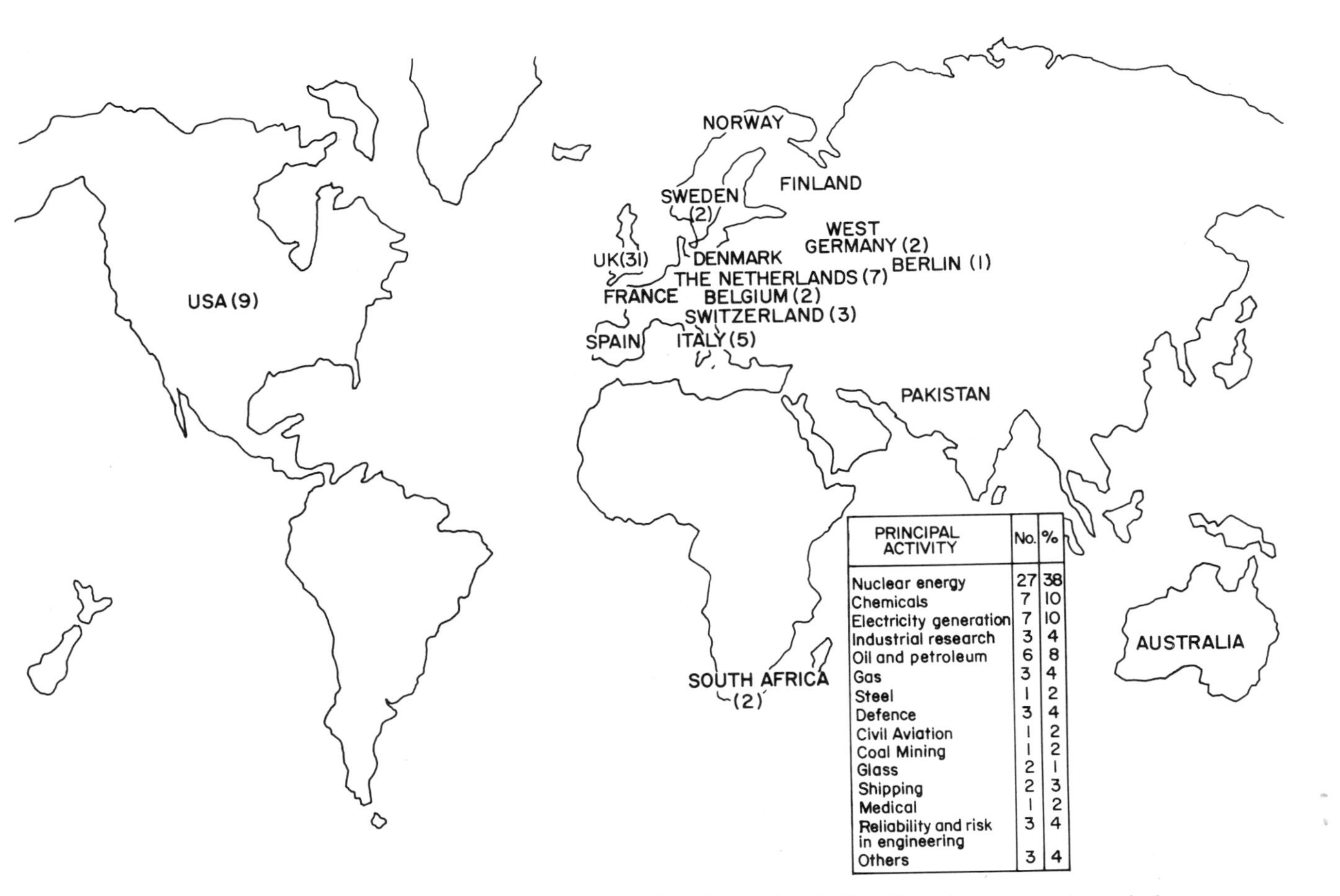

PRINCIPAL ACTIVITY	No.	%
Nuclear energy	27	38
Chemicals	7	10
Electricity generation	7	10
Industrial research	3	4
Oil and petroleum	6	8
Gas	3	4
Steel	1	2
Defence	3	4
Civil Aviation	1	2
Coal Mining	1	2
Glass	2	1
Shipping	2	3
Medical	1	2
Reliability and risk in engineering	3	4
Others	3	4

Fig. 1.7 SRS Associate Members: Geographical location and activities. Based on a sample period

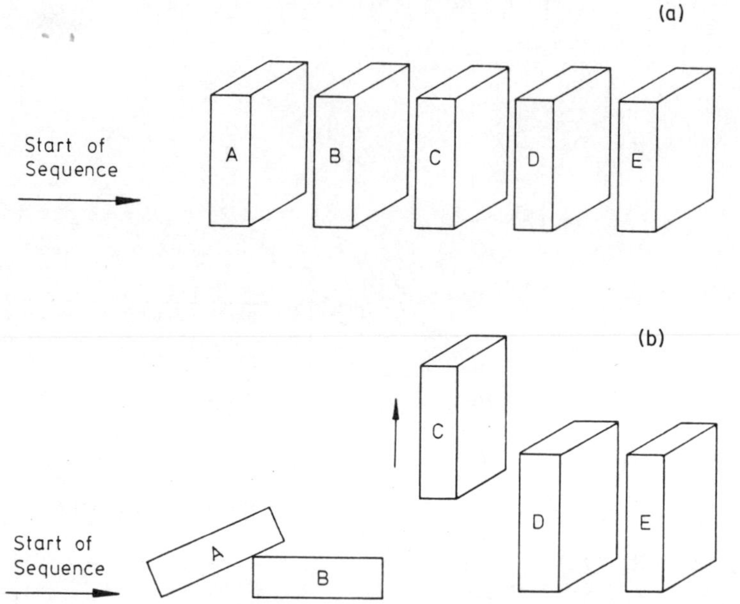

Fig. 1.8 (a) Accident sequence factors. (b) Accident sequence prevented

mechanical or physical failure is a factor then in the conventional factory safety such remedies as machine guards can play an effective part. However, the time for action on behalf of an operative may be small and automatic measures of prevention may be an essential remedy.

In the high-risk type of plant already surveyed the control may be manual or automatic and will normally involve the functional elements of sensing the change in a variable, processing this sensed information and applying it to control the variable. Furthermore, the whole system will have some form of 'feed-back'.

Examples of such systems may be seen in nuclear reactors where there may be general control systems which are designed to maintain the reactor behaviour within required performance limits. Clearly if some part of the control system were to fail then there could be a chance of the process of control completely breaking down. In some cases this may lead to a dangerous situation so that a separate system is introduced purely to protect the reactor by shutting it down. Figure 1.9 shows in simplified form this concept. The various parts of the system will depend to some extent on the particular reactor in question. In a gas-cooled thermal reactor the shutdown systems may consist of solid boron steel rods which drop into the reactor or in some types of water reactor could be liquid 'poison' which is inserted into the reactor to shut it down.

In a chemical plant similar ideas may apply but the shutdown system may consist of mechanical valves which operate to shut-off feeds or vent off the

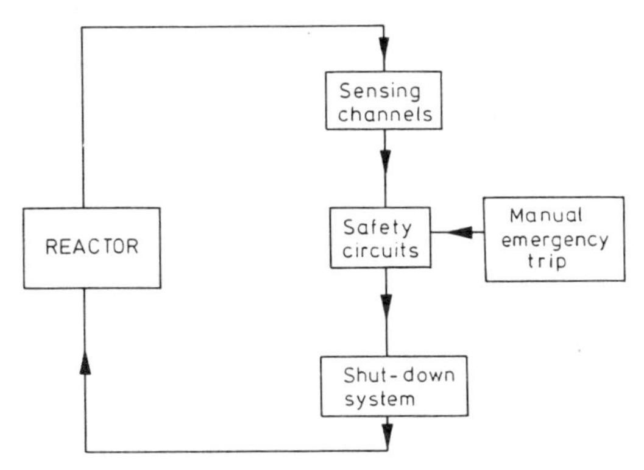

Fig. 1.9 Simplified diagram of nuclear reactor protective system

plant in order to stop the process. This can lead to a generalized protective system which requires to be considered in accident analysis.

1.4 Generalized Protective System

The functional elements of the generalized protective system are basically those for sensing changes in plant variables or states, processing this sensed information and for applying this processed information to control the plant in a predetermined manner. This overall system is automatic in action but usually has provision for manual action as shown in Fig. 1.10.

Normally control systems using 'feed-back' are employed for maintaining the operation of a plant within required performance limits. However, there is a chance of such control processes breaking down and there is a need for an additional safety measure. The generalized protective system meets the

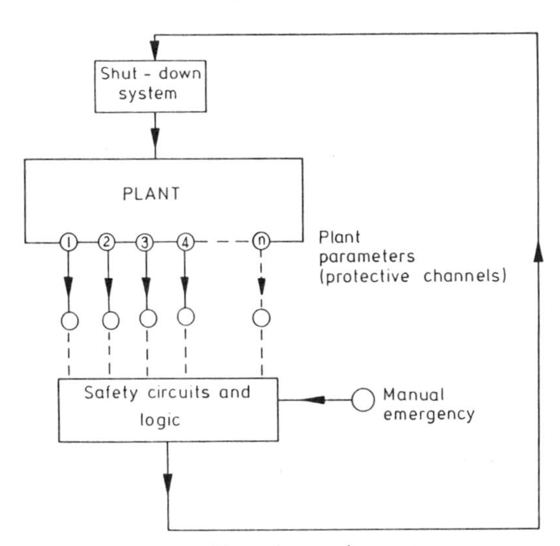

Fig. 1.10 Plant protective system

requirements of this additional safety measure by altering the plant behaviour in one direction only, which is usually to shut down the plant if some operational limits have been exceeded.

Each protective system will depend upon its application for the specific detailed variables to be sensed and the processing of this information for safety control. However, the generalized features are discussed for the basic system functional elements. Aitken and Sayers[34] give a more general introduction to the control and instrumentation for nuclear reactors and discuss the relationship with the protective system.

1.4.1 Protective channels

Plant fault conditions which are to be protected against need to be sensed by detecting abnormal changes in plant parameters in the most relevant way. The measures of these parameters form the basis of the protective channels. The kinetic studies for the plant will have shown the desirability for measuring particular parameters and responding in a given time for a specific shutdown of the plant.

The generalized protective system will have a number of different protective channels. Not all the protective channels will respond to any particular plant condition. Often it will be found that at least two are initiated for any given fault. This provides some diversity in measurement and usually provides a greater defence against common-mode failure. The measurement of each channel is normally at least duplicated to meet the lack of reliability of individual equipment.

The normal mode of operation of the equipment is such that in the main, component failure will take the operation of the system in the direction of safety, i.e. it is 'fail-safe' as opposed to 'fail to danger'. This can, therefore, lead to shutdown action of the plant being initiated entirely due to equipment failure. Such action if occurring frequently could subsequently lead to the system being brought into disrepute hence a majority voting system is employed on an 'm-from-n' basis. An example of this is a two-out-of-three system which requires at least two simultaneous fail-safe faults to initiate plant shutdown. Hence the generalized protective channels will be 'x' in number with each individual channel having measurements on an 'm-from-n' basis. Past experience has shown that frequent equipment failures can lead to financial losses as well as bringing the system into disrepute. In addition, with unreliable equipment there is a natural tendency to disbelieve the equipment if the probability of its failure is greater than the probability of abnormal plant conditions. Protective devices have been known to be made inoperative in these circumstances.

1.4.2 Safety circuits

In these circuits the function of processing the information from the operative channels takes place including any general logic which is introduced by design. An example of such logic is the actual carrying out of logical 'm-from-n' functions to derive signals to pass to the shutdown system. This type of circuit is designed to fail to a safe condition if a fault occurs.

1.4.3 Shutdown system

Depending upon the particular plant so the means of shutdown will be decided. In the case of shutting down a nuclear reactor power plant this may involve inserting sufficient quick-acting negative reactivity by means of safety shutdown mechanisms actuated by the trip signals from the safety circuits. These mechanisms reduce the power of the reactor to a shutdown state. Since a shutdown mechanism could fail due to a physical fault then redundancy mechanisms are employed. The whole system of shutdown mechanisms will normally be divided into groups which may also even extend to diverse groups. This could involve a 'primary' group with a 'secondary' group of completely diverse design.

The methods of operating the shutdown mechanisms would be that in case of failure of signal or supply then the direction of failure would be preferably towards shutting down the reactor plant.

The basic work undertaken in this book has been to research the methods by which the adequacy of such plant protective systems may be assessed to provide plant shutdown when required.

This work has been orientated to provide an integrated reliability approach from the initial design to the ultimate operation of such systems so that they may be analysed and assessed during their lifetimes to establish they are meeting the conceptual reliability requirements. Furthermore to take into account the various factors, such as the hardware and the human elements, which influence the performance reliability achievement and to consider various techniques to cater for the different aspects of the information which may be available.

1.5 Main Objectives of the Study

These objectives are summarized in the following sections; background material related to the formulation of these objectives has been discussed earlier in this chapter.

1.5.1 Historical aspects

To examine in a general fashion the history of certain plant systems related to safety and reliability and to study in particular the assessment of automatic protective systems with reference to present and future developments. This type of protective system has already been defined in Section 1.4. For the purposes of this book the reliability of automatic protective systems as designed and operated on nuclear reactors has been specifically studied. However, similar considerations will also apply to automatic protective systems on many types of chemical plant; this is briefly discussed in Section 1.2.4.

The history leading up to the requirement for the assessment of high-risk plant protective systems has been outlined in Section 1.2. This led to the creation of special high-integrity protective systems which by definition require to have a high reliability.

1.5.2 Reliability methodology

In the particular methods for the reliability evaluation of automatic protective systems have been studied from the point of view of safety assessment. This type of assessment is orientated to demonstrating adequacy of the reliability of the particular automatic protective system being appraised rather than providing the design solution for the particular system.

The methods studied are required to cater for the evaluation of the system not only at the conceptual design stage but throughout the whole safety life cycle. An additional aim has been to research those methods which will permit not only a qualitative evaluation but also a quantitative one, to permit the greatest use to be made of the quantitative type of criterion as shown in Fig. 1.3.

A related objective is the investigation of feasible paradigms for 'updating' reliability predictions from the design stage, which of necessity must have a substantial subjective content, by the addition of operational hard data. A subsidiary objective here is to examine the relevance of Bayesian techniques for this updating procedure.

1.5.3 Validation of applying reliability methodology

An important objective of the study is to establish the validity of reliability methods by comparing derived predictions with results obtained from actual field data. In particular to study at the equipment and system levels for automatic protective systems the correlation between 'hard' or actual data and prediction.

In addition to validating methods the study was also orientated to identify and possibly remedy areas of deficient methodology. A further objective was to investigate the feasibility of using 'soft' or subjective data.

1.5.4 Reliability management and control

The objective here has been to examine the feasibility of determining a reliability parameter to serve as a measure of the efficiency of control and management procedure used to keep the reliability of the system within specified limits. This parameter is to be determined from probabalistic studies over the various phases of the plant constituting its safety life cycle.

1.5.5 Interfaces and human factors

Throughout the whole life cycle interface problems can arise which can lead to the distortion of a true picture as required to be derived in any reliability assessment undertaken for ascertaining safety. These interfaces can be purely technical such as not being able to measure the required quantity or can be created by the human elements which pervade the whole of the protective system design, operation, maintenance and testing. Accordingly, a further basic objective has been to investigate the methods of evaluating some pertinent aspects of human factors and the associated interfaces with the hardware of the automatic protective systems under investigation.

CHAPTER 2

The Initial Approach to Performance Assessment

2.1 Design and Assessment

2.1.1 General process

The various factors previously discussed lead to an overall requirement which lays down the performance and other constraints for the design and ultimate operation for the particular plant concerned. Although it may be the design function to produce a specification and the necessary documentation to permit the procurement of the appropriate system it is considered by the author that a forward design thinking strategy requires to be adopted. Logically this should give the necessary consideration to factors which may arise in the overall process of manufacture, installation and commissioning, operation and maintenance. These various stages of the overall process are shown in Fig. 2.1 but for practical reasons this is not always possible and a total systems engineering approach may not be readily and initially possible. Hence the initial approach may be for the designer to concentrate on producing a design which will have the required functional performance assessment which as shown in Fig. 2.1 can be undertaken stage by stage or in some overall fashion by a comparison of the 'requirement' with the ultimate 'operation'. A stage-by-stage method of assessment can lead to 'interface' problems which do not arise in the overall method. This question of assessment is discussed later in this chapter. Nevertheless the initial approach by the designer may be somewhat deterministic and can in many cases involve engineering judgment.

The protective system has its overall performance defined in the requirement which will include the plant measurements to be made and the response of the overall system. This part of the requirement will have been derived from accident studies looking at the kinetic behaviour of the plant under fault conditions. Limits of operation of the plant will then lead to a specific response requirement.

Another part of the protective requirement is the reliability of the protective system. This may be expressed in different ways such as no single failure

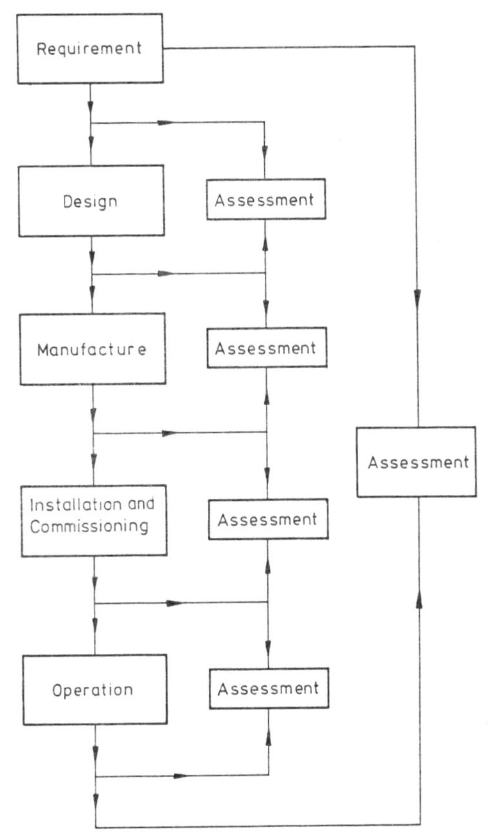

Fig. 2.1 The design and general process of assessment

of a component or equipment shall fail the system or that the system shall be adequate and made from good-quality materials. Initially this may give rise to a starting point for the design and enable a deterministic approach to be taken but ultimately such qualitative statements need to be measured and assessed in some way as discussed later in this book. Nevertheless it must be necessary to have a protective system which has the capability of responding as required assuming no failure and variability. Hence the deterministic evaluation of performance is an important step in the initial design solution and will involve taking into account other factors such as cost which will impose further constraints.

2.1.2 Conceptual design

The protective requirement will lead to a conceptual design being evolved. This design process for the present purposes will be considered as beginning with a need and ending with a set of instructions consisting of drawings (arrangement and detail), schematic diagrams, purchase orders and test schedules. In this process there should be increasing precision and definition with the earlier stages having a greater fluidity than the later ones.

2.1.3 Design selection

Theoretically the problem of design can be stated as a non-linear programme of the following form. Minimize the cost function $C(Q_i)$ subject to a series of constraints $A_r(Q_i) \leq K_r$ where C is cost and Q is the ith parameter the choice of which is decided by the designer to meet the constraints given for K_r. These considerations involve coefficients which are often unknown or only known with relative imprecision. Again the Q_is can be discrete or continuous. In practice the solution of the design problem is by an iterative process until the design requirement has been met within sufficient limits. Therefore a unique solution is not usually readily available in an analytic form and a logical solution is usually only possible by introducing some other techniques of approach.

In the design process various representations and models will emerge and it may be noted that the hypothetical system of the designer will be projected into an actual system. A 'structural model' produced by the designer will demonstrate the elements of the system and the manner in which they are connected. This will also involve certain optimization of various factors which can be controlled at the design stage.

The designer has to arrive at a system solution, however, normally at the start of the design a truly deterministic selection cannot be made due to the limiting of information. Even the basic requirements may be inadequately specified. Hence in general terms a process of partitioning establishes an 'enclosure' in which it is known the system lies. A process of convergence leads to a reduction of the systems in the enclosure and the ultimate choice of a particular system from the design-coordinate space. The present study described has not considered the design process in detail but rather its relationship with safety and reliability assessment. It must be recognized that the particular strategy adopted can lead to different degrees of success or failure.

These factors emerge in the paper by Firth[35] which gives an indication of the process typically adopted within the UKAEA for the design of a prototype nuclear-reactor plant. Clearly the emphasis is of cost control but it is stressed that 'with a new reactor concept, only an internal but at the same time independent safety organization permits a very close collaboration necessary at each stage of development'.

In this design process the design-project office would normally handle a nuclear plant. As would be expected in the conceptual design many uncertainties occur in finalizing a specification because of the prototype nature of the plant. An example of such uncertainties could be in the physics and kinetics of the reactor. However, the reliability conditions for the overall plant and its various systems are directed to the two main areas of achieving certain availability and safety goals.

Experience has shown it is useful to develop the concept of a reference design at a very early stage. Development may then proceed in various areas in parallel without introducing confusion by a review procedure at certain key stages.

This basic design process will be achieved with a staff structure as typically shown in Fig. 2.2. The particular protective systems of interest will be mainly dealt with under control-systems instrumentation and safety. At the same time there is an important relationship between the project design office with other offices and groups as typically shown in Fig. 2.3. The assessment procedures will be mainly focused on the Safety and Reliability Directorate operating independently in carrying out safety assessment.

Although the basic design process described may have some common features with other industries it is increasingly being found that in specific areas of systems dealing with safety that safety goals are assuming paramount importance. This can be seen both in the design of nuclear power plant and chemical plant involving high temperature and pressure chemical engineering.

There is a strong indication that in designing specific systems to protect plant there tends to be similar procedures, e.g. the emergency electrical supply systems or control and instrumentation systems for different plant may be worked on in parallel by the same design office thereby bringing the maximum experience together in finding an adequate design solution. Often this centralization leads to an appropriate optimization. In other words to each system solution there is a cost which is a function of design-coordinate space or a cost function. In the optimization process the preferred solutions will correspond to cost-function minima.

This approach is applied in various forms and Fig. 2.4, which is described by Hill[36], indicates the general approach for generating plant within Ontario Hydro, Canada. The specific aim being to minimize the total life-cycle costs by the appropriate detail.

More recently, Lakner and Anderson[37], discuss an overall programme of work for the design and procurement of National Airspace System equipments within the USA. Here an analytical approach to determining optimum reliability and maintainability requirements is described and it will be found that many of the principles are pertinent to applications other than air traffic control.

In the field of safety and protective systems a relatively small change in a parameter may cause a large increase in system cost. If this increase is extremely large then these high-cost regions will make the system unacceptable. However, if there is a 'cliff-edge' type of discontinuity existing due to the small change in the parameter concerned with grave consequences resulting then it may be necessary to apply the 'domino principle' previously described. This means accepting the various cost-function boundaries in the design-space region under consideration and selecting the system design from the region where the probability of failure meets some acceptably low criterion.

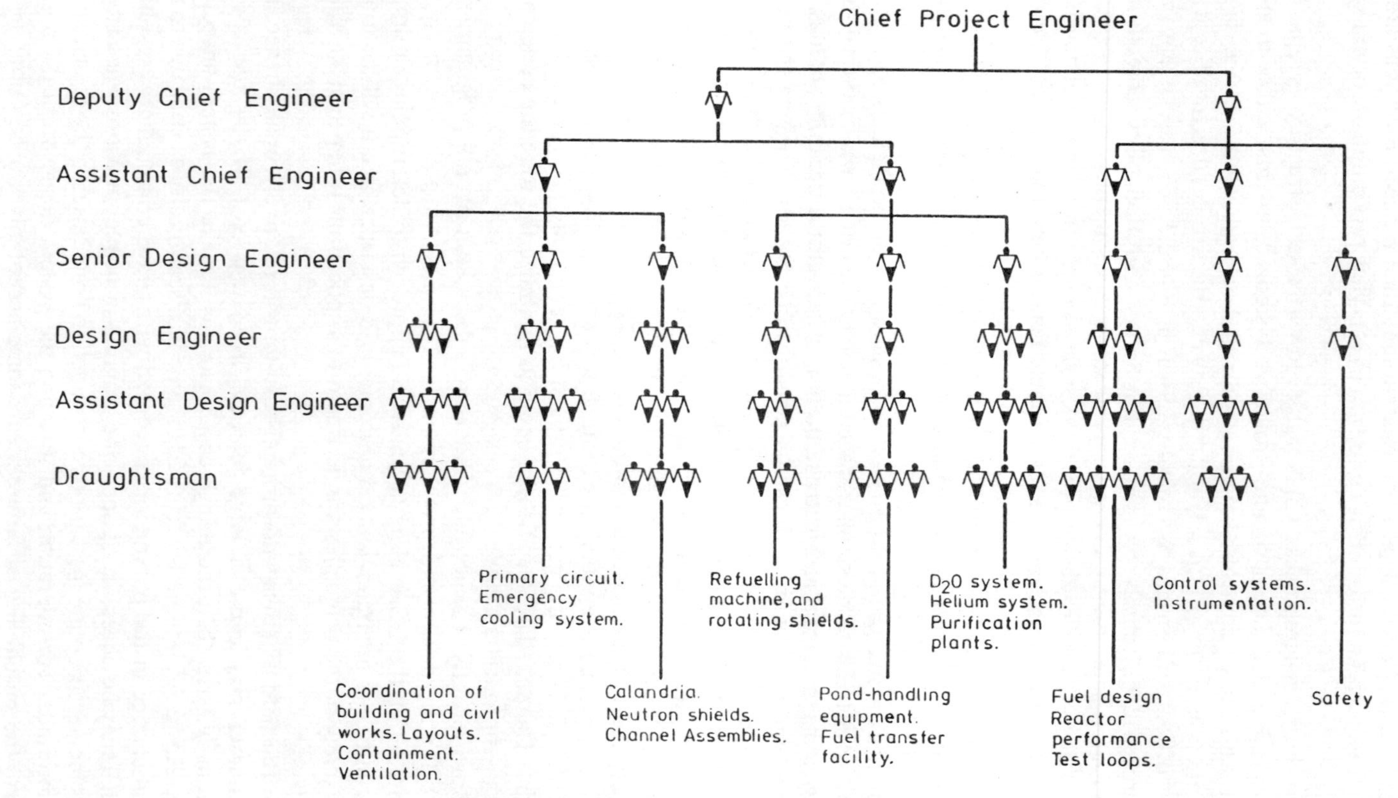

Fig. 2.2 Staff structure of a typical project office

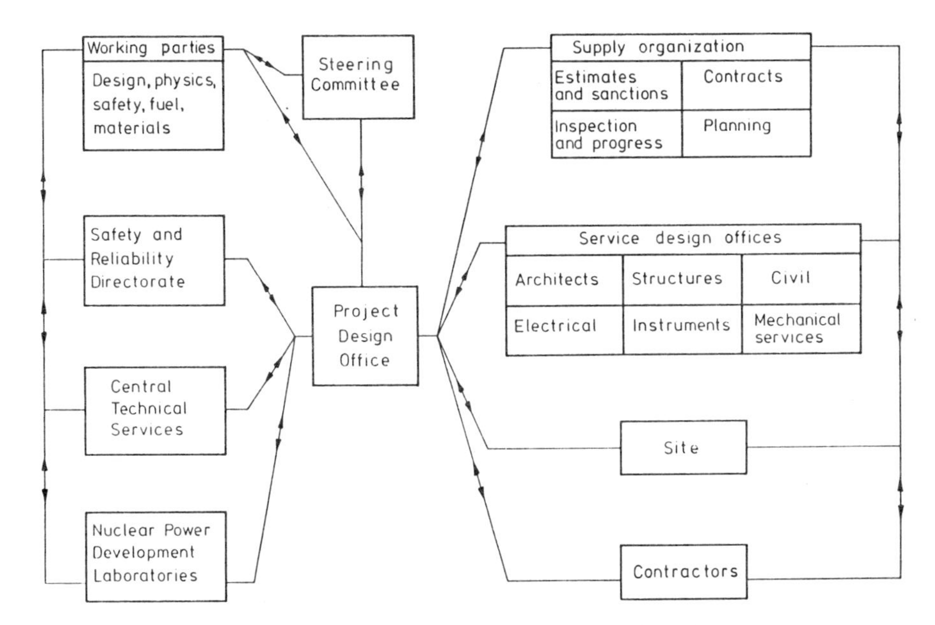

Fig. 2.3 Relationship of project design office with other offices and groups

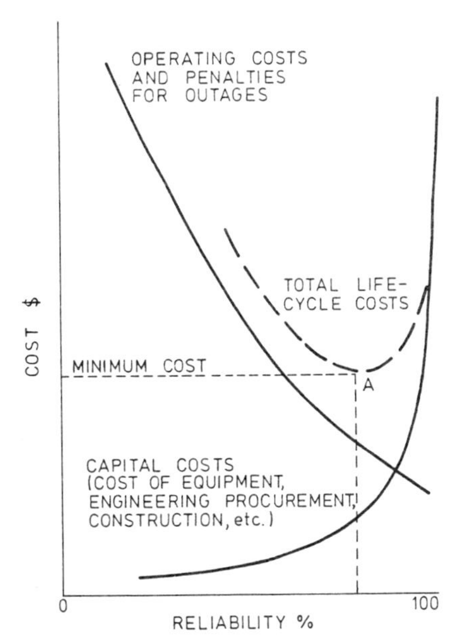

Fig. 2.4 Reliability against life-cycle cost trade-off (relationship of reliability to costs is shown by these curves; point A shows the lowest total cost and the corresponding reliability). (Reproduced from Hill, 1973[36], by permission of A. Hill, Ontario Hydro)

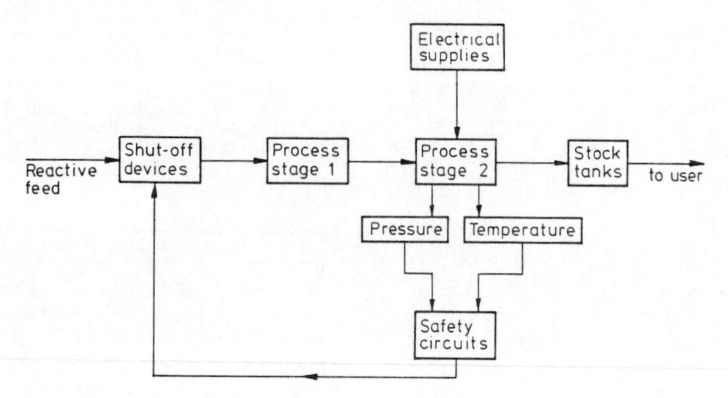

Fig. 2.5 A simpified diagram for a typical process plant

Hence a typical process plant shown in simplified form in Fig. 2.5 may be so dependent upon the electrical supplies that in the event of the loss of these supplies abnormal conditions would exist giving rise for the reactive feed to be automatically shut off by the protective system.

The typical protective system shown in Fig. 2.6 is for a nuclear-power reactor. This automatically initiates shutdown should a reactor fault condition arise. The various reactor parameters are measured by the sensing channels and via the safety circuits action is initiated to shut down the reactor in the event of a fault condition. Depending upon the particular power reactor so the configuration of the system may change but the sensing channels usually

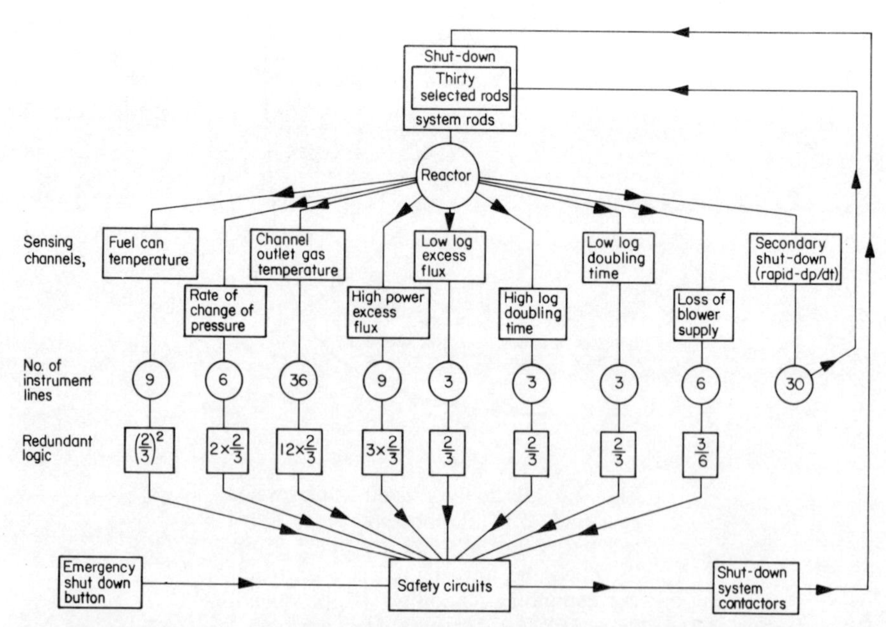

Fig. 2.6 Typical automatic protective system for a gas-cooled power nuclear reactor

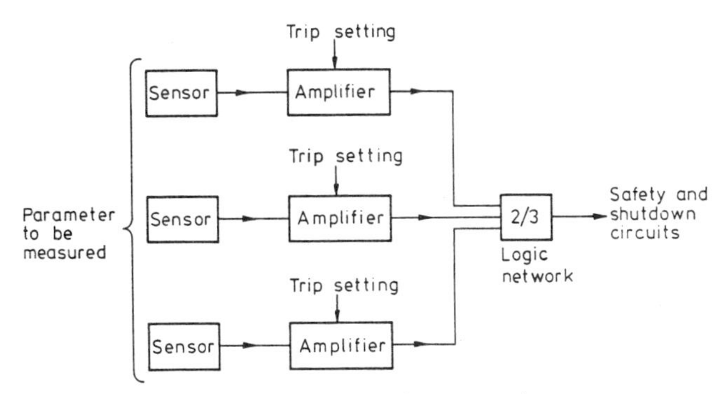

Fig. 2.7 Sensing channel for protective system

include some form of measurement of neutron flux, temperature and coolant flow. In addition, there may be other measurements such as pressure, steam conditions and states of electrical supplies. Usually diverse sensing channels are provided so as to sense each fault condition by changes in at least two parameters.

In principle each sensing line usually consists of a sensor and some form of amplifier feeding a signal into a logic network. A typical sensing channel is shown in block form in Fig. 2.7. It is standard practice to use more than one sensing line to form a complete channel. At some predetermined trip setting, the trip signals from the amplifiers are fed into the logic network, which is often a two-out-of-three configuration (this means that at least two trip signals must be applied to the logic network before the reactor is tripped). This leads to the whole question of the assessment of the adequacy of the particular protective system design.

2.1.4 Safety assessment process

It is the practice within the UKAEA as in many other organizations to carry out an independent safety assessment of the proposed design. This is undertaken by a group of people who have a sufficient multi-disciplinary approach to arrive at a judgement as to the adequacy of the design from the safety point of view. Obviously the design process itself will have already included quite separately feasibility and other critical examinations so that the designer can satisfy himself as to the adequacy of his design.

The independent safety assessment is an attempt to minimize the consequences of failure in systems and equipment. It is directed at detecting errors in thinking and uncertainty in design which could lead to unacceptable risk if the proposed design were adopted. Clearly some standard requires to be adopted and this is briefly discussed in Chapter 1 and will be further considered in later sections.

Although a complete nuclear reactor would be subjected to this independent safety assessment, it is intended to consider a typical automatic protective system for purposes of illustration, as that of Green and Bourne[38]. The purpose of such a protective system may be defined in a general way as all the equipment which automatically senses and produces action to initiate shut-down should a reactor fault condition arise.

Assuming that some standard or criterion exists for the nuclear reactor then two aspects of direct interest to the assessor of the automatic protective system are:

(a) what is the degree of safety against which the adequacy of the automatic protective system is to be judged,
(b) can this degree of safety be reasonably claimed for the protective system in terms of the overall reactor safety problem or other considerations.

The submission by the designer will generally outline in varying degrees of detail, the mechanics, the claimed performance and the intended operation of the system. This will be supported by a safety report with specifications, drawings and other supporting documents and analyses.

The proposed design solution will therefore be substantiated by the designer as having a performance giving adequate protection to the plant. It is the assessor's task to judge whether this particular solution is adequate. Furthermore the assessor is required to judge whether the adequacy of the particular solution has been substantiated and not what the assessor himself might do if he were asked to provide a solution.

Hence the outstanding problem for the assessor is how to prove the case. In the case of the reactor protective system it may be necessary to prove that a district hazard will not arise under reactor fault conditions. Clearly, it is not possible to demonstrate such a case directly on the plant by means of practical tests. This is now the crux of the safety assessment problem. How far can it be proved directly that the design proposal is adequate and what methods can be used for those condtions which cannot be proved directly.

2.1.5 Methods of assessment

The assessor may choose between the following types of assessment; general, sample or detailed.

In the general assessment just the overall concept of the system is examined which may be sufficient under certain circumstances where general principles of safety are solely involved. The sample method involves a detailed assessment of some selected parts of the system. In practice such a sample may not be representative of the whole due to varying qualities of different elements of the system. Hence a compromise between the ideal of doing everything in detail and the sample method is usually adopted. This is to carry out selective sampling on the basis of the consequences of failure and to assess these samples in detail.

The direct method of proof is used if possible to substantiate any required characteristic or quantity by examining the theoretical and practical evidence which has led up to the designer's proposal for the particular quantity.

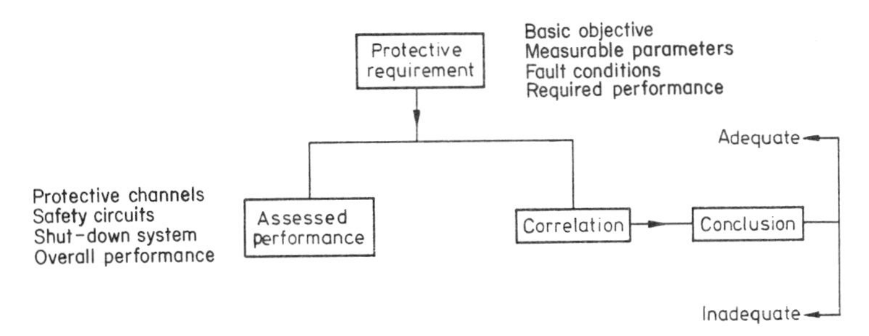

Fig. 2.8 Steps in safety assessment

If the characteristic or quantity is unknown or uncertain then the indirect method of approach will be adopted. Here a boundary value is assigned to the undefined quantity and this now used for the detailed assessment. Usually this boundary value will be set pessimistically and is not necessarily the actual value for the plant. If it is found that this boundary value would make the overall system adequate, then the ultimate proof now depends on showing that the undefined quantity does in fact lie within the assumed boundary value.

The whole process of safety assessment uses various methods of checking the solutions of problems and classifying things and arranging ideas in an orderly fashion. In the case of carrying out a safety assessment of a reactor automatic protective system the steps would be in general terms as shown in Fig. 2.8. The 'Protective Requirement' is, in fact, the complete statement of the problem being solved.

Here is set down in the 'Protective Requirement' the conditions under which they must be met and the required performance of the automatic protective system. The 'Assessed Performance' is the assessors own evaluation of the performance for the system which would appear to be capable of reasonable substantiation under the relevant conditions. A 'Correlation' is then carried out between the required and actual performances and a 'Conclusion' reached and expressed as either 'Adequate' or 'Inadequate'.

2.1.6 Decision-making and problem solving

Reaching a conclusion on adequacy in the assessment effectively involves a process of decision-making in which a problem is being solved. In the investigations carried out by the author it makes a good first approach to do a deterministic method of the solution as the first approximation. This means, of course, approximating quantities which have uncertainty with a single best estimate and examining each particular decision. The investigations for this book show that a question in itself does not necessarily constitute a problem and obviously it is the conditions under which a problem arises which usually exist when difficulty is encountered in finding an answer to a question. When an answer is required involving a difficulty then terms or constraints will be

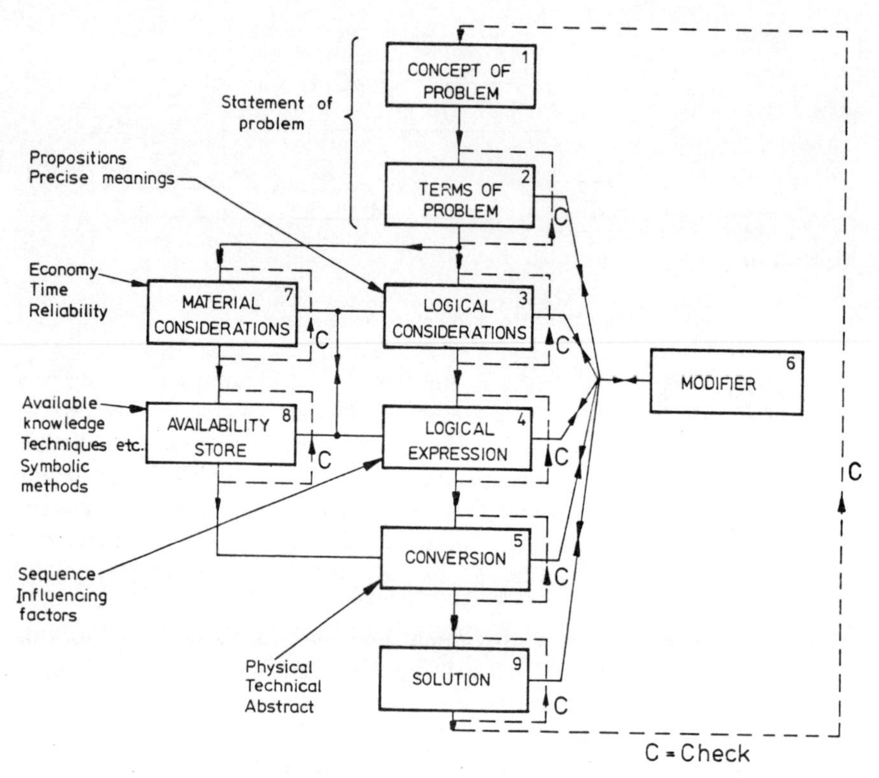

Fig. 2.9 Basic process flow chart

necessary, which together will form the concept and the statement of the problem. In a deterministic sense this leads to a logical method for arriving at a decision based on classical formal logic. From a study of the basic process at arriving at a solution to the problem, Fig. 2.9 represents a basic process flow chart. First there is an awareness of a difficulty not necessarily in words but in concept only which is expressed in some form which shows the concept, represented by block 1. The terms of the problem then constitute a framework in which a solution requires to be found. A check should be made in each stage of the analysis process by indicating a feedback check, as indicated by C in Fig. 2.9. This consists of comparing the output information from a stage with the input information. Otherwise any misunderstanding or misinterpretation which may take place in block 2, may lead to an output outside the terms as originally intended.

The statement of the problem may have every appearance of plausibility but might be unacceptable by a logical solution and must therefore be examined through the processes represented by blocks 2 and 3. This is a most important stage in the analysis of a problem as an attempt is now being made of finding a logical solution without material considerations. In block 3 the propositions are separated and examined for their precise meanings. An attempt is made to identify the elements involved in such a manner as to divorce them from material considerations. It is advisable to do this as experience shows that models having a material suggestion tend to confuse and often bias thinking into set channels. For a clear understanding the first step is to realize what is logically important and express the elements of various events using a symbolic notation as for instance a, b, c etc. A symbolic notation or language permits important and complete appreciation of those considerations from which inference may be drawn. Ordinary language is rarely sufficiently precise in intention and is often misleading being sometimes affected by emphasis which can be misplaced.

Should a difficulty arise at this stage then the considerations are passed through the modifier block, 6, and thence into block 2 for a modification as necessary to the original statement. On re-examination a statement is eventually derived which satisfies the logical considerations but which, at the same time, does not disturb the original concept.

Logical expression is given in block 4 and this in the abstract could for instance take the form of 'all $a < b$ and all $c < a$' from which may be drawn the inference that 'all $c < b$' which is known as a Syllogism and such methods of formal logic are discussed in the general literature, for example by Joyce[39]. It will be noted that at each stage a check back shown by the dotted lines in Fig. 2.9, marked C, should satisfy the reasonableness of the argument. Any inconsistency may necessitate a feedback through the modifier block 6 and re-entry for further consideration into any of the earlier blocks.

The argument is now ready for acceptance as a final conversion 'if necessary' from the abstract into the concrete form. In the case of a problem in formal logic conversion is required but if critical or technical considerations are employed then the conversion must be in the appropriate descriptions of techniques. Critical attention requires to be given to obtaining an accurate statement of any problem which may be the subject of analysis. In block 7 an assessment is carried out of the relative importance of such items as economy, time and reliability, before a decision can be completed and accepted. At this point it may be necessary to feed back via the modifier block and to check the original terms in block 2 to make quite sure that an alternative statement would not perhaps better express the purpose required. To complete the conversion it is necessary to follow through from the material considerations into block 8, 'The Availability Store'. This store is intended to represent and include all available material and the sum total of all available skill, 'know-how' or knowledge.

It may be found that the logical expression may only be converted into a material solution after a hypothetical component is available, obviously it will be necessary to develop such a component. This component would not be readily available in the 'Availability Store' and would generally give rise to another problem.

Too often the problem of logically preparing an 'Availability Store' is ignored and the human memory is left to deal with the problem, or the

'Availability Store' is not in a suitable form for ready access. A typical example of acquiring easy access to information is in research work. Information may be obtained in this case from a library. It is obvious that improving the availability of information can be readily reflected in cost, time and effort. Quite often the application of previously acquired knowledge may yield a solution, but without this knowledge being readily available it would have been necessary to undertake experimental work. However, it should be appreciated that at this stage of the assessment that the decision on adequacy can only solve the problem within the bounds of the 'Availability Store' being available to the assessor. A decision with regard to material considerations is made from the 'Availability Store' which is considered to give the 'best known' answer. After a check back through the various blocks to make sure that the overall structure has not been disturbed the 'best known' method from the 'Availability Store' is applied to reconcile the abstract logical expression with a material solution.

This logical process appears apparently laborious but is followed intuitively in the process of the assessment in some form or other. In arriving at the decisions of adequacy it should be borne in mind, particularly from a safety point of view of distinguishing between a good decision and a safe outcome, this introduces a scale of values associated with certain outcomes. However, as described, decision theory is effectively the application of common sense and the symbolic methods provide a more precise method by which the decision problem may be represented. In the ultimate this representation will involve value, which involves utility theory and information, which also involves probability theory. The basic decision-making process in a deterministic sense may involve apparently great complexity in any systems analysis but in a deterministic and logical framework the problem of decision-making should be reduced to propositions of such a simple nature that they can be answered without difficulty and unambiguously. It is not proposed to develop further ideas on decision-making but important developments have taken place in modern utility theory and have arisen from such earlier classical work as given by Von Neumann and Morgenstern[40].

The inductive use of probability theory for the purposes of reasoning enables information to be considered as to which outcome is likely to occur and is described in the general literature commencing from the time of Laplace[41] and later developments involving both probability and utility theories as given by Savage[42] and Kaufman and Thomas[43].

In resolving uncertainities in the decision-making process it requires to be also recognized that more information can be often gathered at a later period which may serve to reduce the level of uncertainty. Hence initially the deterministic phase of the process will give some insight into the decision and sensitivity analysis can be undertaken by varying the quantities to see how decision may be affected. One of the conditions that would appear necessary in the decision-making for the adequacy of a protective system is that deterministically there is a system which has the capability of performing assuming no failure in variability. However, in the real world there will be failure, variability and growth. Data which are existing or can be found at a later date may be used to impose uncertainty distributions and predict the probability of success in the time domain. The decision-making process of the system can also incorporate at subsequent times remodelling of the system and the use of adaptive analyses to estimate the error from the initial path intended. If need be a remodelling of the system can be undertaken then to reduce errors to acceptable limits but the important question is whether the error is convergent or divergent. Divergent errors would be expected to involve adaptive control with predictations from a growth model of reliability as part of the decision-making process.

2.1.7 Need for a prior

Risk analysis for the plant requires to be undertaken at the initial stages of the design. It is often at this stage where major decisions concerning money and safety are made. For example the overall plant configuration as to whether it is single or multi-stream can be very significant in many cases. With reference to the initiating fault events considered and their outcomes may lead to deriving the probabilities of the protective system or systems functioning and failing to function so as tò prevent the realization of unacceptable consequences.

Detailed assessment of the protective system requires to show that the required performance achievement can be met;

(a) in response time when the system has been newly installed,
(b) during the lifetime over the complete spectrum of operating conditions both normal and abnormal.

Referring to Fig. 2.1 it will be seen that this performance achievement may be assessed at various stages of design, manufacture, commissioning or operation. Furthermore, it may be undertaken stage by stage or in some overall manner. One of the basic principles being advanced in this study is that for high-risk situations involving the use of high-integrity protective systems then a prior estimate is necessary. This permits the 'domino principle' to be employed to the full with a growth programme of posterior estimates being obtained to give progressively more information on the 'tracking' of the trajectory in an ideal space of the performance followed by the actual system design. It requires to be emphasized that the system can be, (a) a paper-work concept, or (b) a hardware concept.

Clearly the important aspect to be considered in the ultimate is the adequacy of the actual hardware system in its operating environment. The complete information on this aspect will be known when the plant has completed its life and the performance adequacy of the protective system can be deterministically assessed in the light of the actual and real demands which have occurred.

Table 2.1 Extract of estimated cost of accidental nuclear releases

Size of release Ci of ^{131}I	Cost of illnesses or casualties (£)	Cost of temporary ban on sale of milk (£)	Loss of GNP (£)	Cost of rehabilitation of people affected (£)	Total cost (£)
10^3	nil	8 000	5 000	25 000	4×10^4
10^4	nil	50 000	200 000	1 000 000	1.25×10^6

In nuclear-power plant and similar process plant experience has shown from accident studies and other considerations such as cost in the widest sense that only a 'forward'-looking strategy can be initially accepted. The lessons and experience of a 'backward'-looking strategy, however, need to be incorporated into the thinking in an adaptive fashion in order to suitably modify the forward-looking strategy.

Some of the estimated costs of accidental nuclear releases from nuclear reactors have been estimated by Beattie and Bell[44]. An extract of some of these costs is shown by Table 2.1.

Further considerations of very extreme cases can lead to total costs of the order of £1000 million. However, it should be noted that high costs have been quoted for the accidents in the chemical industry for example the Flixborough accident which is estimated to have involved an insurance payout of about £36 million. The general literature and experience more than supports having a forward-looking programme. This further stresses the need for having an assessment or prediction carried out before the evidence becomes available which may be properly defined as the 'prior' and the assessor must in some way develop this for the protective system at the paperwork concept. Various techniques of risk assessment formulation can then be applied; Penland[45] shows this where under certain conditions and assumptions that

$$\text{Risk rate} = \sum_i \beta_i \left(\sum_j P_{ij} D_{ij} \right) \tag{2.1}$$

where β_i are statistically distributed with estimates taken at a desired confidence level.

P_{ij} generally involves the reliability of safety systems, e.g. the probabilities of protective system functioning and failure to function and prevent consequences.

D_{ij} are consequences of sequence i–j accident.

With P_{ij} established the assessor can appraise the various co-ordinates of the design space selected for the protective system and derive a prior estimate which can then be compared with P_{ij}. However, this process will be 'information-limited'. A basic problem will be to establish feasibility in the sense that at the early-stage of design a partitioning of a volume in certain hyperspace, i.e. a space involving more than three dimensions, will have taken place. The dimensions of the hyperspace require to be estimated to demonstrate that the protective system has the properties to meet the protective requirement as specified. In practice this partitioning process will lead to some form of bounding or envelope approach which will be initially dictated by the engineering of the particular protective system. The properties of the system will be further established at the various stages of its creation; (a) design, (b) sample testing, and (c) field experience.

Goode and Machol[46] state that experience generally shows that a prior system always exists. The basis for this is that rarely do systems require to satisfy an entirely new human need, and that the need is already partially satisfied by some existing system even though it is of a more primitive nature. Furthermore, it may also be argued that hindsight is a great illuminator. This prior concept is philosophically important to the assessor as it is only possible to arrive at a decision within the framework of the known information. However, it must not be ruled out that information can be collected at a later date. Hence it is considered necessary to have a methodology which permits the decision-making to be modified by new information and Bayes Theorem gives this facility and is discussed later in this book.

Historical information can be considered as being very useful to the assessor technologically, however, often it is the 'rare event' which can be the limiting feature of the assessment of the system. The analysis of rare events in protective systems is discussed in later sections of the text. Unfortunately, such events and their analyses take us to the frontiers of technology. Hence the assessor has to select a strategy which is appropriate for demonstrating the adequacy of the particular protective system. The work involved and the subsequent methods developed as described involve a total technology approach which tends to be of a multi-disciplinary nature applied in a systems-engineering fashion.

2.2 Theoretical Prediction of System Performance

Although as mentioned earlier the design process will have included analyses of the performance of the protective system, the assessor requires to arrive at some independent view of this performance. This will generally be based on a modelling process which will include the dynamic characteristics expressed in some ordered and perhaps abstract form to represent the actual engineered system. The evaluation will of necessity follow an iterative process.

There is an important distinction between the design process and the assessment process as the assessor is dealing with the question of whether or not it is adequate for its purpose whereas the designer is creating the actual design solution. Hence the assessor is interested in deriving performance characteristics which can be readily substantiated. In some cases the substantiated characteristics from a safety stand-point may lead to problems arising in the design optimization process which may require a re-examination of the cost functionals by the designer. Where the designer has critically designed to cost-function minima, the question of 'reliability stability' may arise from a safety point of view. These factors will be considered later, as the first and necessary step is to show that the protective system has the capability of functioning in real time to shut down the plant under the stipulated abnormal conditons.

2.3 Performance Characteristics

2.3.1 Response

In the generalized modelling process of assessment the initial step is to ascertain that the protective system will function as intended. This may be interpreted as assessing deterministically that on the basis of no system failure or variability that on the 'best estimates' the system configuration is correct and that it has adequate response.

The characteristic of response is obviously vital and can include a complete range of input conditions to any of the sensing devices for prescribed abnormal conditions as defined by the protective requirement to a range of equivalent output conditions as given by the shutdown devices. The various equipments comprising the system will be subject to mechanical, electrical or thermal inertias which give rise to some finite time to respond to the particular input stimuli. This gives rise to a lag between the input stimulus and the equipment's output function.

This aspect of response has been well researched in the literature and subject to certain assumptions the response characteristic may be conveniently expressed as a 'transfer function', as described by Chestnut and Mayer[47] and Holbrook[48]. The transfer function expresses the ratio of output signal to input signal in the following mathematical form

$$\Phi(s) = \frac{f_o(s)}{f_i(s)} = \frac{a_0 + a_1 s + a_2 s^2 + \ldots}{b_0 + b_1 s + b_2 s^2 + \ldots} \tag{2.2}$$

where: $\Phi(s)$ = the transfer function
$f_i(s)$ = the input stimulus
$f_o(s)$ = the output characteristic
s = the Laplace operator
a_0, a_1 = constants of a polynomial in s
b_0, b_1 = constants of a polynomial in s

The time lag exhibited may be constant or a changing function dependent upon various quantities. It is important in the assessment to carefully appraise the assumptions made in particular analytical techniques used. As already indicated quite often the assessor does not require to know the exact response but to establish some set of boundary conditions.

Each equipment of the protective system will have a response as shown for example in Fig. 2.10, and in operational form this may be expressed as

$$f_o(s) = \Phi(s) \cdot f_i(s) \tag{2.3}$$

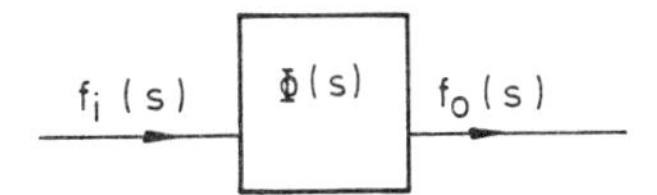

Fig. 2.10 Transfer function block diagram of an equipment

The overall transfer function for the complete system can be derived from the individual equipment transfer functions which for the system of the block diagram Fig. 2.11 would be

$$\Phi(s) = \frac{f_o(s)}{f_i(s)} = \Phi_1(s) \cdot \Phi_2(s) \cdot \Phi_3(s) \tag{2.4}$$

Many devices have a response which for safety assessments can be taken to be exponential in nature and this applies particularly for the envelope approach. The response of an exponential device to an input ramp function is as shown in Fig. 2.12. It will be noted that the output $f_o(t)$ lags behind the input $f_i(t)$ by a time equal to the time constant τ as the output settles to follow the ramp. This form of ramp input is mentioned as it is used in the protective system example given later in the text. Standard tables of transform pairs of functions which are useful in this type of response analysis are available in the literature such as in Green and Bourne[49] and also Chestnut and Mayer[47].

It is not proposed to elaborate further on such techniques for response evaluation since there is a wide literature in existence. However, the assessor would be establishing and seeking substantiation of the actual response techniques from sources other than just the design, such as manufacture and

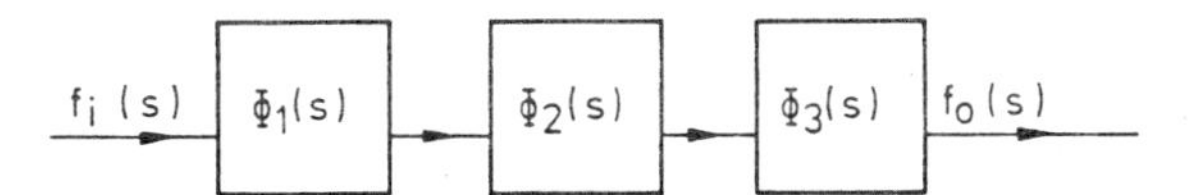

Fig. 2.11 Transfer function block diagram of a system

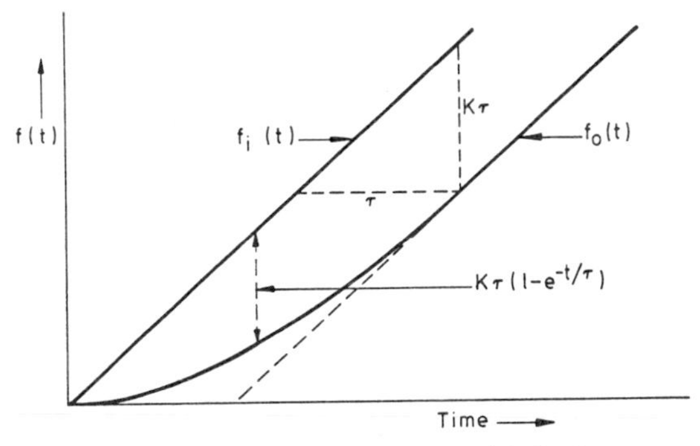

Fig. 2.12 Response of an exponential device to a ramp function

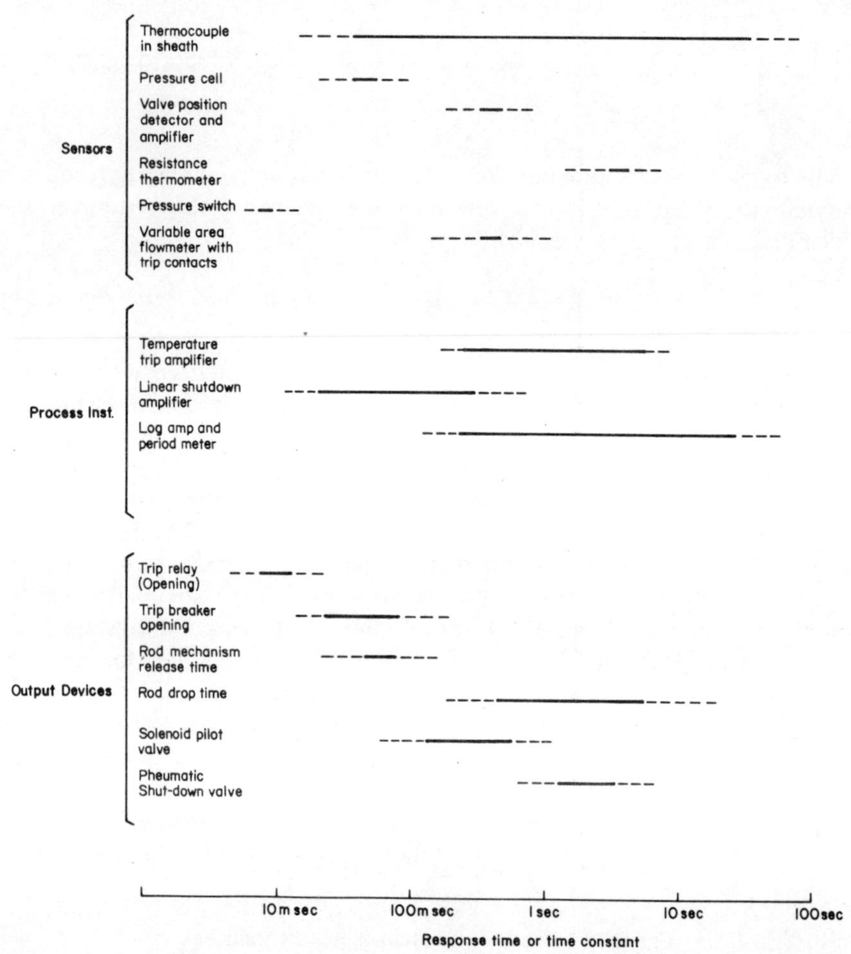

Fig. 2.13 Summary of response times or time constants for items in automatic protective systems

production testing, type testing, field experience and from the intended installation. A review of the general time constants and lags which are associated with the various types of equipment used in automatic protective systems has been summarized in Fig. 2.13.

This type of analysis should give an insight into the sensitivity of the overall response of the system to changes in response times of individual items. Figure 2.14 illustrates the results of the temperature transients in changing the response time of the automatic protective system in tripping a specific type of nuclear reactor under particular fault conditions and this type of information helps in developing an understanding of the sensitivity involved. Furthermore applying a maximum limit to the temperature involved enables a deterministic boundary or envelope approach to be adopted as an initial approach.

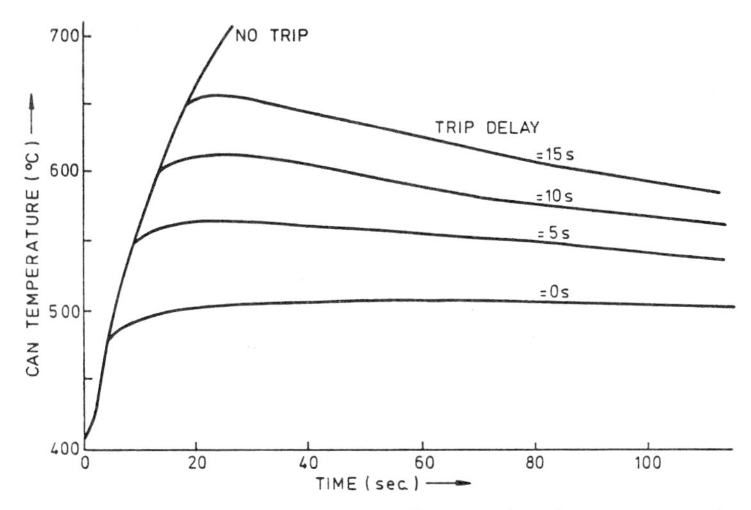

Fig. 2.14 Temperature on a specific type of nuclear reactor

The type of chart shown in Fig. 2.13 can be extended to include ranges of response times for different types and designs of various items of equipment. This can also form a basis for initial assessments in the early stages of design where the 'partitioning' process is being adopted in the conceptual design as previously discussed. In the initial assessment from a deterministic point of view the lessons of the past will be also taken into consideration. As for example ensuring that certain basic principles are being followed such as 'there should be adequate provisions for proof testing and maintaining the system at regular intervals and such provisions should not reduce the degree of protection below that required'. These basic principles are given in Table 3.4 and discussed by Eames[50]. Deterministically this should lead to the adequacy being determined with reference to the protective system having the 'capability' to respond as required.

2.3.2 General transfer characteristic

The output of the generalized protective system will be dependent on the various device characteristics already mentioned. These characteristics will vary in space and time domains and the response will also depend upon the input conditions of the system. Typically the input I_q can be written in the form

$$I_q = f_i(x_j \mid t, z) \tag{2.5}$$

where x_j represents a range of input variables and t and z represent the general time and space conditions on which they are dependent. The equivalent required output conditions, O_q, can be represented by

$$O_q = f_o(y_i \mid t, z) \tag{2.6}$$

where y_i represents a range of relevant output variables in the space and time domains z and t.

Hence the complete required conversion or transfer characteristic between all the relevant output functions and all the relevant input functions may be represented by a single complex matrix function of the type given by

$$\frac{O_q}{I_q} = [\phi_{ij} \mid t, z]_q \tag{2.7}$$

In general terms, the transfer characteristic for the protective systems are made up of elements, $\Phi_{ij}(s)$, which are each in the form of a transfer function. As described by Green and Bourne[49], the general matrix is of the form

$$\frac{L[O_q]}{L[I_q]} = \lfloor \Phi_{ij}(s) \mid t, z \rfloor_q \tag{2.8}$$

which contains $\Phi_{ij}(s)$ which are usually polynomials or ratios of polynomials in the complex Laplacian operators.

It is obvious that in assessing the adequacy of the response achieved the assessor is effectively taking the required transfer characteristic O_q/I_q and comparing it with the achieved transfer characteristic O_h/I_q.

This results in the transfer characteristic matrix being achieved as a result of the assessor's analysis in the form

$$\frac{O_h}{I_q} = [\phi_{ij} \mid t, z]_h \tag{2.9}$$

or in operational form

$$\frac{L[O_h]}{L[I_q]} = [\Phi_{ij}(s) \mid t, z]_h \tag{2.10}$$

where the elements $\Phi_{ij}(s)$ are set up by the protective system configuration and definition under the required conditions.

2.4 Interface Problems

The general engineering analysis of response and performance can be deterministically undertaken and is well described in the general literature. In practice there are problems in defining the system boundaries and various assumptions creep in. Often in post-accident analysis it has been shown that an assumption made in good faith may have led to the particular accident condition. The validation of such assumptions would sometimes require questions being asked as to why such simple questions were not asked (perhaps because of possible embarrassment) or so subtle that the problems were not foreseen. It is therefore worthwhile to consider the implications of Fig. 2.15.

A salient feature of a viable assessment process is ensuring that no 'gap' exists in the examination and appraisal of the system from the required input to the required output. The interfaces are shown in Fig. 2.15 in dotted lines

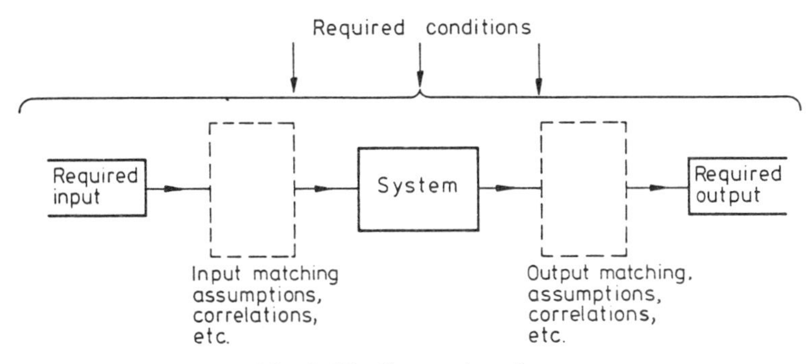

Fig. 2.15 System interfaces

and involve assumptions. As an example a particular plant may have the requirement to prevent conditions leading to an explosion. However, due to practical or other difficulties it may be impossible to measure the explosion conditions themselves and other quantities related to these conditions have to be measured. This leads to a set of assumptions and the creation of an interface problem.

In other words it is necessary to establish the exact relationships and it is necessary for the assessor to be carefully recording the various assumptions arising. A typical relationship diagram shown in simplified form is given in Fig. 2.16 for the heat input and output to fuel in a nuclear reactor for which there would normally be a balance. The heat input will be derived from the nuclear energy of fission and chemical energy of combustion. The heat removal from the fuel will be dependent upon the geometry of the core, the temperature of each constituent part, the coolant mass flow and the various appropriate heat transfer coefficients. It requires a carefully controlled analysis to obtain a full understanding under both static and dynamic conditions. Obviously the plant kinetics will depend upon the particular plant

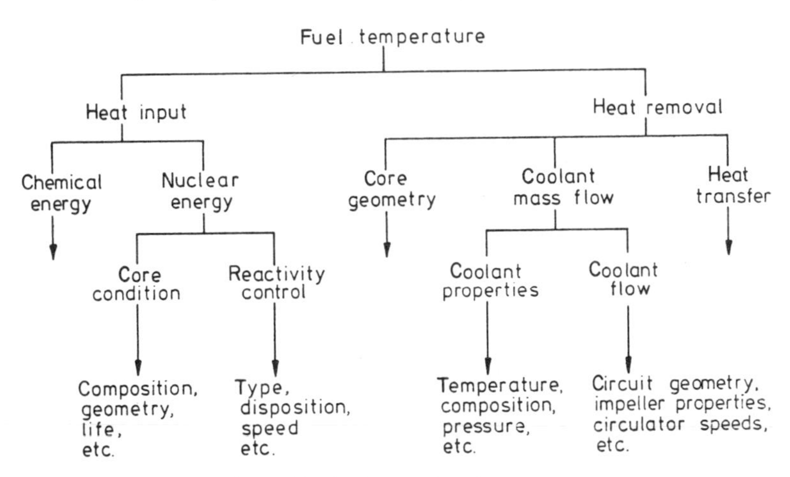

Fig. 2.16 Relationship diagram

processes involved and there is a wide general literature available on plant behaviour and relationships under various conditions.

2.5 Information and Communication

The adequacy of the assessment will be obviously related to the quality of the information on which it is based. Ideally the safety assessor would require complete information on every aspect of his problem. Obviously this is not possible nor practical, therefore the strategy set by the assessor should take into account what basic information is likely to exist and how it may be used in the most appropriate way for the assessment. It must be appreciated that essentially the process of assessment is one of logical and critical thinking. This commences with the concept of the problem, works through the various conditions and considerations and finally arrives at a reasoned solution.

However, the information has to be communicated and there exists a wide literature on this topic. First, the assessor needs to acquire, from the creator of the actual data, information on which to base his assessment. Secondly, the assessor is required to impart to some authorized recipient the final conclusions of his assessment. A process of information transfer is involved which may be subjected to numerous types of distorting influences. A simplified block diagram is shown in Fig. 2.17. The person who is in command of all the facts may not be able to present them in a way as to be understood or convincing. This may lead to some inadvisable solution being adopted. Therefore, the assessment may be of little value unless proper consideration is given to the presentation of the conclusions, as described by Deverell[51].

If this question of information and communication is not properly organized with a technique for the acquisition and organization of information and its circulation then effectively further interface problems (see Section 2.4) are being created. In fact an important keynote to a proper assessment is understanding and is also a general theme of safety. Therefore, concepts should be arranged in an orderly fashion so that any inconsistencies and uncertainties may be recognized and justification of the adequacy of the protective system obtained in a logical manner. This general topic is considered in more detail by Green and Bourne[38] and Ablitt[52], and will be considered as appropriate in other sections of this book.

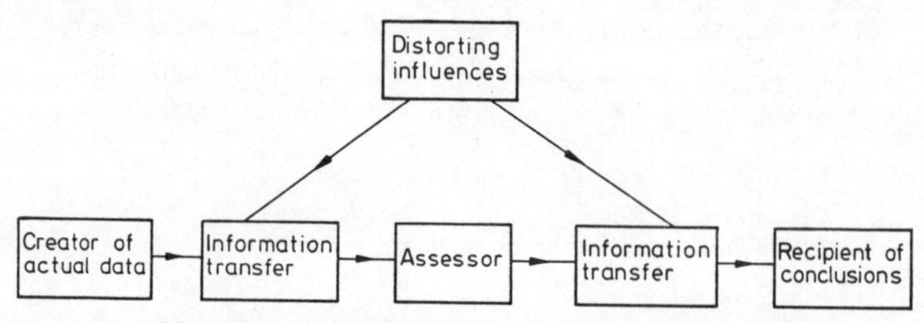

Fig. 2.17 Simplified block diagram of communication

CHAPTER 3
Variability in Performance and the Assessment of Reliability

3.1 Variability of Performance

The basic requirements considered in the previous sections will in practice be subject to variation both in space and time. Similarly the functional capability of the protective system will not be just represented by say an n-dimensional matrix with points fixed at some nominal or singularly representative value of system performance within the appropriate dimensions of the protective system. In general it will be an n-dimensional matrix which can be regarded as an n-dimensional surface which is changing systematically and randomly in time and space. This achieved performance may be represented by a function which will be called the H function.

It is this aspect which represents the general problems in extending the modelling assessment process from that of a type which is deterministic to one which is probabilistic. Hence if the protective requirement which has been discussed previously can be represented by a function called Q then it may be equally regarded as some dimensional surface which is changing systematically in space and time. The assessor has therefore the problem of comparing these two functions Q and H and measuring their correlation in some way which involves probabilistic evaluation.

As an example, if the protective requirement has a Q function which has simply two n-dimensional surfaces, one of which represents a lower limit Q_l and the other an upper limit Q_u, then it can be compared with the assessed performance achievement H as shown simply in Fig. 3.1.

The probability of the H surface remaining within the required limits of the Q surface is a measure of the reliability R of the performance of the protective system and can be generally represented functionally (Green and Bourne[49]) as

$$R = f(Q, H) \tag{3.1}$$

This process is therefore an extension of the previously mentioned 'Deterministic Assessment' which would have included a full engineering appraisal.

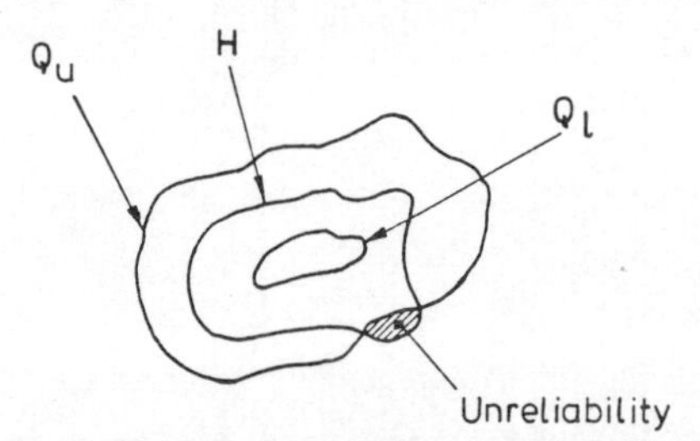

Fig. 3.1 Representation of performance requirement and achievement in two-dimensional form

Such aspects may havc also included environmental testing and other evidence. Its conclusions may even be subjective but it must be stressed that before entering into a fully quantified reliability assessment some reference point requires to be established that under some known set of conditions that there is a capability of functioning.

3.2 Performance Spectrum Method

In general terms an attempt is made by the assessor to make a probabilistic statement on the variability and complete failure of the equipment and of the overall protective system. The 'performance spectrum method' has been studied and used as a basic tool in this book as one approach to this assessment problem and employs the following model

$$\text{Probability of failing} = p + (1 - p)\int_{Q}^{\infty} f(x)\,\mathrm{d}x \tag{3.2}$$

where $f(x)$ = pdf for the variability of performance of the equipment, e.g. in time.

Q = required performance level which should not be exceeded.

p = probability of complete failure of the equipment.

This model is illustrated in simplified form in Fig. 3.2. It is assumed that some design estimate of performance exists about which there is variability extending over the whole spectrum of performance. In a protective system a variate which is of interest is time of response, typically variations are often more likely to occur near the design estimate or in the catastrophic failure region. For example, the time of operation may be increased or complete catastrophic failure take place which permits no operation at all. Superimposed upon this pattern is a requirement shown as Q which as indicated previously may also be distributed in space and time.

It has already been discussed that in safety assessment work often a boundary or envelope approach is taken which leads to a demonstration that some value is unlikely to be exceeded. Therefore, often it is not always necessary to know a specific value for a performance parameter of interest. It is often found that where the requirement is well removed from the design estimate

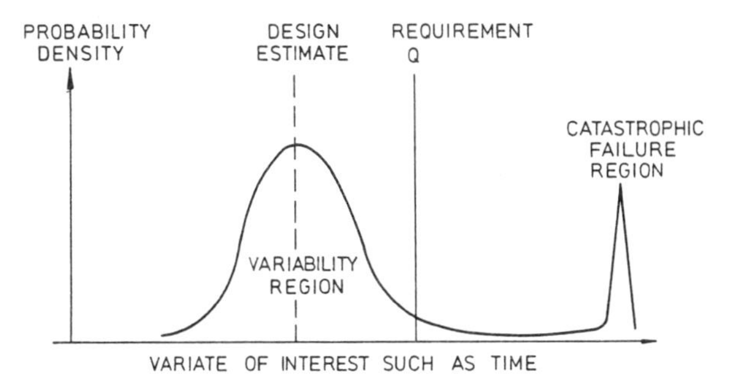

Fig. 3.2 Performance, achievement, and requirement

that the variability region is not very significant and a catastrophic failure model may be quite adequate for the purpose. Conversely, where the requirement is close to or overlapping the design estimate, this may lead to 'marginal' operation and cause variability in this region to be very relevant. In some cases such variability may perhaps be more relevant than catastrophic modes of failure, described by Green.[53]

In practice the selection of the type of model and the areas for critical examination depend on various factors including the particular historical experience involved, the techniques and presentations favoured by the regulatory body when a licence application is required. Normally the risk and consequences which are undertaken assist in directing attention in particular areas. However, surveying the various approaches used in different organizations and countries it is found that the automatic protective systems for

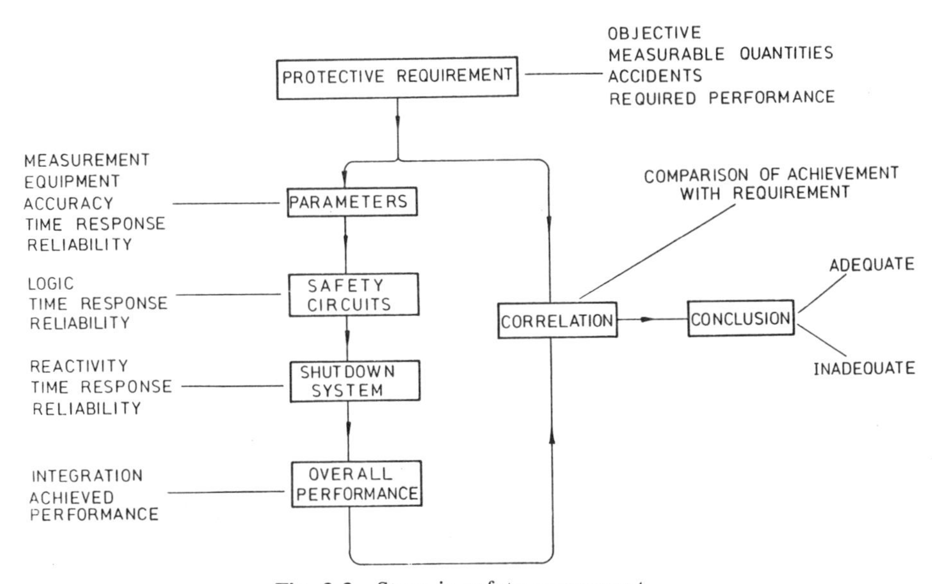

Fig. 3.3 Steps in safety assessment

shutting down nuclear reactors normally receive a fully detailed assessment. The steps for the safety and reliability assessment for an automatic protective system as typically undertaken in the UKAEA are shown in Fig. 3.3 and are described by Green and Bourne[38]

3.3 Variability Region

The variability region mentioned in the section 3.2 is dependent on the exactness of the output quantity of the device in relation to some desired value. These deviations in the device arise from inaccuracies giving changes in characteristics. Such deviations may be systematic or non-systematic in nature. Obviously, the systematic deviations will follow some known or perhaps ascertainable law and may be catered for in a variety of ways, e.g. by the use of particular operating procedures. This then can be arranged to cater for the 'bias' introduced by these systematic deviations. The non-systematic deviations which are essentially 'unbiased' or 'random' in nature cannot be readily predicted or defined individually but the deviations in the mass can be understood and calculated by resorting to probabilistic techniques. This type of deviation has been subjected to much research and is described in the general literature and also by Topping[54]. Figure 3.4 shows the types of deviations for some quantity of interest.

It can be shown that the 'normal' probability function may be taken to represent a typical distribution of random deviations. The normal density function is given by

$$f(x) = \frac{1}{\sigma\sqrt{2\pi}} e^{-(x-\mu)^2/2\sigma^2} \qquad (3.3)$$

where μ = mean
σ^2 = variance
σ = standard deviation

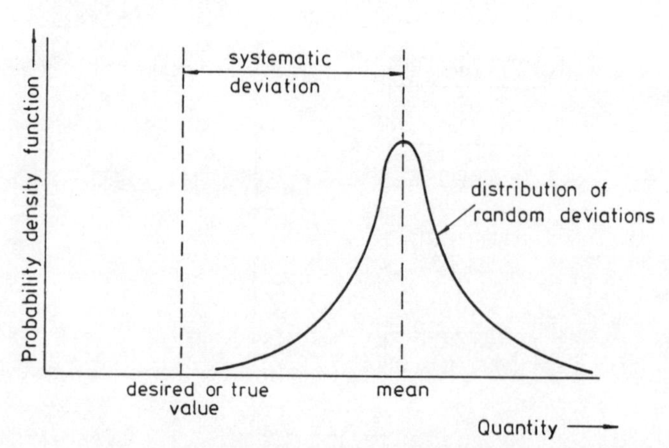

Fig. 3.4 Variability diagram

The probability that the random deviation is less than or equal to some particular value x_0 is given by

$$p(x_0) = \frac{1}{\sigma\sqrt{2\pi}} \int_{-\infty}^{x_0} e^{-(x-\mu)^2/2\sigma^2} \cdot dx \tag{3.4}$$

Values of this integral may be obtained from suitable tables, for example those given by the *Tables of Normal Probability Functions*[55].

The general situation with regard to deviations is that the deviation characteristic of a component is better known than that of the equipment of which it is a part and the deviation characteristic of an equipment is better known than that of the corresponding system. A process of deviation prediction is therefore possible if component deviations are known and methods of combining component deviations to give the equivalent equipment or system deviation are available. Suppose that the random deviations for each component can be expressed by a normal distribution with a mean value of μ_j and a standard deviation of σ_j. Then, in the general case, the mean characteristic of the equipment, μ, will be some function of all the individual values of μ_j, i.e.

$$\begin{aligned} \mu &= f(\mu_j) \\ &= f(\mu_1, \mu_2, \mu_3 \ldots) \end{aligned} \tag{3.5}$$

and it can be shown approximating to the first order of small quantities that the variance of the equipment deviation is then

$$\sigma^2 = \left(\frac{\partial\mu}{\partial\mu_1}\right)^2 \sigma_1^2 + \left(\frac{\partial\mu}{\partial\mu_2}\right)^2 \sigma_2^2 + \left(\frac{\partial\mu}{\partial\mu_3}\right)^2 \sigma_3^2 + \ldots \tag{3.6}$$

Two particular cases are of interest since they represent most situations met in practical equipments. The first case is where the components operate in an additive fashion so that the mean equipment value is simply the sum of the mean component values, i.e.

$$\mu = \mu_1 + \mu_2 + \mu_3 + \ldots \tag{3.7}$$

In this case, the overall variance is simply

$$\sigma^2 = \sigma_1^2 + \sigma_2^2 + \sigma_3^2 + \ldots \tag{3.8}$$

The second case is where the mean equipment value is a function of the product of the mean component values. So that

$$\mu = \mu_1 \times \mu_2 \times \mu_3 \times \ldots \tag{3.9}$$

It can be shown by Keller[56] for this case that the overall variance is given by

$$\left(1 + \frac{\sigma^2}{\mu^2}\right) = \left(1 + \frac{\sigma_1^2}{\mu_1^2}\right)\left(1 + \frac{\sigma_2^2}{\mu_2^2}\right)\left(1 + \frac{\sigma_3^3}{\mu_3^3}\right) \times \ldots \tag{3.10}$$

This for $(\sigma_i^2/\mu_i^2) \ll 1$ reduces to

$$\frac{\sigma^2}{\mu^2} = \frac{\sigma_1^2}{\mu_1^2} + \frac{\sigma_2^2}{\mu_2^2} + \frac{\sigma_3^2}{\mu_3^2} + \ldots \tag{3.11}$$

in accordance with equation 3.6.

It can also be shown that similar techniques may be applied to component deviation distributions which are not of the 'normal' form.

In many cases it is obvious that the variability region cannot be just a normal distribution which implies the quantity as varying from $-\infty$ to some particular value x_o as shown in equation 3.4. An example of this would be the time lag of a protective device which at its minimum would be zero. However, in certain cases the normal distribution may be a sufficient approximation or it may be necessary to use some skewed type of distribution such as that of a lognormal distribution. The various distributions which can be applied are well described in the statistical literature, such as that described by Kendal and Stuart[57].

3.4 Reliability Model

In order to complete a reliability evaluation the model will be based on reliability parameters such as the mean rate at which the protective system may fail. This may in practice be more complex and require a statement being made as a function of time or environment such as

$$f(t) = \theta(t) \exp\left[-\int \theta(t)\, dt\right] \tag{3.12}$$

where $f(t)$ = the probability density function for the events in the time domain,

$\theta(t)$ = the function describing the rate of occurrence of events with respect to time sometimes known as the event rate function.

Equation 3.12 yields the well-known exponential form when $\theta(t)$ is constant of value θ and gives

$$f(t) = \theta \exp(-\theta t) \tag{3.13}$$

The probability of failure up to any given time t is given by the cumulative probability function, $p_f(t)$ that is

$$p_f(t) = \int_0^t f(t)\, dt \tag{3.14}$$

When the event-rate function cannot be considered as constant then the mean time to the first event or the first failure is very pertinent. This mean value μ is given by

$$\mu = \int_0^\infty t f(t)\, dt \tag{3.15}$$

or

$$\mu = \int_0^\infty [1 - p_f(t)]\,dt \tag{3.16}$$

where $p_f(t)$ is the cumulative probability function.

In protective systems such as those for automatically operating to shut down a nuclear reactor they are only operational when some abnormal event or demand arises. The testing of such a system may occur periodically at time intervals τ to prove that it has the capability of functionally performing. If a failure is found then a repair process is put into operation to restore the system to full working order. However, if the system is observed to have failed at any other time then a process of breakdown maintenance will take place. This failure, testing and repair process is shown in a simplified form in Fig. 3.5.

Combinations of failures may take place, for example, a failure may occur and stay in this particular state for a period of time during which a second failure occurs. Hence it is necessary to consider restoration or repair time which will include the repair, replacement or restoring the failed device. The characteristic of repair time can be represented in a similar way to that described by replacing $\theta(t)$ with a repair-rate function.

The complement of availability $A(t)$ is the fractional dead time $D(t)$ which is very important in assessing automatic protective systems. The changes in the system due to random failures are broadly of two types, those which are 'revealed' and whose presence are known immediately and those which are 'unrevealed' and whose presence will not be known until some routine test procedure takes place.

Generally in assessing the automatic protective systems for safety it is the 'unrevealed failure' which puts the system in a dangerous state and it is the calculation of the mean fractional dead time $D(t)$ which is of interest and is given by

$$D = \frac{\mu_r}{\mu_r + \mu_f} \tag{3.17}$$

where μ_f and μ_r are the mean values of the relevant failure and repair distributions. This is subject to certain assumptions as given by Bourne[58] where it is also shown that the mean fractional dead time is given by

$$D = \frac{1}{\tau_c}\int_0^{\tau_c} p_f(t)\,dt \tag{3.18}$$

where τ_c is the time between the checks or testing of the system.

In the reliability analysis of a system various parameters of the type described may enter into the mathematical model and the calculations. Logical flow diagrams may be prepared which enable the probability of failure or success of each item in the system to be represented and the results combined in some logical fashion.

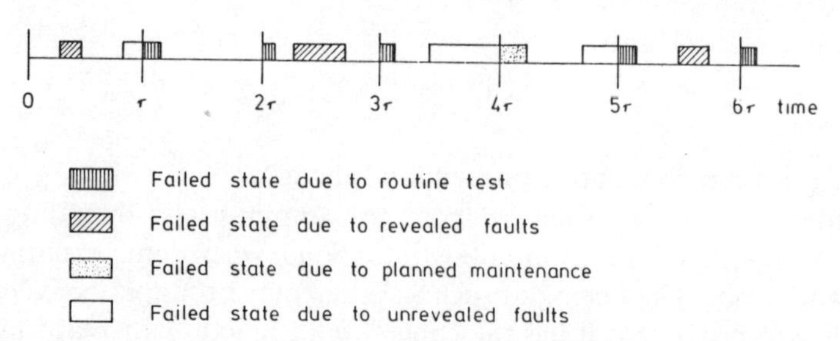

Fig. 3.5 Typical patterns of two-state behaviour of a system

This question of modelling in general and the problems involved in dealing with complex systems has impinged to some extent upon the investigations undertaken by the author. The indications are that the methods tend to revolve around the historical evolution, the data available and the particular system in question. Usually some form of synthesis is developed in the case of high-reliability protective systems as the data are to be found at the lower hierarchical levels of the system. This is combined with various analytical techniques and the combinational process involves problems which revolve round the independence and dependence of the analysed items of the system. Figure 3.6 gives in a generalized form these ideas and later sections describe the application of these ideas. It will be seen that from Fig. 3.6 basically the analysis may be considered as proceeding from the direction of the apex towards the base and the synthesis from the base towards the apex of the diagram.

However, it should be noted that there is a whole subject of fault-tree modelling and other methods which arose from the aerospace industry in the USA during the early 1960s. These are reviewed by Gangloff[59] with Barlow and Fussell[60] and Dhillon and Singh[99] giving references for further reading on

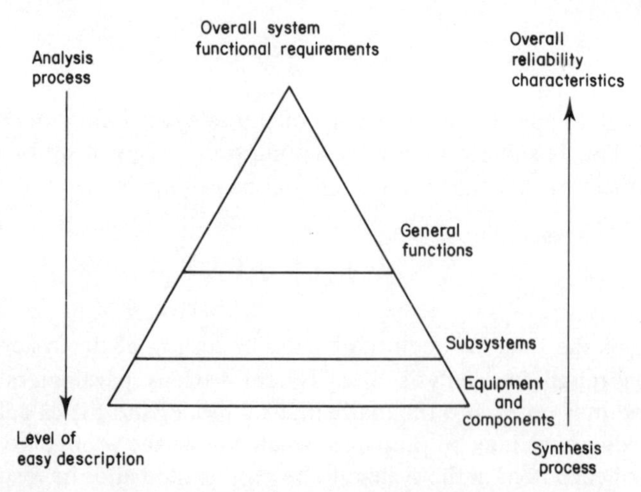

Fig. 3.6 Illustration of basic system analysis and synthesis

both the theoretical and applied aspects of fault-tree analysis. Since fault-tree techniques have been used extensively for the evaluation of protective systems in the Reactor Safety Study[21], and are mentioned later in the text a brief description of them is given in Appendix 2 with illustrations of their applications in Appendices 7 and 8.

Various methods exist for 'tearing-up' complex systems using topological models and previous work has considered to some extent the application of this process known as 'diakoptics'. For example, Kontoleon *et al.*[61,62] give an account of some of these diakoptic processes which have been researched and applied over a number of years. Decomposition techniques of analysis have been appropriately developed NATO Advanced Study Institute[63] for solving large-scale problems by finding means of interconnecting partial solutions of large systems. The overall strategy is that the complete system may be 'torn up' at any section and if any changes are made in one section of the system then it may be easily programmed in the process of solution. This type of work being of a topological nature as a generic problem dealing with, for example, nodes, meshes, branches and planes, is discussed by Balasubrannanian *et al.*[64]. Figure 3.7 shows a simplified diagram of tensor analysis developed by Kron for large electrical and electronic systems of the Lagrangian type. Here Kron's approach was to tear up the model rather than the equations in order not to lose the interconnecting constraints between the torn-up sequences (sections of the system). Examples of these modelling techniques for electrical machines and networks are given by Kron[65] and Lynn[66].

In a number of different methods considered it has been found that basic methods are often applicable to the smaller type of system. 'Tearing-up' and

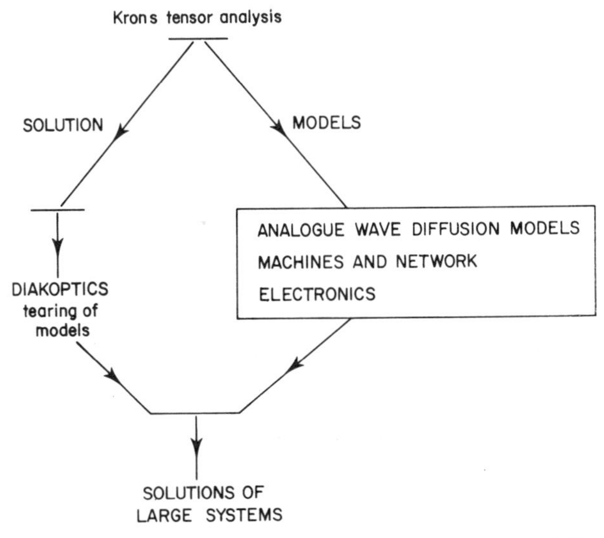

Fig. 3.7 Large electrical and electronic systems of the Lagrangian type

synthesis methods have been applied successfully provided that the non-linearities which appear at the system level can be readily handled and where computer techniques are used that sufficient capacity exists.

3.5 Hazard Probability

In applying the reliability assessment so far described the ultimate objective in the case of a plant shut-down protective system is to assess the probability of a hazard arising due to the failure of the protective system. This means that an abnormal condition has occurred on the plant at the same time as the protective system is in the completely failed or 'dead' state. This is illustrated in Fig. 3.8 where the protective system is tested periodically at time, T, and the shaded areas show that due to failure the protective system is in the 'dead' state. The demands for a protective system are shown by X and Y. It is obvious then when demand X arises the system is not in a 'dead' state and there is no hazard but there is a hazard when demand Y arises. Therefore the probability of a hazard occurring depends upon the fractional 'dead' time of the protective system and the probability of the demand arising subject to certain assumptions (Bourne[58]), is

$$P_h(T) = P_d(T) \times D \tag{3.19}$$

where $P_h(T)$ = probability of a hazard arising
$P_d(T)$ = probability of the demand arising
D = mean fractional 'dead' time
= mean time failed/total time of interest
T = time of interest

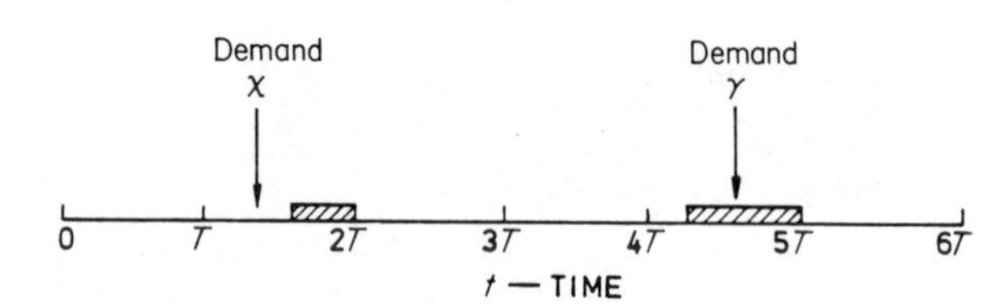

Period between tests is T

Failed state of system is indicated by [hatched bar]

Fig. 3.8 Illustration of hazard occurrence related to dead time for system

3.6 Application of Reliability to Example Protective System

The protective systems described in the previous sections can be typically represented in simplified ideas as shown in Fig. 3.9. The general function of this type of system has already been described and it is assumed that a capability functional assessment has been carried out.

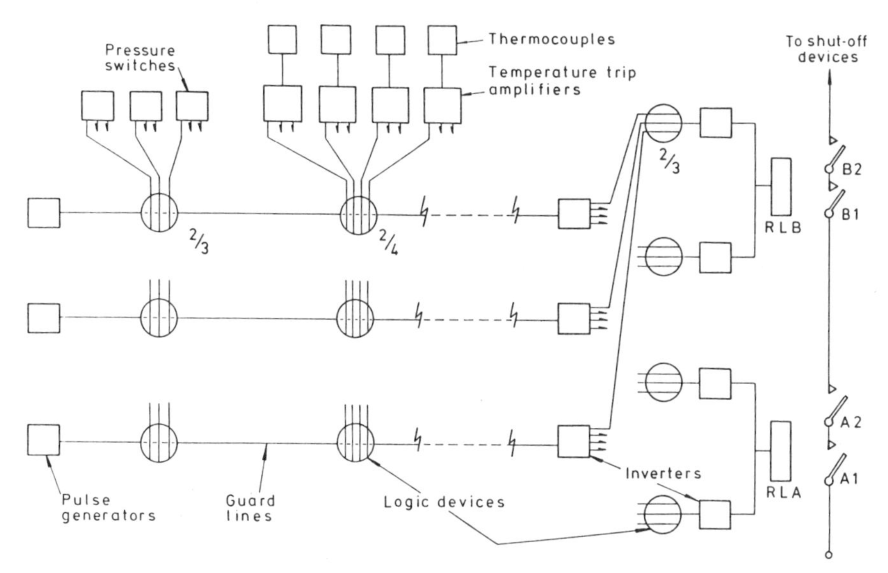

Fig. 3.9 Example protective system

From the general literature, Green and Bourne[49] or by the use of appropriate data from a data bank, the individual component failure rates can be derived. Assuming it can be considered that all these failure rates are independent of one another then the overall mean failure rate for each equipment can be derived.

If the equipment has n components working in their useful life phase and the failure of one component causes the particular equipment to fail, then the overall mean failure rate for the equipment is given by

$$\theta = \sum_{i=1}^{i=n} \theta_i \tag{3.20}$$

where θ_i is the mean failure rate of the ith component. The usual life phase for each component assumes a constant failure rate θ which gives a pdf as defined by equation 3.13.

In studying the failures it is necessary to consider the various modes of failure and their effects from a safety point of view. The effect of each component part fault requires to be analysed in sufficient detail to enable it to be fully categorized. In Appendix 3, Sections A.3.1 and A.3.2, two samples of failure mode and effects analysis are given for a differential pressure device which is of the mechanical type and for an electronic gamma radiation monitor. This type of approach is assumed and further applied for investigations undertaken later in this book. The simple terminology of fail-safe or fail-dangerous requires to be more fully defined and a four-character code, as given in Appendix 3 Section A.3.3, has been found to be of great assistance in this type of analysis which is described by Green[67].

Table 3.1 Typical characteristics

Item	θ_c (f/y)	τ_c (y)	θ_r (f/y)	τ_r (y)	Φ_s transfer function	μ_τ mean (s)	σ_τ standard deviation (s)
Pressure	0.025	0.25	0.075	10^{-4}	$\frac{s\tau}{1+s\tau}$	39.7	8.0
Temperature	0.05	0.25	0.37	10^{-4}	$\frac{s\tau}{1+s\tau}$	24.3	4.5
Logic device	2×10^{-4}	2.0	—	—	—	—	—
Inverter	10^{-3}	2.0	—	—	—	—	—
Contactor	10^{-3}	2.0	—	—	$e^{-s\tau}$	0.2	0.04

Where θ_c = mean rate of occurrence of unrevealed faults in units of faults per year
τ_c = time in years, between the routine test procedures which correct unrevealed faults.
θ_r = mean rate of occurrence of revealed faults in units of faults per year.
τ_r = mean repair time, in years, for the repair of unrevealed faults.
Φ_s = transfer function representing the operational delay time of the protective system.
μ_τ = mean value, in seconds, of the appropriate time constant.
σ_τ = standard deviation, in seconds, of the appropriate time constant.

Standard procedures and methods have been established and are described by Eames[68] and Green[69] for deriving the failure rates and also taking into account the distribution of response time. It may be noted that the variations or deviations for the response time tend to follow a lognormal distribution. There are indications, however that in certain types of shutdown equipment for nuclear reactor protective systems that the standard deviation of such distributions do not normally exceed about 10 per cent of the mean value. Hence, quite often the assessor can take a normal distribution as a good approximation over certain ranges of working. In the example protective system means of manual shutdown by an emergency trip facility have not been included. However, it will be discussed later in the text under human factors as this requires an evaluation of the response of the human operator.

Testing and maintenance has already been discussed and it is assumed that the assessor would have considered and appraised the times specified for periodic testing and the time to repair. This would lead to a complete set of characteristics being derived for the system as shown in an abbreviated form in Table 3.1.

The foregoing analysis would permit the preparation of a logical flow diagram as shown in Fig. 3.10 giving the various signal paths and logical operations which are indicated by circles. The characteristics defined in Table 3.1 explain the boxes shown in the logical flow diagram Fig. 3.10. The overall system performance may then be synthesized and calculations undertaken as described by Green and Bourne[49], using the techniques of logical analysis and probability theory. With a specific form of input into the protective system via

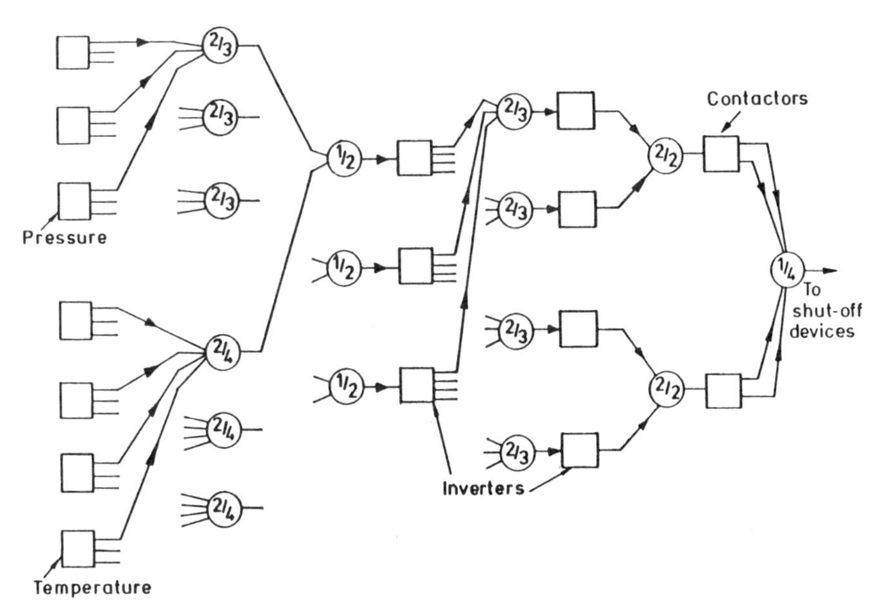

Fig. 3.10 Logic diagram for example protective system

the protective channel sensors, then the probability of the failure of the system can be plotted against time after the demand arising for protective action, as shown in Fig. 3.11.

From the form of the plot in Fig. 3.11 it may be noted that the particular

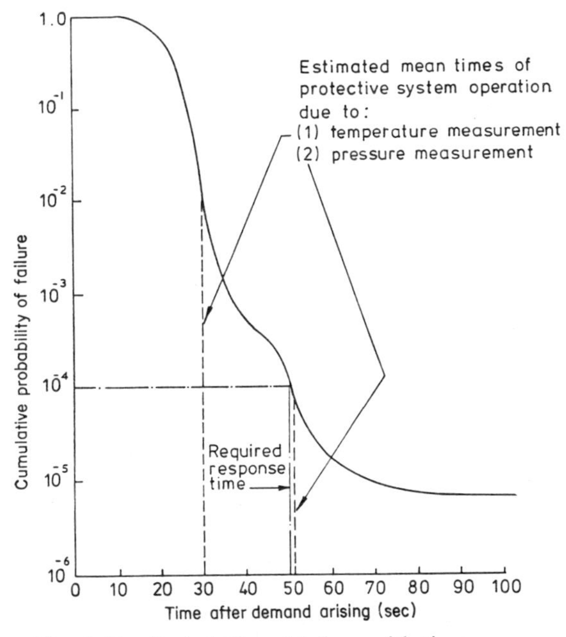

Fig. 3.11 Probability of failure with time

shape arises from there being two independent subsystems, i.e. pressure and temperature, which are both sensing the same demand for protective action. However, there is a time lag in each subsystem which has variability associated with the time constants involved. These variations in the analysis have been approximated by a normal distribution; in addition there is a further time lag which can be attributed to any catastrophic failure during the lifetime of the system. A particular life-time has been chosen so that the probabilities of the two independent subsystems can then be logically combined according to equation 3.2. A family of plots can also be produced for different lifetimes if required.

It will be noted from Fig. 3.11 that on average the temperature measurement would operate the system before the pressure measurement. However, the complete spectrum of performance is taken into account by the curve describing the probability of the complete system failing to operate taking into consideration both measurements. If the required performance time is known, for example 50 sec, then this would correspond with a 10^{-4} probability of failure. As mentioned previously, a family of curves can be determined for the particular application involved.

3.7 The Human Factor

In the assessment process so far investigated and discussed it is evident that the assessor is dealing with a 'man-machine' system. The human element will enter into the basic specification of the protective system requirement and its specification, through the design process, manufacturing, commissioning, operating, testing and maintenance. Since the purpose of the assessment is to evaluate the performance of the system with the objective of assessing its adequacy then since human factors exist they require to be given consideration. Often the assessor will attempt to evolve a strategy of placing the minimum reliance on having to evaluate the 'man' and place more reliance on evaluating the 'machine'. Hence the automatic functions of the system may be given initial priority in the evaluation and it may be that the effects of human factors are investigated to show that they are at such an hierarchical level in the overall system that the consequences of failure on behalf of these factors are not significant.

This strategy of placing reliance on the 'machine' is often reflected in safety criteria issued by regulatory bodies in different countries and often involves a time of response. As an example, H. M. Nuclear Installations Inspectorate of the Health and Safety Executive[70] under the heading of 'principles for the protection system', quotes

> The protection system should be automatically initiated. No operator action should be necessary in a timescale of approximately 30 min. The design should however be such that an operator can initiate protection system functions and can perform necessary actions to deal with circumstances which might prejudice the maintenance of the plant in a safe state but cannot negate correct protection system action at any time.

The whole question of operator response to various stimuli is complex as is the effectiveness of the operator in a testing role. In some situations the assessor finds it necessary to assess the response of the operator to an alarm or in taking a particular action even to operating a manual shutdown facility. Hence at the design stage of the protective system there may be a need for the assessor to 'model' the response of the operator. This topic has been researched and is still the subject of development in different organizations.

These developments and considerations can be seen from the literature for example in the UK chemical industry. Kletz[71] and Lawley[72] indicate some approaches by operability studies and hazard analysis. It has entered into the arguments for particular types of landing systems for aircraft, as described by Adkins *et al.*[73], and in most applications where man can be a controlling element of the system under consideration.

3.8 Operator Response

In the protective system for automatically shutting down nuclear reactors in the UK a manual facility is provided in that the operator may impart his own logic by taking action to shutdown. The operator, in effect, becomes a component of the 'loop' and it is useful to have an ergonomic appraisal of the quantification of his expected performance. In the same way as protective equipment is appraised for variability and failure so the assessor may require to involve himself in similar estimates for the operator. It would be true to say that the general literature on 'human engineering' shows that the assessor is facing a problem which is complex and that physiological and psychological aspects are important.

However, it may be concluded from assessing operator performance that although fault or accident kinetics are calculated knowing that it is an accident being studied the operator does not know that an accident is about to begin. Therefore, if he has been performing a series of operational procedures and the information being sensed by him is normal then it can be argued that his reactions at this point could be normal. It is only some time later when the operator knows that abnormal and perhaps full-scale accident conditions are in being that his reactions would be expected to change from the norm. Hence a first approach to the assessment of the operator response is to approach the initial transient accident phase using normal procedural modelling and data. Nevertheless it must be appreciated that this may involve many assumptions as can be seen in the post investigations of accidents for example in the collision of ships at sea, outlined by Mackworth[74].

The ergonomics of the situation need to be studied which involve the structure of the human body so that the assessor should ensure that the equipment provided caters for varying factors in this structure. In simple terms it should be ensured that the equipment fits the man who may vary in his bodily make-up. Such aspects as the bodily mechanisms involving the physiology and the capability of handling the available information may involve psychology. Reviewing the general literature available indicates the

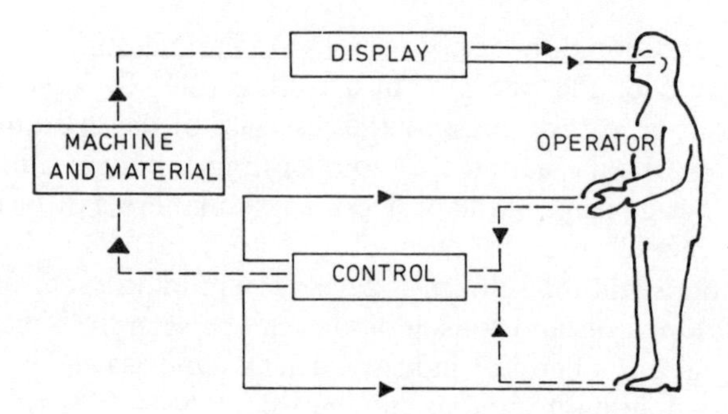

Fig. 3.12 The flow of information between operator and machine

need for having the assessment of the ergonomics carried out in the same way as previously described for the engineering of hardware. It being a first priority that the whole system including the man–machine relationship is appraised. Figure 3.12 illustrates in principle the general flow of information between the operator and machine.

These various facets of ergonomics are being and will continue to be researched and developed. Embrey[75] gives an introduction to some of the considerations and further detailed discussion, particularly for large-scale man–machine systems, has been outlined by Singleton *et al.*[76], McCormick[77], De Greene[78] and Meister[79]. Human detection and the diagnosis of system failures are covered in Rasmussen and Rouse[158].

The basic example of a protective system which is being considered is typically that used for shutting down a nuclear reactor if normal conditions do not prevail. From the preceding discussion it is essential to view the operator in the first instance as a 'component' of the overall system. Generally, this requires having an overall appreciation of the reactor control and instrumentation which is typically shown in Fig. 3.13. It will be seen that information

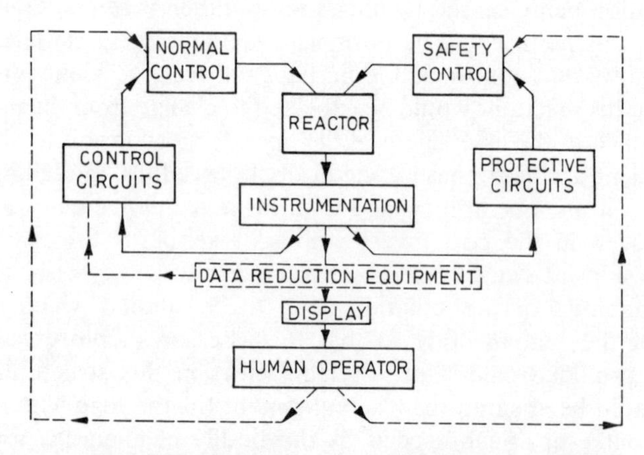

Fig. 3.13 Reactor control and instrumentation

concerning the state of the reactor is used basically in three ways;

(a) in the automatic control circuits it is used to maintain automatically the reactor in a predetermined state
(b) it is used via automatic protective circuits to protect the reactor from a safety point of view
(c) additionally the information is displayed to the operator for purposes of manual control of the reactor both from a normal operation and for the purposes of abnormal control which may involve the operator taking action to manually shut down the reactor.

3.8.1 Control-system approach

Ideally it would be most useful to be able to establish a transfer function for the operator placed in the control system. The work reviewed shows difficulty in this approach and is dealt with in the general literature as outlined for example in early work by Sheridan[80] and in Morgan *et al.*[81], McRuer *et al.*[189], Birmingham and Taylor[190]. Here the operator is considered as a 'black box' in the particular control loop in a similar way to that previously described for protective equipment. Hence the output quantity would be related to its input quantity by a transfer function is a similar way to any element in the control loop. If the ratio is output to input for a human operator is given as F_H in terms of the Laplace transform variable then

$$F_H = \frac{K \cdot (T_2S + 1)\exp(-T_dS)}{(T_1S + 1)(T_NS + 1)} \tag{3.21}$$

The value of the time constants T depend upon the differences between operators and also upon the dynamics of the loop being controlled but generally they would appear to lie in the following ranges

T_d about 0.2 sec with very small range
T_1 between 0.2 and 30 sec
T_2 generally between 0.3 and 2 sec—sometimes zero
T_N between 0.1 and 0.7 sec—sometimes zero

Operators have an optimizing ability such that the loop gain is adjusted until the overall control response is that of a marginally stable system. The delay T_d is considered to be the reaction time of the central nervous system.

In the very early work carried out by Green *et al.*[82] this type of control-system approach appeared to be most relevant to describing the performance of an operator engaged in a continuous task, e.g. some tracking operation. However, it was not clear that it could be readily extended to situations where there are long periods of inactivity followed by a sudden burst of activity. Whilst it appeared that the equations could be used the choice of coefficients would appear to be obscure and there may be non-linear effects which could predominate in the particular role of the operator taking emergency action to shut-down a nuclear reactor. However, more sophisticated control system methods have been applied which are referenced for example by Embrey[87], Regulinski[91], Dhillon and Singh[99], and Rasmussen and Rouse[158].

The appropriate techniques of the evaluation of the operator when considered as a component of a protective system would appear to be still very much in the development stage. It is difficult to know whether general data on the response of subjects to stimuli may be directly applicable to the reactor situation. There is a wide range of data available in the response of subjects to stimuli which may not be directly applicable to the prediction of reactor-operator responses to reactor-fault indications. First, the subjects may not have been tested in the environment of a nuclear-reactor control room. Secondly, the rate at which artificial stimuli have been introduced has often been many orders higher than the expected rate of reactor-fault indications. Hence the basic strategy has been developed along the lines of breaking down the overall operator function into elements and then developing methods of synthesizing the overall response.

3.8.2 Methods-time study approach

It was appreciated that information may be derived from the field of production methods as this had been already investigated for determining the terms necessary for an operator to move switches, press buttons etc. Data for this type of motion for repetitive tasks is already available in the literature for example as described by Maynard *et al.*[83]. Hence some simple experimental work was carried out using this approach, as given by Green *et al.*[82] and Ablitt[84].

Obviously in this experimental work it was appreciated that a reactor operator would not be working under the production conditions. His role first being to realize that action is required, then to decide upon the action and finally to perform the action in the environment of a reactor control room. It was felt that some information could be derived on the time of the final action although for a more detailed investigation the fatigue factor, motivation, etc, should be considered.

A typical test carried out was for the operator to be sitting in a chair and the time required to stand up, reach forward and press a button after lifting a covering flap was estimated using the method-time study approach as follows

Time to stand up	43.4 TMU
Reach forward 30 in	17.5 TMU
Lift flap 1 in	0.9 TMU
Apply pressure	16.2 TMU
Total	79.7 TMU
which equals	2.85 sec

It may be noted that a Time Measurement Unit (TMU) is 0.036 sec. This was found to compare with a measured mean time for the complete test in an actual control room of 3 sec. Although in practice there will be variations for example the individual motions may not occur separately, in sequence, it indicates some estimate of the minimum time to be expected for this elementary operation.

3.8.3 The vigilance approach

The more direct approach is to measure the time elapse between the appearance of the particular phenomenon of interest and the response of a human subject. This approach has been attempted by a number of workers in the field. In the assessment of the operator response it is difficult to know whether the general data on the response to stimuli may be directly applicable to the particular nuclear-reactor situation. Apart from the various data which may not have been collected in a reactor control room, the stimuli may be of a duration and at a rate not expected to apply to that of reactor-fault alarm indications. On the other hand it could be anticipated that certain data may apply.

Tests have been therefore undertaken to obtain appropriate data on operators under laboratory conditions and in the actual reactor environment. These have been reported on by Green[58]. An automatic equipment known as HORATIO (Human Operator Response Analyser and Timer for Infrequent Occurrences) was developed for the experimental purposes and is described by Ablitt[84] in the appendix. This gave the facilities for the timing of operator

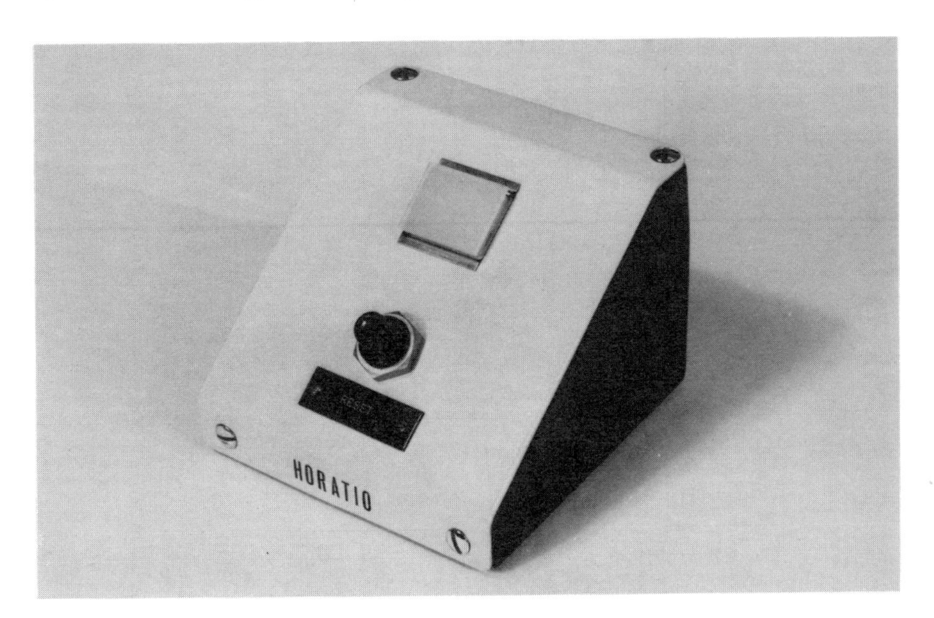

Fig. 3.14 Display and control unit

responses to stimuli which could be programmed at mean rates varying between ten events every 1 hr to one event every 10 000 hr.

Signals were produced by the HORATIO equipment which were reasonably representative both in form and in time distribution of certain types of reactor-fault indications. Measurement of the time interval between the onset of the signal and the operator's acceptance or cancellation of the signal was also incorporated in the equipment. Although the actual instruments, alarms or controls in the control room could be used for display it was found advantageous for experimental purposes to use an independent alarm and push-button unit, as shown in Fig. 3.14. Other facilities in HORATIO gave the opportunity for presenting visual or audible alarms and for the programming of thc signals to be either systematic or random. If there were no operator response after a predetermined duration of signal then this was recorded and the equipment automatically prepared for the next programmed signal.

The laboratory results of tests for simple situations involving visual indications and audible alarms and the pressing of a push-button are shown in Fig. 3.15. The results given in Fig. 3.16 were obtained under actual reactor-control-room conditions. Figure 3.17 shows a typical control-room situation involved in the experimental work; on the control desk can be seen the

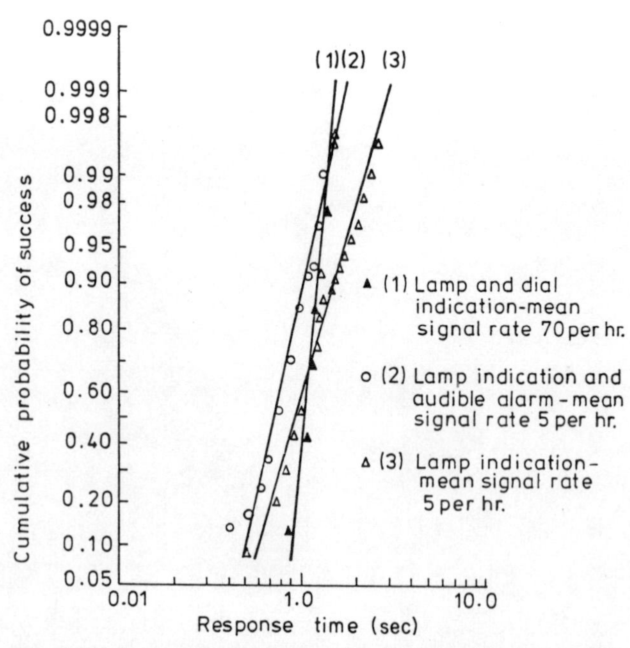

Fig. 3.15 Response results for visual indications and audible alarms in the laboratory

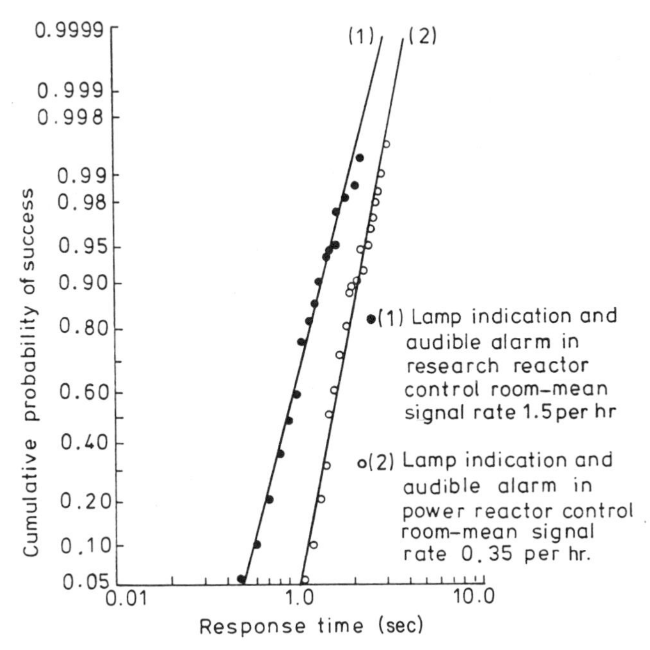

Fig. 3.16 Response results for visual indications and audible alarms in reactor control rooms

HORATIO alarm and push-button unit. In Fig. 3.16 test one is for a small research reactor with day working only and test two is for a research power reactor with full three-shift working.

The results obtained in these tests are important as they are providing basic information for considering the operator as a component in the loop. The distribution of response times has been found to be approximately lognormal.

It is of interest to note that similar tests were carried out by the use of an equipment called HAMLET, described by Rasmussen and Timmerman[86], which used the instrumentation itself and added no extra equipment to the control console. The test equipment was installed in the control room of an operating reactor and a collection of response time information was undertaken in a similar way to that in the HORATIO tests. The distribution of operator response time was found to be lognormal in a similar way to that on HORATIO but with slightly different mean response times, which could be due to differences in the distances for the reset buttons or due to the special significance of the HORATIO signal, Fig. 3.18. This information now permits the quanitification of variability to be made in a similar way to that described in Section 3.3 for the 'variability region' as discussed for protective equipment.

Fig. 3.17 Reactor control-room test

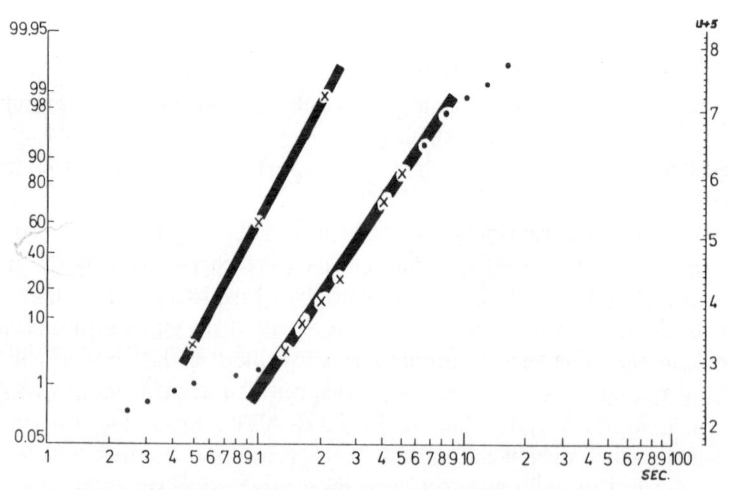

Fig. 3.18 The log normal distribution of operator response times for HAMLET and HORATIO. •, Response times HAMLET from bell-to-bell reset 15/4–71–23/5–71; ×, response times HORATIO

3.9 Human Operator Reliability

It is obvious from the foregoing discussion that it is not easy in any assessment to isolate the human being from the machine as there is a man–machine interaction. The human factor involved is really an inter-disciplinary activity and depends upon the skills and training involved, the particular operating procedures, the design of the equipment and the particular environment. These factors and others tend to lead to a greater variability than in equipment giving in general a wide range of variables and the performance shaping factors can have greater sensitivity. There exists a wide literature on these factors, see Rasmussen and Rouse[158] and Green[159] leading to the assessment of the reliability of a human being carrying out various tasks and Embrey[87] gives an overview of human reliability in complex systems. This also illustrates the extent of the literature available and outlines many of the problems involved.

However, developments have been undertaken in trying to 'model' the reliability of the operator in specific tasks and have been researched particularly in the nuclear field with reference to highly reliable protective systems. Quite often in carrying out a safety assessment it helps to highlight limitations which may be required to be placed upon the operator in arriving at an adequate system. An important point which occurs in the assessment of a protective system is that the assessor has before him a specification which defines the equipment whereas in the case of the operator it is not normally possible to know who this will be. When assessing a system for a life of 20 years or more there may have been a few generations of operators and other personnel who will have moved on. Whereas one can visit a reactor after 20 or more years' operation and still see the same equipment as originally supplied. On the other hand it will have been subject to testing and maintenance and even modification which in some ways can be an easier problem to deal with.

It is beyond the scope of the present study to describe the considerable volume of literature and the various approaches which have been tried by many workers in the field of human factors. However, it was considered to be constructive to concentrate on some quantitative assessment techniques in connection with the human operator and the protective systems under discussion. In the field of reliability analysis the methods of synthesizing equipment and system reliability have formed a useful foundation on which to build the techniques for predicting operator reliability. Basically this has tended to follow the basic steps as follows

1. Define the system which is to be analysed with the basic tasks broken down into elements such as hearing an alarm, reading a dial and presssing a trip button. All these human operations are carefully identified and listed with reference to the particular system functions.
2. Investigate the various modes of failures and their effects and categorize them in some manner.

3. Assign failure probabilities to the various elementary tasks and their particular modes. This requires human error data to be derived by some technique or taken from a data bank specific for the purpose and may need an extrapolation process.
4. Synthesize the overall probability of failure or success for the operator in undertaking the tasks.

An early approach using an *Index of Electronic Equipment Operability*[88] has been investigated by the American Institute for Research, so that each operator action, e.g. an instrument check, is broken down into three stages; the input, the meditating, and the output stages. The input may be defined as written instruction and a time and reliability is assigned to reading the instruction prior to carrying out an 'output' operation. The 'thinking' time between the input and output is then defined as the meditating state. The actual action of reading a dial or moving a switch is then defined as the output. In this approach a data store in the *Index of Electronic Equipment Operability*[88] quotes a 'Base Time' for each operation. Each parameter of the dial or instrument being read, such as scale size, brightness etc, requires to be taken into account which affects the operation and/or reliability. The additional parameter times are added to the 'Base Time' to give the overall time for the instrument reading. The individual reliabilities associated with the individual parameters are multiplied together to obtain the overall reliability. This method is illustrated for the reading of a dial as shown in Table 3.2. The first column in the Table gives the parameters of linear-scale instruments as listed in the 'Data Store', column two gives the sub-classification of these parameters, column three lists the time to be added and column four the associated reliabilities for completing the tasks.

The investigations of this method on a typical reactor installation with a multitude of operations sequences, dial readings etc, was found cumbersome.

Table 3.2 Time and reliability for an 'output stage' of reading a dial (Base time 1 sec)

Parameter	Dimension indication	Time added (secs)	Reliability
Scale size	6-in scale length	0	0.9998
Scale style	Moving pointer	0.15	0.9976
Scale layout	Horizontal	0	0.9998
Mark spacing	0.1–0.25 in	0.75	0.9992
Resolution	Good	0	0.9998
Number of marks	50–100	0	0.9998
Scale increase	Left to right	0	0.9998
Proportion of marks numbered	1 : 2	0	0.9999
Viewing time	Indefinite	0	0.9999
Scale illumination	Easily perceptible	0	0.9999
Totals		1.9	0.9954

Furthermore, in principle, the basic strategy of the method is trying to take in all the factors but it is really doubtful that the elemental data required can be readily derived and made available to the assessor. Nevertheless, this method brings out the basic idea of breaking down into elements and synthesizing the overall probability of success or failure required. Obviously there is a basic problem emerging here, that in general terms the operator error rates will depend upon the equipment, and the failure of the equipment will depend upon the human handling.

From some points of view it may be easier to carry out an evaluation of the human errors or deviations in a more independent fashion using a technique involving a tree logic such as THERP (Technique for the Human Error Rate Prediction), as described by Swain[89,90]. THERP is useful for estimating human error for incorporating in the reliability analytical techniques already described in preparing the various reliability diagrams or for use with other methods such as fault-tree analysis.

The question of data and the monitoring of reliability in a general sense will be considered as appropriate later in this book. However, it may be noted that the operator may completely fail, in the first instance, to perform the action as required and data indicate that for simple operations the error rates involved are approximately in the range of 10^{-2}–10^{-4} per operation. Obviously from the foregoing discussion it is necessary to break down the system into elements but ideally these elements should be compatible with the data which are available on actual plants. Clearly if good data are available on complete sequences of operation applicable to the plant system under investigation then these make for ease in the general safety and reliability assessment process.

3.10 Human Factors in Maintenance and Testing

In the general design of a system, decisions will have been made as to the questions of having disposability which may be defined as a 'characteristic of design and installation which is expressed as the probability that a system will fulfil its design function over a specified life-time with a given reliability without maintenance'. On the other hand it may have been decided to depend upon the strategy of maintainability which may be defined as a 'characteristic of design and installation which is expressed as the probability that an item will conform to specified conditions within a given period of time when maintenance is performed in accordance with prescribed procedures and resources'. There are many issues involved in the question of maintainability versus disposability, some are easily quantified, others are more nebulous and the choice between design for maintainability or disposability can sometimes hinge on tradition or prejudice. However, depending upon the particular application then various factors may be more significant than others as discussed by Green[160]. The general factors involved in maintenance and disposal are summarized in Table 3.3.

Table 3.3 Factors in favour of maintenance and disposal

Maintenance	Disposal
1 A very high ratio of cost of product/cost of maintenance.	1 Larger manufactured quantities leading to lower unit costs.
2 The furtherance of engineering skills and knowledge.	2 Simpler design less development on features associated with maintenance (accessibility).
3 The feedback of data to designers, researchers. reliability analysts.	3 Elimination of failures associated with maintenance errors.
4 Less ecological damage conservation of raw materials less rubbish less effluents from production processes.	4 Elimination of all costs associated with maintenance activities.
	5 Environmental or operational conditions which make maintenance difficult and/or costly.

As discussed earlier a system may enter into a failed state because of some planned maintenance or some fault occurring. Usually planned maintenance procedures tend to be systematic whereas the occurrence of faults tends to be random. The changes due to random faults will be broadly of two types as previously discussed, those that reveal themselves so that repair or renewal can take place immediately and those that remain unrevealed until some later routine procedure.

In considering maintainability versus disposability it is necessary to differentiate between the problem on the component or small subassembly level and at some higher system level. Disposability is much more common on the component/sub-assembly level and can be looked on as a form of 'maintenance' when viewed from a system level. On large plants or systems, where availability is all important, the choice whether to scrap or maintain in various ways can be very significant. From the plant point of view immediate replacement will be required, but at this point a choice will have been made by the designer or even by the operational staff, that the item is to be designated for maintaining or for disposal.

There is often a tendency to associate disposability with cheap mass-produced products; and although this may often by true, it is interesting to note there is an increasing number of cases where the opposite is true. For example, the most advanced highly reliable products are designed to be maintenance free, such as those in space applications or on nuclear reactors where valves may be welded permanently *in situ* in active regions. Some of the most advanced technology developments are centred around the design of long-life, maintenance-free plant.

It is vitally important in the case of high integrity protective systems to make the correct decision in design philosophy and the assessment processes and to know the likely consequences of taking a particular decision.

Present problems of this kind in the form of 'models' may involve a number of different specialists such as Terotechnologists, System Engineers and Reliability Engineers. It may be argued that a 'maintained design' may be less subject to outside influences such as escalation costs or changes in basic characteristics. On the other hand it is necessary to live with the particular design system as it cannot be disposed of.

The choice of modelling techniques is largely predetermined by; (a) the conceptual requirements that the model needs to satisfy, (b) the amount and quality of the data available, and (c) the level of accuracy required.

Whatever choice is made and however weak the numerical data may be, the greatest merit of such techniques is to provide new insights into the workings and performance of the system involved. It is normal practice in the protective systems involved to follow a rigorously laid down procedure on maintenance both 'breakdown' and 'preventative' and testing procedures to prove the functional capability of the system. In addition there may be throw-away module strategies adopted in the maintenance policy. Clearly running right through this topic is the human-influence factor and in any assessment procedure it is an important aspect to be considered.

Furthermore the method of proof testing the system is an important philosophical consideration particularly as to whether it tests the system right through from the initial sensor to the ultimate output and shutdown action of the system or whether it is a stage-by-stage check of each part of the system. Clearly at a base-load power station, it is not acceptable to shutdown the station to prove the protective system by initiating an abnormal input to the sensing elements. Hence it is important to take into account the human influence factors and the assumptions created by the various interfaces of testing procedures so that proof testing of the protective system to operate in the correct time can be undertaken and ascertained. A further point which may be considered in the maintained and tested system is the psychological factor of having something in which confidence is building up because the item is believed to be known. The example discussed in the next section brings out some of these factors.

3.10.1 Analysis of example channel trip test system

It has been discussed in Section 3.7 that protective systems are often designed and operated so that they are largely automatic in operation. They have inbuilt redundancy and for some reactor-fault conditions require to operate on too short a time-scale to depend on the operator taking manual emergency shutdown action. However, these systems still involve the operator and maintenance personnel at different hierarchical levels in such functions as testing and maintenance of specified items of equipment.

As an example of human action in the testing of a protective system it is assumed for the purposes of illustration that a two-out-of-three set of

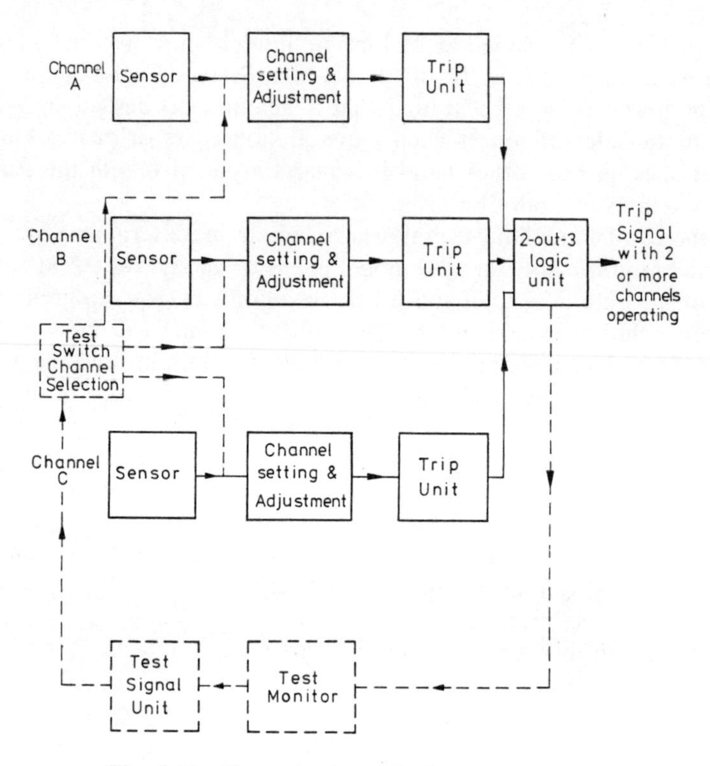

Fig. 3.19 Example channel trip-test system

protective-system channels are to be tested to show that they are functioning correctly. It is normal procedure never to deliberately put a two-out-of-three into a two-out-of-two operating state as this reduces its reliability or it may be viewed as increasing the fractional dead-time of the system. Basically the system is shown in Fig. 3.19 with three channels, A, B and C, and in dotted lines the testing system. The test signal unit generates in appropriate form a signal which can be set up by a calibration procedure to simulate a real signal coming from the sensor.

In Table 3.4 is shown an assumed test procedure which the operator should follow. Using the THERP technique for example it should be possible to derive the human error in calibrating the test signal unit incorrectly so that if a channel is adjusted by using this test signal unit it would be unable to trip at the required setting. Furthermore, it could be that the human error involved causes an incorrect calibration such that the sensor output would have to be so high as to represent a really dangerous fault condition on the reactor plant. It is obvious that if two of the three channels were adjusted with incorrect calibration this could lead to a common cause for each channel to be inoperative and would represent a common-mode failure of the system. This type of failure event is further considered later in this book as it can place limitations on the reliability achieved by the use of redundant sub-systems.

The analysis carried out for the human influence factors of the example channel trip test system is shown in Fig. 3.20 and is presented in the form of a 'matrix logic' array. This condensed form of analysis links dots vertically for the 'probabilistic AND' combinations (individual event chains, $p_a = p_1 \times p_2 \times \ldots$ etc.) and link dots horizontally for the 'probabilistic OR' combinations (accumulation of the event chains, $p_T = p_a + p_b + \ldots$ etc., assuming small probability values). The triangles modify the 'dot' probabilities by a half since only high errors as opposed to high and low errors are being considered.

Table 3.4 Assumed Trip Test Procedure

1	Calibrate test signal unit.
2	Operate test switch channel selector to required channel.
3	Check all channels are in a normal untripped state.
4	Operate test signal unit.
5	Observe response from trip unit and two-out-of-three logic unit.
6	If no trip unit output, check test signal unit calibration.
7	If test signal unit calibration is satisfactory follow maintenance procedure required for corrective action and put trip unit into operation on one of the other channels.
8	After completion of maintenance repeat steps 1 to 5.

In the analysis is shown the collection of what are judged to be the most significant event chains—many possible higher-order chains are omitted which have associated probabilities of low significance. The total quantification predicts a mean error probability per calibration (for the three channels taken as a whole) as 0.002. In other words on an average of 1 in 500 occasions the trip levels will be set and left over high. If the channels are calibrated monthly, a dangerous calibration error will occur with a mean frequency of approximately once per 40 years.

This example can be used to see the interrelationship between the operator, maintenance personnel, and the system and can become an important factor which requires careful consideration. In Fig. 3.13 the use of data reduction equipment is indicated in dotted lines which can become part of the display. The basic element of such equipment is a digital computer which gives the operator the facility to select information he requires according to preset programmes and furthermore provides facilities for automatic logging of plant variables. In certain complex plant this man–machine relationship involving computers and the appropriate consideration of the human factors is important and is a problem which is under continual review. A set of relevant papers relating to human performance reliability is contained in Regulinski[91] with a review by the editor.

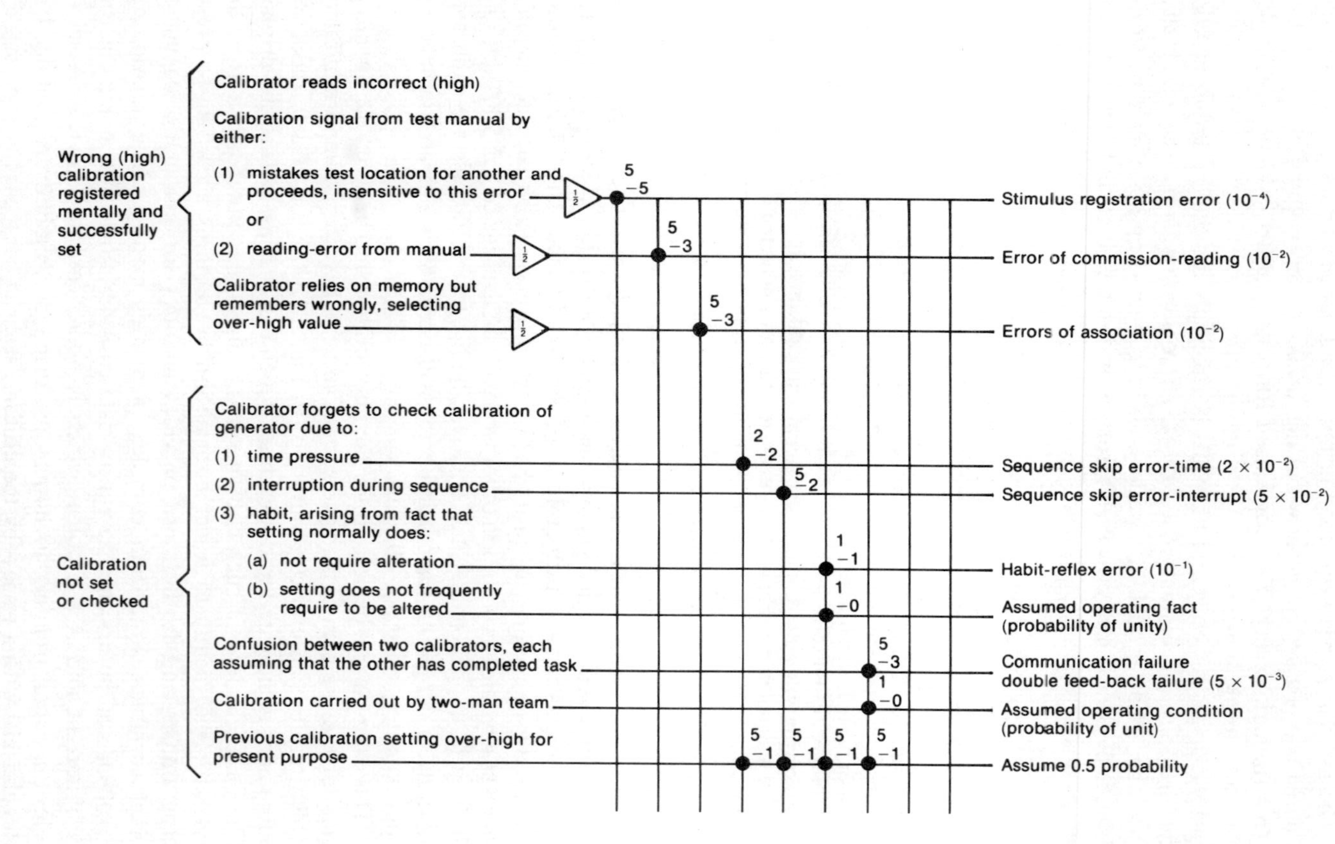
Wrong (high) calibration registered mentally and successfully set
Calibrator reads incorrect (high)
Calibration signal from test manual by either:
(1) mistakes test location for another and proceeds, insensitive to this error
or
(2) reading-error from manual
Calibrator relies on memory but remembers wrongly, selecting over-high value
Calibration not set or checked
Calibrator forgets to check calibration of generator due to:
(1) time pressure
(2) interruption during sequence
(3) habit, arising from fact that setting normally does:
(a) not require alteration
(b) setting does not frequently require to be altered
Confusion between two calibrators, each assuming that the other has completed task
Calibration carried out by two-man team
Previous calibration setting over-high for present purpose
½
Stimulus registration error (10^{-4})
Error of commission-reading (10^{-2})
Errors of association (10^{-2})
Sequence skip error-time (2×10^{-2})
Sequence skip error-interrupt (5×10^{-2})
Habit-reflex error (10^{-1})
Assumed operating fact (probability of unity)
Communication failure double feed-back failure (5×10^{-3})
Assumed operating condition (probability of unit)
Assume 0.5 probability
5 −5
5 −3
5 −3
2 −2
5 −2
1 −1
1 −0
5 −3
1 −0
5 −1
5 −1
5 −1
5 −1

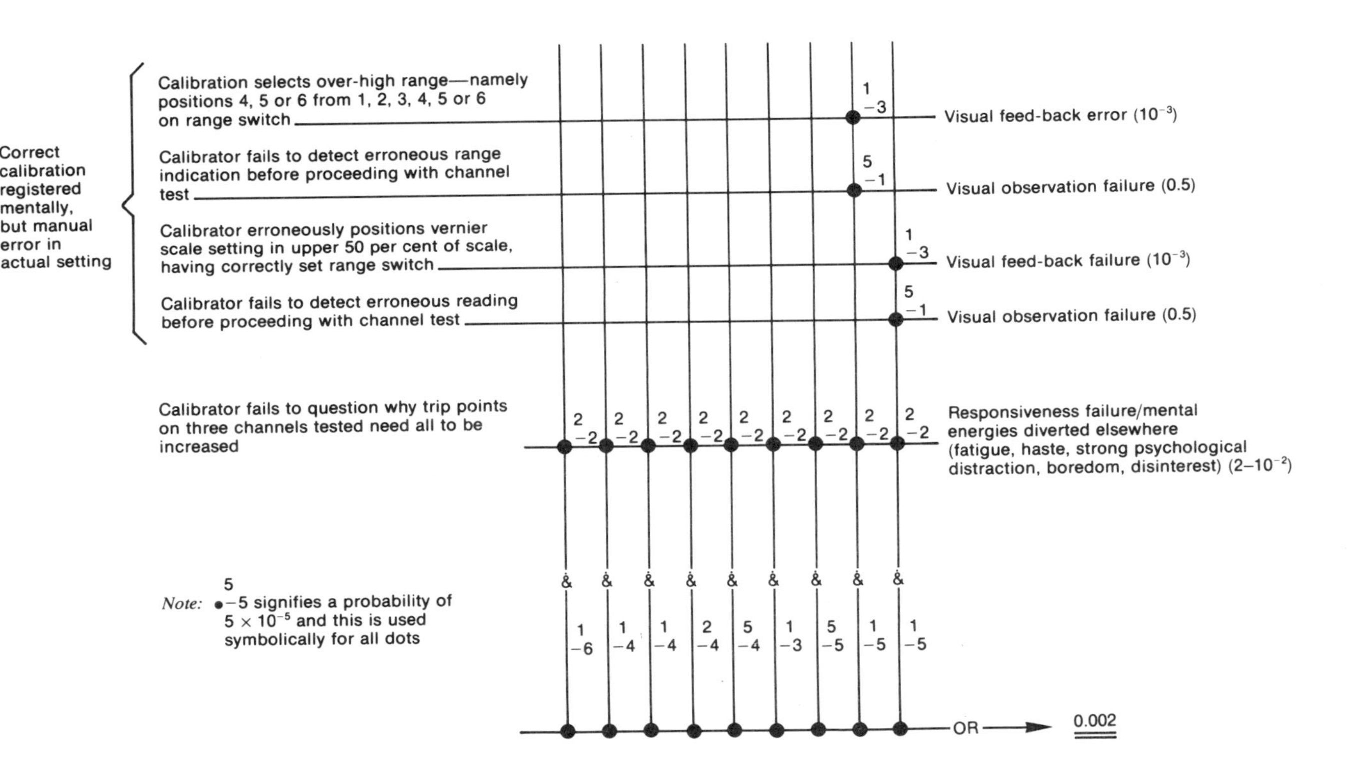

Fig. 3.20 Analysis of channel trip-test system with reference to human factors

3.10.2 Task counting method

A similar approach to that already used for assessing failure rates of electronic and other equipment using a simple fault mode and effect analysis (FMEA) has been used as described by the CNSI Group[92]. It is assumed in the approach that the individual tasks of the procedure are independent and a failure rate is attached to each individual task and the fail to danger ratio is assessed. In the particular example given by the CNSI Group it was carried out on an actual plant and a basic fault rate of 10^{-2} per task was adopted. The result of this analysis was a fail to danger rate of 0.83 for the particular task sequences.

It is interesting to observe that a simplified method which considerably reduces the required effort is by counting the tasks of which there were 186 and adopting a standard task error rate of 10^{-2} which gives an error rate of

$$186 \text{ (task)} \times 10^{-2} \text{ (task error rate)} \times 0.5 = 0.93$$

This figure assumes that the tasks are independent and does not take into account the operator's self monitoring and error correction function. Consequently the figure of 0.93 appears to be unrealistically high. This is discussed further in Appendix 8 on page 249. However, this error rate which is derived represents the error committed in not following the procedure rather than the rate of errors induced into the channel. From this interpretation it is evident that the task counting methods gives quickly an estimate which is 0.93 and is little different from that found from the laborious method first described which had the result of 0.83. Furthermore, it is interesting to note that the figure chosen for the task error rate which was 10^{-2} implies for 186 tasks an overall rate of just less than two errors (1.86) could be expected. An actual test was carried out on the plant for which this analysis was undertaken and it was observed that two errors were actually made.

3.10.3 Matrix approach to the evaluation of human factors and the system

In order to present some methods by which human factors may be assessed, the work undertaken in a comparative study of protective system designs for a particular boiler plant has been reviewed. In this study it was necessary to compare designs with various degrees of automation for the protective system.

This requirement leads to simultaneously considering the effects of the human factors and the system hardware as an integrated whole. Therefore the approach used in the analysis of the example channel trip test system may be more generally considered as a logic matrix strategy involving 'multiple event chains'. This requires defining the various conditions which can arise for the different protective system designs taking into account the probability of the actual hardware failing as well as the failure of standard operational sequences involving the human element. In order to illustrate more completely the application of this technique a case history from a real plant has been extracted from Hunns[93]. The details of this analysis are summarized in Appendix 4.

3.11 Implementation of Design and its Assessment

In the foregoing sections of this book the investigations and discussions have been concerned with establishing on some prior basis the assessment of the system design. The various techniques for assessing the generalized protective system to be found on a nuclear reactor or perhaps a chemical plant have been orientated to demonstrate a functional capability and to appraise the probability of this capability being available when required. Employing a quantification of these various characteristics imposes a discipline on the assessment process and the assessor himself. In many ways the techniques which involve a modelling process will have had a certain abstractness and boundary assumptions. However, the protective system itself will be materialistic and a conversion takes into account all the logical design considerations from which the designer will have progressed from the conceptual design to his final set of drawings to permit the progression to the ultimate operation of the system installed in the plant.

Since the ultimate objective is to establish an adequately reliable protective system it is found in practice that the design process will be overlapping into other stages of the 'growth' of the actual material system.

The procurement of the equipment and its general manufacture will be required to meet specific standards and methods of quality assurance will be extended to the installation, construction and commissioning. Commissioning, operational and maintenance manuals will obviously involve the designer and other personnel. This overall process is one which is well documentated in the general literature and will include various trials, tests and evaluations. In addition, particularly in the case of nuclear plant where licensing procedures are involved, there will be imposed additional requirements over and above those on conventional plant of a non-nuclear nature.

In most countries including the UK these safety requirements are set out so that the designer/operator has a clear knowledge of what is required to be met. This question of regulatory procedures is outside the present scope of this book but Fichtner *et al.*[94] give a catalogue and classification of technical safety standards, rules and regulations for nuclear-power reactors and nuclear-fuel cycle facilities. This catalogue contains references to about 2800 standards drawn from 145 organizations from 31 countries and it may be noted that Section 7 of Fichiner *et al.*[94] deals with 'Plant and Reactor Protection'.

It is of interest to note the *Safety Codes and Guide*[95], issued by the Institut für Reaktorsicherheit der Technischen, for compiling the information on the reactor protection system required for examination purposes in the licensing procedures of nuclear plants. This safety code and guide is for plant in the

Federal Republic of Germany and an abstract is given in Appendix 3, Section A.3.4, to illustrate some of the phases which have already been discussed. There are common points which arise in connection with meeting certain policies for ensuring adequate reliability. Similar requirements may be found in the UK regulatory documents such as that issued by the Health and Safety Executive[96].

In addition the assessor will have prepared a safety assessment report with all the prior findings and recommendations which should have resulted from the modelling and appraisal of the design. If an appropriate dialogue has been maintained between the designer, operator, the assessor and other interested parties then it would be assumed that if the assessment prior findings can be subsequently demonstrated then the protective system should be adequate for its purpose. Hence the problem arises of searching for techniques to show that the 'materialistic' or actual protective system is adequate.

There are interface problems which obviously exist in the general design, manufacturing, installation, commissioning and operations processes but there is an important interface problem between the assessor's prior findings in the assessment and the actual system. The general process for the provision of the protective system will provide the facilities for additional information which requires to be analysed in such a form as to permit proper decision-making on the adequacy of the protective system.

3.12 The Safety Assessment Report

In order to establish a prior starting point for the ultimate demonstration of the adequacy of the protective system the assessor should have prepared with great care an appropriate report. From reviewing assessments carried out on various systems over a number of years it is essential that the report covers various facets both technically and from the point of view of presentation. This has been reported upon by Green and Bourne[38].

Not only should the report state the aspects of the proposals which have been demonstrated and found to be inadequate, but also those which have been appraised and found to be adequate. It is important that there should be no doubt as to the areas of the assessment covered and the degree to which they have been investigated. In this respect the document should be self-explanatory and contain full references to the information used. Quite often this is the only real way by which the reactor actually assessed may be defined. Also, in the event of a reactor accident arising in an area of acceptance, it should be possible to deduce where the assessment failed. This can make for an added contribution to future safety.

It is found convenient and helpful to introduce the main body of the assessment report with three types of introductory material. These may be classified under the following headings; Summary, Contents, Guide to the Assessment.

The Summary is normally designed for the reader who requires a quick appraisal of the situation and who may not have a detailed knowledge of the proposed scheme and equipment. To this end it is useful to have a few pages of summarizing material leading from a brief description of the scheme (including block schematic diagrams) to outstanding issues and recommendations as seen by the assessor. The overall conclusions are preferably given with sufficient explanation without too much technicality.

The Contents should list the main section headings, together with the headings of the sub-sections where appropriate. Numbering of these sections and sub-sections should, if possible, be derived from the original reference numbers planned at the stage of preparing the file. This makes for easy cross reference from the report to the file when investigating various points at a later stage. A typical contents list, based on the section headings used throughout this report may be as follows:

Guide to the Assessment

Section 1 Protective Requirement
- 1.1 Basic objective
- 1.2 Reactor properties affecting the basic objective
- 1.3 Limits on measurable parameters
- 1.4 Reactor fault conditions
- 1.5 Reactor fault analysis
- 1.6 Required performance
- 1.7 Conditions of operation

Section 2 Protective Channels
- 2.1 Fuel-can temperature
- 2.2 Rate-of-change of pressure
- 2.3 High-power excess flux, etc.

Section 3 Safety Circuits
- 3.1 Principles
- 3.2 System and components
- 3.3 Integrated performance

Section 4 Shutdown System
- 4.1 System
- 4.2 Components
- 4.3 Integrated performance

Section 5 Overall Performance
- 5.1 Lines of protection
- 5.2 Reactor faults
- 5.3 Conclusions

References

Drawings

Figures

Following the Contents the body of the document should be introduced with a Guide to the Assessment. Such a guide may take, typically, the following form:

In order to deal with the information, and the corresponding appraisal, in a systematic manner, the assessment is divided into five main sections.

Section 1 considers what the protective system is required to do and how this requirement is related to the reactor behaviour and the various operating conditions.

Section 2 assesses the proposal for each automatic protective channel in detail from input sensed quantity to output trip signal with reference to each instrument line.

Section 3 evaluates the safety circuits which process the output trip signals from each protective channel.

Section 4 considers the operation of the shutdown system which is designed to shut down the reactor on the basis of the output information from the safety circuits.

Section 5 assesses the integrated performance of the complete protective system in relationship to the reactor fault conditions against which protection is required.

Where appropriate in the text, various aspects of the assessment are summarized in the form of 'notes' or 'recommendations'. The notes deal with points on which agreement has been reached, intended action has been expressed or further assessment is required. The recommendations represent facets of the system which, in the opinion of the assessors, are deemed to be inadequate. In order to assist in appreciating the relative importance of each recommendation, they are classified into three categories, A, B and C.

A These are recommendations which arise from points leading to an acceptable overall performance and which, in the opinion of the assessors, should be definitely implemented.

B These recommendations which, in the light of further knowledge, may become graded as either A or C. However, failing such final resolution, they should, in the opinion of the assessors, be implemented as in A.

C These are recommendations which, in the opinion of the assessor, should be implemented on the grounds of 'good practice'.

It should be noted that where, due to lack of information, boundary performance figures have had to be estimated or assumed by the assessors, such figures are shown in **BOLD** type.

The preceding examples of layout and wording can lead to a uniform presentation which permits a ready understanding of the points covered and the assumptions made. It also facilitates easy reference to similar facets from different assessments by whoever they are prepared or read. Once a format has been logically prepared and developed, the ideas may be arranged in an orderly fashion so that any inconsistencies and uncertainties may be recognized and the appropriate justification of the system obtained.

One final aspect connected with the layout of the report is the method of keeping the assessment up to date. If the conclusions are to remain of value over the years, then the report must be in a format which permits it to be maintained current. During the course of the assessment various questions and answers will have been posed and the information should have been stored such that an investigation can be readily carried out as to whether a particular question has been raised and as to whether an answer has been given. This is most important at a modification stage of the system. The previous questions should be investigated and the validity of the answers should be established with the new modified condition. Failing this, it may be possible to find a position where the answers were correct but do not apply to the reactor in its current form. The assessor's final task, therefore, is to establish that the means by which the reactor will be operated, together with the committees and procedures to be set up to feedback the information and maintain a current assessment are in fact valid.

3.13 The Need for a Posterior

Irrespective of the equipment and the system there will be variability in performance characteristics and in some cases failure. The whole complex procedure from design concept to the ultimate operation requires to be monitored in some way in order to provide posterior information to establish the safety and reliability assessment under discussion. This information should preferably be derived not just qualitatively but quantitatively and requires to be used for updating the priors estimated in the original assessment. In fact one of the considerations in the strategy undertaken by the assessor should have taken into account the posterior information and its form which would be expected to be generated by the equipment and the system as it 'grows' into its ultimate 'materialistic' form.

The assessment models should be able to accept this posterior information and to enable decisions to be made in trends. In a natural fashion certain time lags, e.g. for manufacturing and testing etc., occur and the demonstration of the growth of reliability cannot be very fast. Dependent upon whether the failures occur early or late in the overall growth programme for the protective system so depends the decisions which can be made.

This leads to the assessment strategy of the allocation of the requirements against which the estimated priors will be initially allocated and to be subsequently updated by the posterior information. Hence the assessment is a continuous process which starts life as a prediction, continues through various 'sample testing' stages and ends up with the confirmed results of the plant system's performance at the end of its life. This implies that some form of control for the implementation of the overall assessment is essential especially in the case of highly reliable systems such as those providing protection for a nuclear plant or hazardous chemical plant. In the context of the real world the protective system is dynamic in a reliability sense due to the variability which can exist.

3.13.1 Allocation of system targets

The importance of considering the allocation of target probabilities is best given by a specific example, as those of Aitken[97] and Green[98], of the automatic

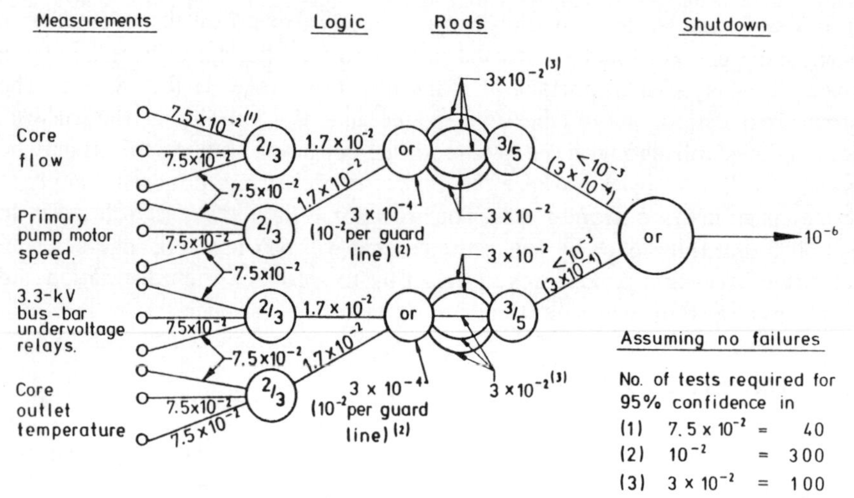

Fig. 3.21 Logic diagram showing reliability and test requirements (loss of electric power)

protective system designed for the prototype fast reactor, PFR, which is a liquid-metal-cooled fast-breeder reactor (LMFBR) at Dounreay. Considering one specific fault condition, i.e. loss of electric power as an initiating event this had a frequency of one per year. From the Farmer-type criterion discussed in Chapter 1, Section 1.2.3, a target was indicated of the probability of failure to shutdown on demand against loss of coolant flow due to loss of electric supply as 10^{-6}. As described by Aitken[97] the assessment strategy was evolved in the early design concept of the protective system so that the allocation of the target probabilities for the assessment took into account the feasibility of the posterior information which could be obtained. Figure 3.21 shows the allocations made in the form of a logic diagram showing the requirements for the particular case of loss of electric power.

Figure 3.21 shows two independent systems, one based on the measurements core flow and primary pump motor speed, and the other based on the measurements 3.3 kV bus-bar under voltage relays and core outlet temperature. Each measurement is carried out on a two-out-of-three majority voting logic and there are five rods for each independent system of which three operating are sufficient to shut-down the reactor. Either of the two independent systems has the capability of shutting-down the reactor thereby meeting the overall protective system. The probability targets for each part of the overall system configuration are shown and the figures in parenthesis, i.e. (1), (2) and (3), indicate the number of tests required for 95 per cent confidence to be undertaken assuming no failures. Furthermore, the response and accuracy requirements for the automatic protective system were ensured not to be onerous.

This example quite clearly raises the question of the assessor setting up the assessment strategy as in general terms it is taking part in a 'statistical experiment' and such an experiment requires to be designed. However, this is going to require careful consideration of such aspects as the frequency of the events, the sample size and 'rate' at which the growth of reliability is expected without leading into 'reliability instability' giving rise to safety problems.

3.14 Implementing the Reliability Policy

The implementation of the reliability policy which is contained within the framework of the safety and reliability assessment will obviously vary according to the motivation. In the design office there may be the principle of not to 'tread new ground' unless forced to do so. If it is necessary to tread new ground then it may be necessary to find out from other people their experiences in such a procedure. When there is no information or experience available the starting point as already discussed commences by putting trust in the design team and picking the staff with the appropriate characteristics to see the job through as intended.

In the design process if complete generations of plant designs were reviewed, e.g. nuclear-power stations, chemical plant, etc., it would be anticipated that there would be limitations set by human endeavour. Hence a strategy based on fundamèntally putting right design process defects may involve many generations of people. On the other hand, there may be latent defects which would still remain.

In reviewing various types and configurations of protective systems it is clear that as equipments, sub-systems and cables are brought closer together in a physical sense then the probability of the occurrence of common faults becomes more significant. This question of common faults will be discussed in a later section, however, in properly segregated designs the probabilities of common faults tend to put limitations on the overall system probability failure of less than the order of 10^{-6}. It is found that this figure also applies to other types of 'rare events'. Usually it becomes difficult to substantiate a probability of failure of less than 10^{-4} when equipment is installed in close proximity or for example in one cubicle. Various investigators in this field have discovered similar limitations, as have Eames[50], Green[100], and Epler[101], and other more detailed work on these limitations is considered later in this section.

The foregoing type of consideration has led to guide lines for the prediction of protective-system design and assessment. From a probabilistic point of view the experience is summarized in Fig. 3.22 for several types of system. A simple single-system 'A' would consist of just a single safety-monitoring assembly. System 'B' is a simple redundant system containing a number of single safety-monitoring assemblies operating with some redundant logic, e.g. two-out-of-three logic. The partly diverse system 'C' would have at least two separate and physically different parameters which are used to detect or sense

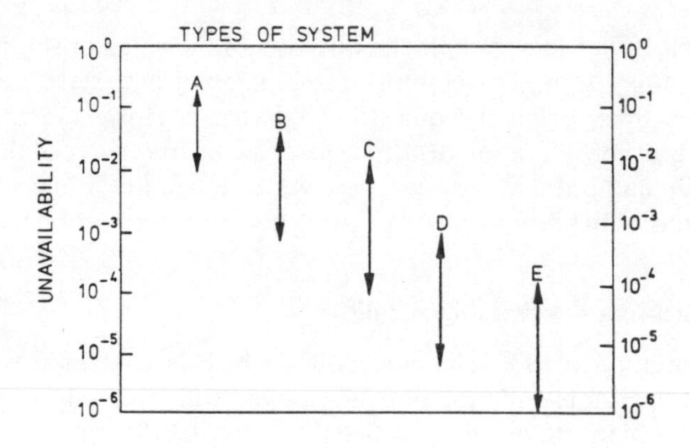

Fig. 3.22 A guide to possible reliability ranges for various system types. Types of system: A. Simple single system; B. simple redundant system; C. partly diverse system; D. fully diverse system; E. two separate systems (each diverse)

a particular plant condition. Furthermore, physically different types of measuring instruments and sensors would be used for this type of system 'C'. Normally the partial aspects of the diversity arise because the outputs from the two parameter measuring systems feed into a common shutdown system. In the case of the fully diverse protective system 'D' this is a further extension of system 'C' where the instrument outputs are fed into two separate guard-line systems, both being physically different and separate. However, they then feed into a common shutdown system.

The information given for two separate diverse systems implies that each system is similar to the 'C' system but are quite different from one another. This means that the quantities being measured will be different such as flow and temperature for one system and for the other say neutron flux and pressure. Furthermore, the whole configuration from the sensors to the shutdown equipment will be designed and will operate on different principles in two separate systems. This may even mean using two separate design and construction teams in order to limit common cause failures and to ensure complete techniques of segregation.

The various ranges given for the unavailabilities of the system will be clearly dependent on various factors such as proof testing and its periodicity. It is found there are limitations to the extent to which the proof-testing frequency may be increased without getting diminishing returns.

Other factors enter into these limitations and are summarized in the case of common-mode failure in Fig. 3.23 as opposed to independent failure modes.

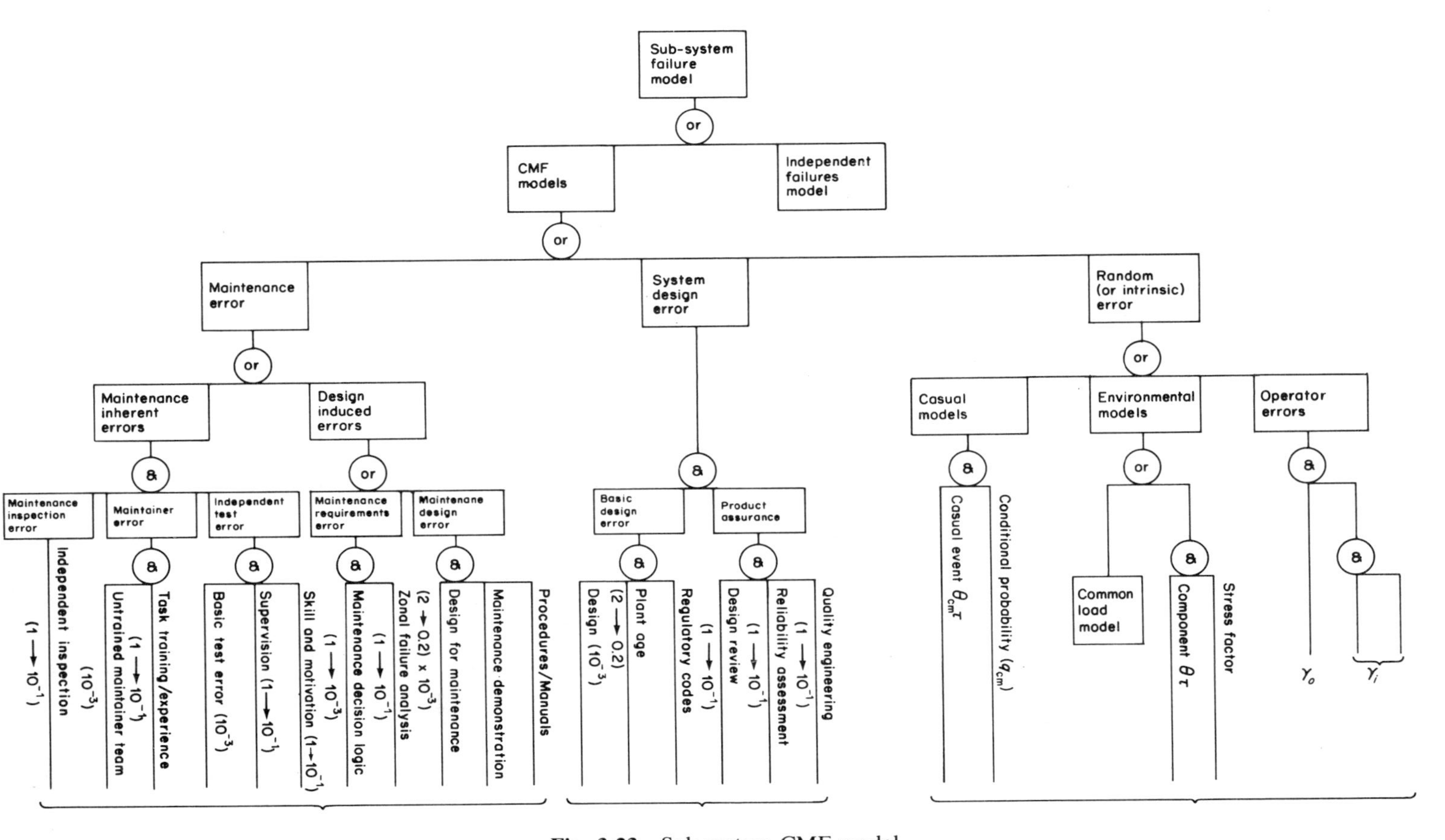

Fig. 3.23 Sub-system CMF model

Table 3.5 Some basic principles for the assessment of protective systems

1 No credible common system-fault condition affecting the protection system should prevent the plant from being safely shut down when required.
2 No credible combination of two equipment faults should inhibit automatic protective action under opreational conditions requiring such action.
3 There should be adequate provisions for proof-testing and maintenance of the system at regular intervals and such provisions should not reduce the degree of protection below that required.
4 No anticipated credible human action should prevent the plant from being safely shut down when required.
5 The protection system should make use of measurements related as closely as possible to the effect guarded against and its causes.
6 Designs should be such that a high proportion of faults will result in the state of the system being moved in the safe direction. Fail to danger fault rates should be as low as possible.
7 The design of plant should be such that the rate of demand made on the protection system is kept as low as practicable. The protection system should cause the minimum number of spurious shutdowns, trips, or alarms.
8 The protection-system design should meet the reliability requirements as defined by the overall safety criterion.

The components of errors for maintenance, system design and random (or intrinsic) causes which enter into these system reliability limitations are illustrated in Fig. 3.23 together with the ranges of factors involved. However, it will be observed how the maintenance error can be a dominating factor in the assessment and realization of highly reliable automatic protective systems. Hence depending upon the particular system and the various conditions involved the ranges given as a guide in Fig. 3.22 could have a wider variation. This general problem is discussed in more detail by Watson and Edwards[102] and also by the CNSI Task Force[103].

It may be noted from Fig. 3.21 and the allocation of probability targets that the guides given in Fig. 3.22 have formed a basis for consideration. Nevertheless, there is another important implementation of reliability policy which has emerged by having a set of precepts or principles on which to design, assess and operate the system. Experience over many years has produced such principles which often give a practical bias to the system incorporating prior experience which has arisen from routes and different system approaches to reliability implementation. Table 3.5 gives a typical list of such precepts or principles which have in practice been used as a foundation upon which to base practical design and operating details as by Aitken[97]. These principles are somewhat idealistic but should lead to the implementation of high reliability in a protective system if good quality components are used. However, the use of the word 'credible' introduces subjective elements which lead back to the quantification of reliability.

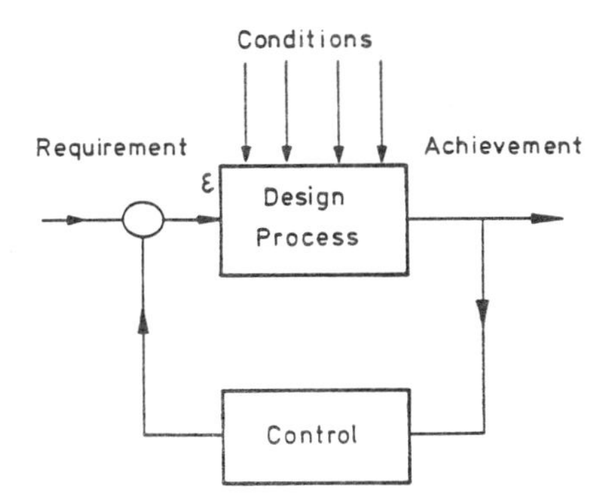

Fig. 3.24 Design process control diagram

3.14.1 Reliability control phases

Another approach is to consider the basic design process and procedure as evolved and to impose on it a control to reduce error and its consequences to be acceptable as shown in Fig. 3.24. This general type of task error model with feedback and monitoring has been discussed in the study undertaken on common-mode failure and reported upon by the CNSI Task Force[103]. In addition Rasmussen, of the Danish Atomic Energy Commission, Risø, is also investigating this type of model empirically in the case of maintenance and some aspects are discussed in CNSI Task Force report[104]. The important conclusion from this work is that it is essential to cover design and maintenance task errors specifically. This would be envisaged as leading to the requirement for the engineering and operational processes and tasks or organizations that produce the systems to being assessed in some way and may be considered as a generic problem in reliability management.

The control shown in Fig. 3.24 requires to be predictive so that the feedback process of modification can take place as early as possible in the design. If large lags are introduced then the consequences of failing to produce the desired achievement may be costly in all senses of the word. The 'control' requires to have the characteristics of predicting and give 'phase advance' in the design procedure. On the other hand it requires to have an 'integral' action so an error integral will also result in a change in the design procedure.

The conditions under which the design is evolved may also be changing particularly as new information is created as time passes. Furthermore the design which represents 'software' requires to be converted into hardware. In this conversion process it is necessary to consider the manufacturing phase which involves communication as shown simply in Fig. 3.25. This can produce distortion and error can result so that control is necessary to reduce this error. The techniques of communication are an important feature in conveying the exact meanings involved in the 'software descriptions'. Communication tech-

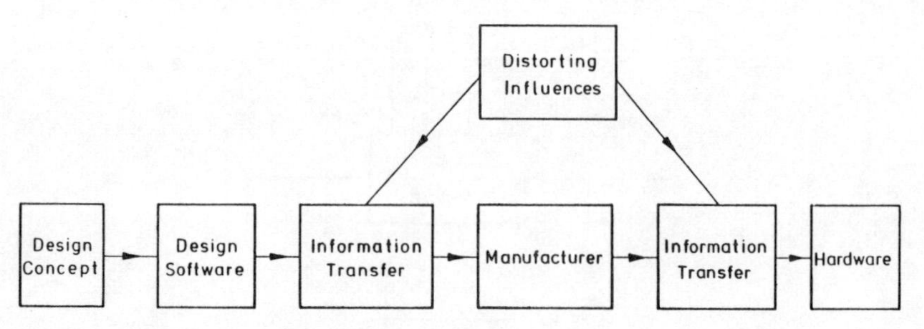

Fig. 3.25 Basic diagram for communication in manufacture

niques involved in reliability and probabilistic terms have been the subject of an investigation by the CNSI Task Force[105]. In the general review work undertaken by the author it is considered that not sufficient attention has been given to this problem.

Further sources of error arise where the manufacturer becomes involved in the installation phase with a view to realizing the design concept and error can result in this phase.

Commissioning requires the actual demonstration of the capability of the system and equipment as expressed in the design concept. It would appear that in this phase, control in the adaptive sense is necessary to dictate the procedures and thereby to converge on to the designed capability and general indications of reliability. Finally there is the ultimate phase of operation which involves the actual working of the system as originally specified and this requires feedback for adaptive control and also feed-forward for future system improvements.

3.14.2 Overall control of reliability

Clearly from the foregoing discussion some basic overall control is required to evolve as shown in Fig. 3.26, where the complete design and operation pro-

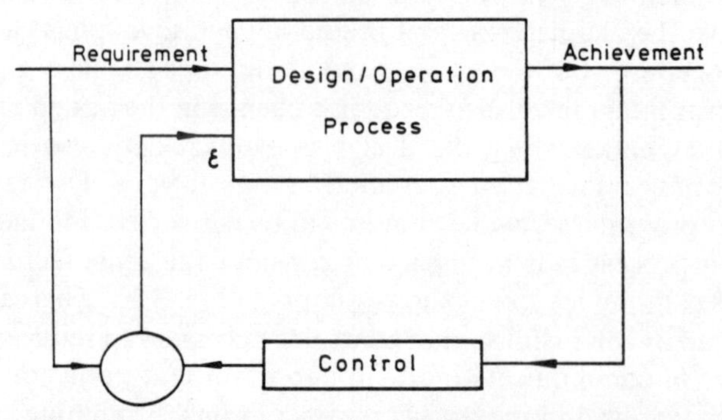

Fig. 3.26 Design/operation simplified control diagram

cess through all its stages is considered as a whole. This then leads to the error becoming more and more practical and real as the design/operation process permits the protective system to 'grow' to its full operational form on the plant.

The functions of such a control system for reliability will obviously require some measuring system for determining the values of the various reliability characteristics. Since the protective system is growing then a dynamical model of the system should be available to give physical predictions of the performance behaviour of the protective system and its equipment. Clearly there must be observations made available in the appropriate form of the protective system and its equipment. These ideas are expressed in simplified form in Fig. 3.27. The satisfactory evolution of dynamical system models of the basic type proposed is an iterative procedure and is described in the literature, as by McFarlane[106].

Although various types of control models could be evolved it is necessary that they meet the requirements of the whole design/operator chain and at the same time have data input which are compatible with the practicalities of the whole process of manufacture, sample testing and operational procedures.

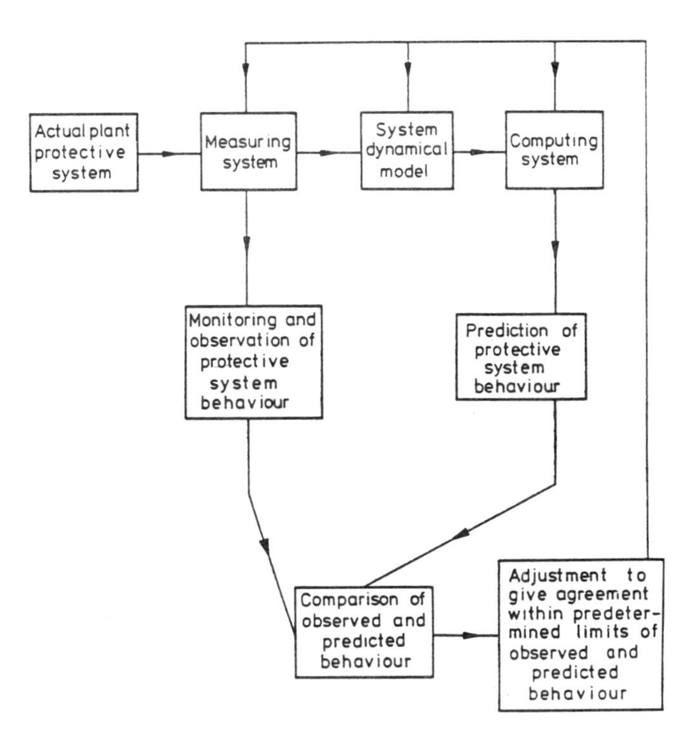

Fig. 3.27 Basic functional diagram of reliability characteristic control

Furthermore having the appropriate technique for estimating the various reliability characteristics at any particular time is another important attribute. Therefore this question of monitoring and estimating reliability characteristics at the equipment and system levels both in the sample testing phases and the operational field experience has been investigated and is discussed in the next chapter.

CHAPTER 4

Reliability Estimating, Monitoring and the Investigation of Operating Plant Systems

4.1 Monitoring and Estimating Reliability Characteristics

A basic ingredient in reliability analysis is deriving the appropriate data and this topic is one which is occupying research and development by many workers in the field which has been commented upon for example by Keller[107]. However, there are two fundamental questions to be asked: (a) Are the data wanted and why? and (b) In what form should they be?

Once these questions have been satisfactorily answered then it is possible to make a start on the appropriate methodology to be used. However, this is quite logical but it is a rather simplistic approach as the whole problem is complex.

Sufficient has already been said in previous sections to justify the need for having data for substantiating the adequacy of highly reliable systems such as protective systems for shutting down nuclear reactors under plant-fault conditions. It enables an appreciation to be gained of the significance and impact of applying the 'domino effect' mentioned in Chapter 1, Section 1.3, in terminating a condition which if allowed to continue may lead to extreme and hazardous conditions. Furthermore, the data are an important part of the synthesis process usually adopted in the safety and reliability assessment.

Another important aspect of question (a) is that technology is increasing at such a rate that workers in various fields of application find there is a lag between the present-day techniques and those which are required for the future. In the missile field for example Robert Lüsser who was a pioneer in reliability techniques emphasized over two decades ago in 1956[108], that there was a need for greater simplicity and, furthermore, there was a need for a greater emphasis on reliability concepts where simplicity is not reasonably possible. The recent trends today in plant systems show an increasing complexity and in particular this is seen in the protective systems which have been discussed. Hence there is a danger of complexity increasing more rapidly than the ability to design and substantiate high reliability. Therefore, it is essential to overcome this interface problem involving these limitations by having an in-built learning technique as part of the developing technology and this mainly revolves round having data made available on the engineered endeavours. In other words not only should the plant and its protective systems be let to 'speak for themselves' but it is necessary to 'listen' to them in the appropriate way by using the right form of language to enable this communication and understanding to take place. Even if the technology has matured there is still a requirement to concentrate on reliability particularly with reference to safety. However, there are other motivations which involve general availability and economics which should give encouragement to take full advantage of the full extent of the technological capabilities.

It may be assumed that in some form or other it would be generally agreed that the answer to question (a) is in the affirmative. This leads to question (b) as to the form of the data and by implication the extent of the data. There is an increasing cry in most countries from designers and assessors of highly reliable systems, and other associated parties, that the appropriate data are not available. This whole question has been the subject of research and development. The answer extends from requiring just crude 'ball-park' data to requiring every aspect in detail of the plant system both when it is performing correctly and extending to every aspect in detail to events which cause the system to fail completely in its performance. It may be noted that the overall design/operation of the protective system which involves some complexity by its very nature automatically generates data by its own documentation and procedures. Unfortunately, these may not be in the form required by the assessor or it may be so difficult and time-consuming as not to permit the extraction of the appropriate form of data.

It would appear to be most constructive to review some of the methods and approaches which have been developed and are still being researched in some cases. The opportunity will be taken to point out the form of the data for particular requirements. There is a general controversy over the use of subjective (or soft) data and so-called 'hard' data which is considered as real data taken from the system environmental conditions. However, the general objective which is being attempted in this book is to derive the relevant reliability parameters as opposed to the data which are required for accounting purposes and administration involving cost, materials, manpower, etc.

4.2 Subjective Data

Reference has already been made to considering experience or subjective judgement at various stages of the overall design/operation process for the protective system. In the early stages of analysis of the reliability characteristics of the system data may not be available. However, there may be prior experience which may be purely qualitative of similar situations, equipment or systems. As an example, in the early stages of design the engineer has often built up a subjective feel for the likelihood of a system having a feasibility of performing in the manner desired. From the general configuration and other factors it may well be that a highly experienced engineer has formed the opinion that the system is 'highly reliable and made of good quality materials' which may have been the intent for the system. Furthermore, such an engineer may have taken a consensus from various views expressed by specialists on various aspects of the particular system in question. Similarly the assessor takes a similar approach independently but clearly it is preferable if such subjective judgements can be quantified.

The technological method must make the best use of the experience that is available to predict the future. In some cases the data are sparse or non-existent for the particular equipment under review. This may be for a variety of reasons such as the equipment is new and has no history or that data has never been collected on the reliability of the equipment. There are other factors connected with the operation of very high reliability so that by its very nature it does not give large sample sizes for analysis purposes. In the case of rare events by their very definition may never be expected even on a world-wide basis to give a large sample size. It can be argued that even if the sample size is very small or just the benefits of subjective judgements are available this permits making an initial start in arriving at some prior quantification of a reliability characteristic. The philosophy of such arguments will be considered as appropriate in later sections.

4.2.1 Method of estimating by consensus

Although methods of consensus had been used informally it was decided in 1966 to conduct an experiment in the UK on the failure-rate estimation by consensus and this is described by Green[69]. It is a well-known technique in many walks of life to put a question to a group of people and from their answers to obtain a consensus. This technique was used to obtain the failure-rate of equipment.

At a private meeting lasting two days on the subject of reliability, sixteen equipments of the type found in nuclear protective systems were displayed. Each equipment was exhibited with the appropriate diagrams and operating manual as shown in Fig. 4.1. All the equipments were of the electronic and light mechanical types and the delegates to the meeting knew that the equipment environment was defined to be that of certain nuclear-plant applications. A total of seventy-three delegates participated, representing a cross-section of engineers both electronic and mechanical, physicists, chemists, statisticians, and plant operating staff. Each delegate was asked to give his best estimate of the failure-rate for each equipment but by severely limiting the time available, this could only be given as an option.

It may be noted that all the equipment exhibited had been observed in the field with good sample sizes and operating experience.

The results of these opinions were analysed with reference to the ratio of the estimate to an average failure-rate observed in practice under normal

Fig. 4.1 Display of differential pressure transmitter

operating conditions. The results obtained are given in Table 4.1. A ratio less than unity indicates optimism and one greater than unity indicates pessimism. Table 4.1 also includes the average ratio, the highest and lowest ratios, and an idea of the distribution for each of the equipments. It will be seen that typically one-third of the estimates for each equipment are within a factor of two, but for the different equipments it was a different group of delegates that comprised this third. The overall result is pessimistic, but for some equipments most individuals are optimistic. Additionally it may be noted that most individual averages are inflated by one or two excessively pessimistic estimates. In general, it is of interest to note that delegates fared better with electronic equipments, and for these there were forty-four cases where the averages were closer to unity than the averages for mechanical equipments from the same delegate. Furthermore, it was observed in the analysis of the results that no one was consistently good, but it was noted that the very optimistic were generally optimistic on all items.

As part of the present study programme the consensus experiment was repeated at the 1977 Reliability and Maintainability Symposium held in Philadelphia USA. In this experiment different aircraft navigation equipments were displayed, i.e. warning equipment and inertial navigation

Table 4.1 Summary of ratios of estimated to observed average failure-rates

Description of equipment	Ratio average	Ratio range		Ratio distribution			
		Lowest	Highest	<0.5	0.5 to 1.0	1.0 to 2.0	>2.0
Electronic							
1. Temperature trip amplifier	0.53	0.01	3.85	44	17	6	2
2. Shutdown amplifier	0.61	0.01	5.13	36	22	8	4
3. Gamma monitor head amplifier	2.32	0.01	21.05	15	9	17	29
4. Gamma monitor	2.64	0.01	17.65	18	10	14	29
5. Gamma monitor	3.34	0.10	29.44	12	10	13	36
6. Gamma monitor	5.28	0.05	68.25	12	20	9	29
7. Pulse head amplifier	0.73	0.01	4.00	35	24	6	6
8. Pulse main amplifier	0.52	0.004	4.00	49	15	5	2
9. Pulse discriminator	0.51	0.002	4.22	47	13	10	1
10. Pulse logarithmic ratemeter	0.95	0.01	6.06	30	18	14	9
11. Period meter	1.34	0.01	5.56	21	14	17	18
Mechanical and pneumatic							
12. Pneumatic controller	2.69	0.03	18.95	16	13	13	29
13. Pressure reducer and relay unit	8.57	0.12	215	7	10	10	44
14. Differential pressure transmitter	1.28	0.01	7.89	27	12	15	17
15. Pressure switch	5.49	0.07	45.0	14	11	8	38
16. Pneumatic transmitting flowmeter	2.61	0.02	23.81	26	14	10	21
Delegates' means	2.46	0.10	13.61	6	16	15	34
Electronic means	1.71	0.04	10.68	13	20	20	18
Mechanical means	4.13	0.09	43.13	5	10	20	36

Table 4.2 List of actual operational failure rates and estimates

Equipment	Actual failure rate per 10^6 hr	Number of estimates	Estimated failure rates					
			Mean	Standard deviation	Skewness	Range	Median	Geometric mean
Inertial navigation system								
1. Gimbal assembly W/servo cards Stable element	113	54	578	1031	2.55	0.05–5000	120	117
2. Quantizer	28	53	126	253	3.90	0.05–1500	35	40
3 Computer cards (4 cards) Computer interface cards (3 cards) Program memory (2 cards) Random access memory Gyro bias memory	49	53	271	436	3.01	1.0–2400	100	66
4 RX/TX (2 cards)	6.4	53	73	95	1.66	0.05–400	30	30
5 D/A card A/D converter/A/D mult (2 cards)	29	54	100	165	3.54	0.05–1000	40	41
6 CDU (less cards) CDU cards (4 cards)	67	53	187	280	2.05	0.1–1100	55	58
7 Complete system	292	61	1331	1861	1.94	1.3–7200	485	455
Ground proximity warning system								
8 Complete system	111	57	309	422	2.21	0.5–2000	133	126

Table 4.3 Distribution Fitting

Equipment	Distribution of failure rates per 10^6 hr				
	Best fit	Mode	Median	Mean	95% band
1	Lognormal	0.18	105	2518	0.75–14680
2	Weibull	0.00	39	112	0.13–672
3	Weibull	0.00	118	250	1.09–1263
4	Weibull	0.00	32	81	0.17–455
5	Weibull	0.00	41	107	0.19–610
6	Lognormal	1.20	57	390	1.21–2660
7	Lognormal	35.7	513	1947	20.9–12600
8	Lognormal	12.8	130	417	6.6–2591

equipment which were mainly electronic. Estimates of failure-rates were obtained from sixty-two participants and other questions on extreme limits were asked which are not discussed here (see Green[98]).

Table 4.2 shows the actual failure-rates derived from airline operating experience in the USA. It may be noted that not all the sixty-two participants gave an estimate for each equipment. Statistical tests were carried out with probability distribution fitting and the lognormal and Weibull distributions gave the best fit and the results are shown in Table 4.3. The particular type of fit reflected the extreme skewness of the data. In general the results are in keeping with other experience that people tend to estimate logarithmically. It was shown that the logarithms of failure-rate followed a normal distribution and Table 4.4 shows the results for the means of the fitted distribution obtained for each equipment together with the best estimates of failure-rates derived from these means. It may be noted in passing that the choice of a suitable statistic is simplified since the mode, median and mean coincide.

Table 4.4 Best estimates of failure rates per 10^6 hr

Equipment	Mode/median/mean of distribution of LN (failure rate)	Best estimate of failure rate
1	4.82	124
2	3.81	45
3	4.54	93
4	3.42	31
5	3.83	46
6	4.08	59
7	6.24	513
8	4.89	133

Table 4.5 Comparison of the ratios of estimated to actual failure rates

Equipment	Ratio best estimate/ actual failure rate	Distribution of individual estimates: Ratio estimate/Actual failure rate				No. of estimates
		<0.33	0.33–1.0	1.0–3.0	>3.0	
1	1.1	16	11	10	17	54
2	1.6	13	10	10	20	53
3	1.9	12	4	16	21	53
4	4.8	7	3	15	28	53
5	1.6	11	11	12	20	54
6	0.89	17	11	12	13	53
7	1.8	11	9	18	23	61
8	1.2	12	15	14	16	57

Table 4.5 shows comparisons of the failures rates derived from the means of the normal distribution of the logarithm of failure-rates with actual failures found in practice by a number of USA airlines. It will be seen that for each equipment at least one-third of the estimates lie within a factor of three of the actual value. Furthermore, the estimated values were within a factor of five of the actual values. It should be borne in mind that there was a severe time limit imposed during which these estimates were made. A good agreement was obtained between the estimated and the actual failure-rates and it could be noted that deriving the mean of the particular normal distribution used here was equivalent to taking the geometric mean of the individual estimated failure-rates.

This type of technique extending to modified Delphi techniques have been applied to basic data and equipment, by Shooman and Sinkar[110] and by Booth *et al.*[111]. Furthermore a guide giving reliability data derived by the use of these techniques and applicable to nuclear-power generating stations has been published as a result of the IEEE Project 500[112] which was set up in the USA for this purpose.

It will be observed from the original UK experiment in 1966 and the USA experiment in 1977 that the comparisons indicate that this technique can form a basis where there is a lack of data. Informally, the technique has been used on other occasions for various reliability characteristics in reliability assessment work and, in general, has been quite useful. Often in the initial stages of an assessment very 'coarse' data are often sufficient for sensitivity types of analyses. Usually this type of prior estimate using expert opinion can be surprisingly good when posterior information is used to up-date the prior. From an engineering point it can be a useful tool if the limitations and assumptions are given careful consideration. However, there can be controversy on the use of such methods as they appear to be somewhat novel and more conventional methods using data banks have been researched and developed which are now discussed.

4.3 Reliability Data Banks

A data bank may be defined as a comprehensive collection of libraries of data and usually it is centralized although this is not an absolute essential. The rate of increase of knowledge has been so great that in most industries greater attention has been given on plant and throughout whole organizations to the collection, storage and analysis of data.

In the 'real world' these data come from manufacturer's tests or from the actual operation of the plant. Most reliability workers like in the ultimate to have data from the system operating under the plant environment. However, the user of the data may find that there is not the luxury of time to read all the reports on the various experience and the actual events which have taken place. Therefore, there is a need to analyse the data in the form necessary for the synthesis of the system's reliability or perhaps its correlation with predictions which have already been carried out. Obviously there are other functions to be met by the data bank under the headings of plant management and reliability management. Figure 4.2 shows some management activities utilizing data.

Over the past two decades there have been intensive efforts to research and develop data banks. Often the data bank in its ultimate form has needed government backing and takes on a national character. The nuclear industry have been actively engaged in bringing into being the appropriate type of reliability banks. Goarin *et al.*[113], gives a description of an electronic data

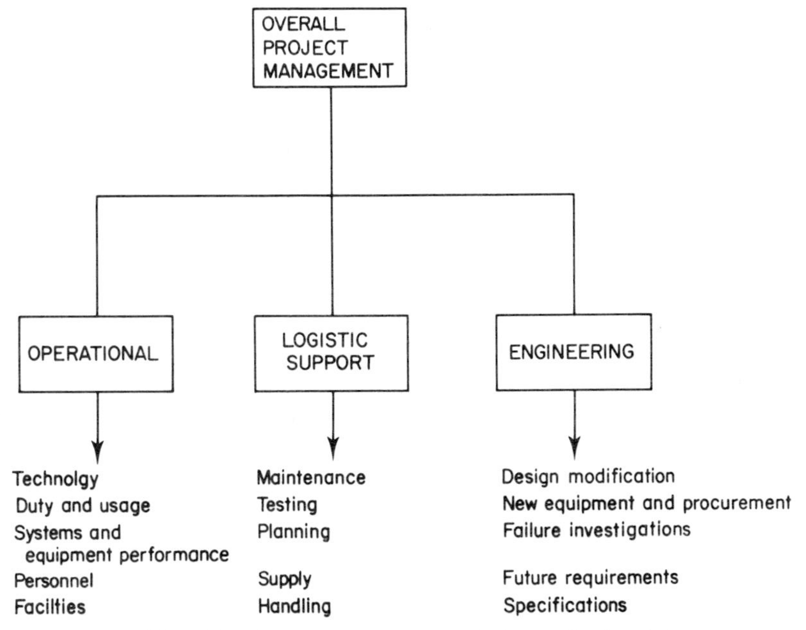

Fig. 4.2 Some management activities utilizing data

bank in France with an example of the use of the data applied in assessing the reliability of the protective system of a pressurized water reactor. Since the topic of data banks is so wide it is proposed to concentrate on the SYREL data bank developed in the UKAEA under the auspices of the National Centre of Systems Reliability. This data bank will serve to bring out the various facets of the techniques involved.

4.3.1 SYREL data bank

Arising from the need to report abnormal events or incidents on nuclear plants within the UKAEA there was a formalized system set up in 1961. Obviously this was orientated to qualitative information rather than quantitative which was the requirement of the reliability assessor. Hence, at this time, each analysis of reliability data was undertaken mainly by the endeavours of the reliability and safety assessor in conjunction with plant personnel. Certain classical features emerged for plant performance in the field, for example the performance of research reactors at the UKAEA, Harwell, was established by using a system of data collection as from 1960. Moore[114] describes the type of data scheme which was used for this purpose and comments on the quantitative nature of the resulting reliability and availability analyses. After intensive reviewing and a pilot exercise carried out in conjunction with the Safety and Reliability Directorate (formerly Health and Safety Branch) of the UKAEA, the CEGB, SSEB and UKAEA operators, a specification emerged for a practical system for the collection, coding and storage of reactor plant

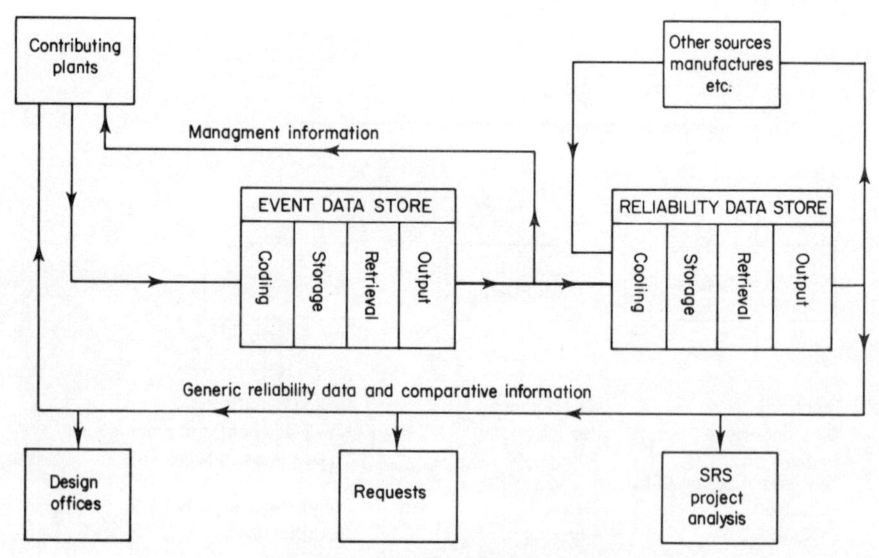

Fig. 4.3 The SYREL data bank

data, described by Eames[115]. This work in about 1967 gave a foundation for the System Reliability Service Data Bank (SYREL), outlined by Ablitt[116].

SYREL is a comprehensive reliability bank and has been an essential part of the process of applying reliability technology by the Systems Reliability Service (SRS). Typically, SYREL, is shown in Fig. 4.3. The bank consists of the event-data store and the reliability-data store. It is feasible for automatic processing of the raw data in the event-data store to be carried out periodically to provide an input into the reliability-data store. However, this can cause problems and manual checks are carried out during event-data analysis before transfer to the reliability-data store.

The event data is collected from the plant and coded in accordance with an agreed convention. It is then stored in computer files. There is a high degree of output flexibility from this store which can be, for example, just a routine feedback for the guidance of plant management, or the derived generic reliability which is one of several inputs to the reliability data store.

For different items of information it is necessary to draw up an inventory of technical data of a type shown in Table 4.6. Next it is important to have a report from which event data may be derived as shown in Tables 4.7 and 4.8. These event data are obtained from documentation such as job cards or work orders which are commonly used in industrial plants. The data are then stored in computer files and analysed according to the particular system and installation.

Reliability data can be derived from the event-data store as shown, for example, in Table 4.9 for a nuclear reactor period meter which could be an

Table 4.6 Item inventory data

Inventory number
Item description (words)
Identity (mark, type)
IDEP code number
Manufacturer
Designer
Serial number
Design year
Commissioning date
Range/size
Quality
Circuit type
Material of composition
Environment
Application
Maintenance interval
Operating-time function
Operating-cycles function
Population
Entry date into system

Table 4.7 Event data

Event number
Type of event
Time
Date
Installation
System
Items faulty
Installation effect
Effect on system
Operating state (installation)
Shutdown information (installation)
Outage hours
Fault importance
Site reference number
Skilled labour—man hours
General labour—man hours
Materials and parts cost—£
Other descriptive information

item used in a protective system. Such data would be stored in the reliability-data store.

Furthermore, generic data can be derived so that failure-rates on various items can be built up for use by the reliability assessor. Such information

Table 4.8 Faulty item data

Event number
Inventory number of item
Description of item (SRS code)
Failure mode
Cause of failure
Category of fault
Operating parameters of item:
Media handled
Temperature
Pressure
Fluid flow
Speed rpm
Position in installation
Hours outage
Active repair time
End of life (date)
Replacement item inventory number
Details of repair (replaced parts, etc.)
Measured or estimated time since previous event

Table 4.9 Example of reliability data for a nuclear reactor period meter with trip facility

Location	'X'	
Manufacturer	'C'	
Identity	'Y'	
History time	26 000 hr	
Mean operating time	11 400 hr	
Items × time	205 000 item hr	
Number of items	18	
Number of faults	22	
Operating failure rate	Mean value	107 faults/10^6 hr
	Lower confidence limit	67 faults/10^6 hr
	Upper confidence limit	162 faults/10^6 hr
Confidence band	95.0%	
Distribution assumed	Exponential	

derived from the event-data schemes can provide high quality reliability information for the various reliability assessors and for other uses such as maintenance on the plant.

Another source of input information to the reliability-data store is from reports in the open literature, special studies carried out on plant including student exercises, reliability assessment projects which have been undertaken by SRS project engineers, and data exchange schemes. The general flow of this reliability input information is shown in Fig. 4.4.

A useful illustration of the banking and analysis of information from different

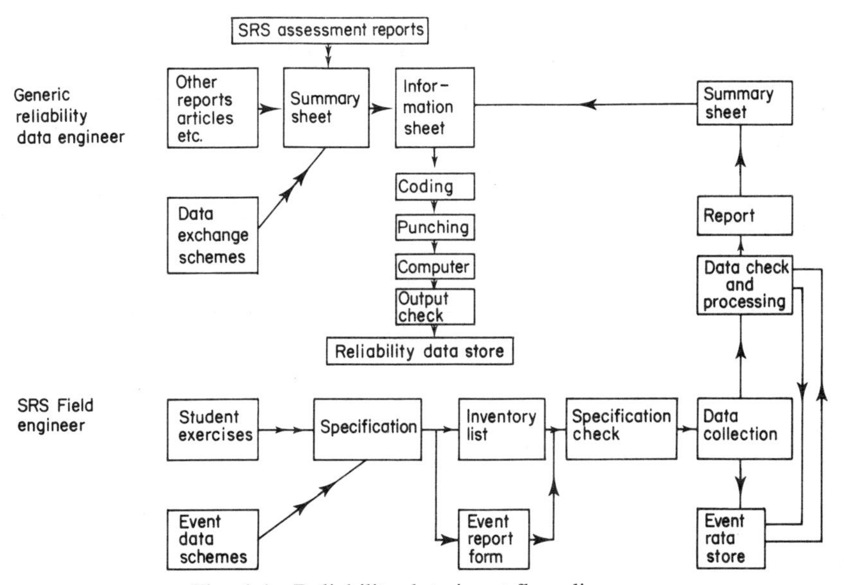

Fig. 4.4 Reliability data input flow diagram

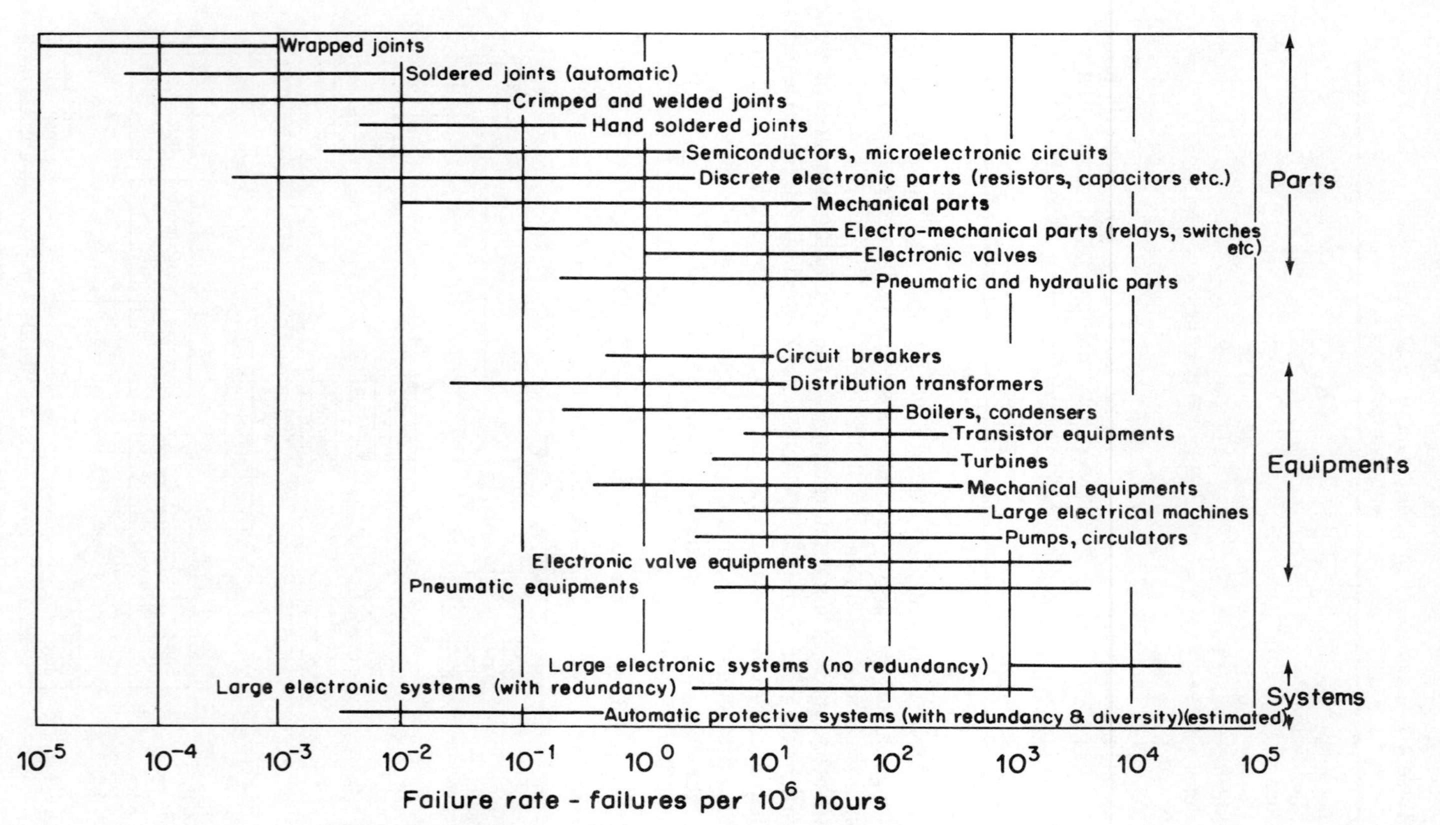

Fig. 4.5 Failure rates for parts, equipments and systems

sources is to permit the study of the same item used under different stress and environmental conditions which can be shown in general to give correspondingly different failure-rates. Hence, if there is sufficient statistical evidence, a series of failure-rate curves for each condition of stress and environment can be prepared. Figure 4.5 shows, as an example, the spread of some items from selected fields of experience. Usually a basic failure-rate θ_o under some standard conditions is used and κ factors are tabulated to allow for varying stress and environmental conditions. This enables a failure-rate θ_a to be derived for the actual conditions, where

$$\theta_a = \theta_o \prod_{i=1}^{i=n} \kappa_i \tag{4.1}$$

Therefore θ_a will be an actual weighted failure-rate for each item in its application. The author has reviewed in more detail the failure-rate available for various items of plant automatic protective systems and the summarized results are given in Appendix 5.

More detailed information on the analysis and presentation of derived reliability data is given in Fothergill[117] and some reliability characteristics for operating plant are given by Eames and Fothergill[118]. Developments in the SRS Reliability Data Bank are further described by Moss[119] and Daniels[159].

From the survey carried out of the data banks such as SYREL it is obvious that experience has shown that well-structured data-bank systems exist to permit the use of modern data-base techniques to be applied. However, an important point emerges where data is sparse and the frequency of events is such that it raises questions as to the methods for deriving the necessary posterior information on the reliability characteristics in the required form.

4.4 The Correlation Between Predicted and Practical Results

In the preceding chapters various methods of modelling, analysis and synthesis of reliability have been investigated and the developments discussed. Clearly variance in the prediction can result due to the various assumptions made in the mathematical model and the errors in the data and its application. Equally the methods of data collection and its recording as discussed in the previous sections can introduce variability. This can arise due to the practical difficulties in detecting, classifying, timing and recording all the changes and fault modes of the actual system behaviour.

It is, therefore, pertinent to investigate the precision of the theoretically predicted values of the various reliability characteristics to compare them with the reliability actually achieved in practice for the same system over a suitably long period of operation. This is a typical investigation which has been undertaken for a range of systems and their elements. Usually the predictions are carried out at the design or production stage and the reliability performance is subsequently monitored over a reasonable sample period of practical operation. Two investigations which have been undertaken are given, one for protective system trip initiators and the other for a diverse range of equipments and systems.

4.4.1 Plant protective system trip initiators

A reliability assessment was carried out in 1969 on a chemical plant in order to predict the reliability of protective channels containing special circuits of electrical, electronic and pneumatic components. Several of these predictions were based on failure modes and effect analysis of the equipments and the allocation of failure-rates for various components, thereby enabling the overall predicted failure-rate of an equipment to be derived. Thus each equipment can be deemed to represent a small system comprising many elements.

In the subsequent 7 years which elapsed since the report was written, field data were accumulated on specific equipments. A comparison of the mean observed values with the original predictions is contained in Table 4.10. Column 3 represents observed data supplied by the plant operator from their operational records. Following the receipt of these data various discrepancies and lack of understanding of data interpretation necessitated further investigation of the data and the relevant records. A summary of the field failure-rate data for the same equipment are recorded in column 4 of Table 4.10. It will be noted that there is not a great difference. Nevertheless the separate investigation gave confirmation of and confidence in the field data collected by the operator. In addition, further information was obtained such as causes of faults, faults due to maloperation, common-mode faults affecting redundant units, e.g. fire, wrong setting up, which provides further explanation for the deviation for predicted and observed values.

Table 4.10 Comparison of predicted with observed failure rate data for chemical plant protective channel trip initiators

Equipment (1)	Predicted faults/year (2)	Direct from plant reporting systems faults/year (3)	Investigation of plant records faults/year (4)	Population for Col. data (5)	Ratio O/P (6)
1 Pressure switch	0.14	0.5	0.7	57	3.57
2 d/p transmitter and power supply	0.58	1.34			2.3
3 Trip amplifier	0.39	0.92	1.13	39	2.36
4 Level switch	0.22	0.64	0.62	6	2.9
5 O_2 analyser plus trip amplifier	1.9	1.2	1.3	15	0.63
6 Pneumatic differential pressure transmitter	0.59	0.94		12	1.6
7 Pressure gauge	0.088		0.032	54	0.36

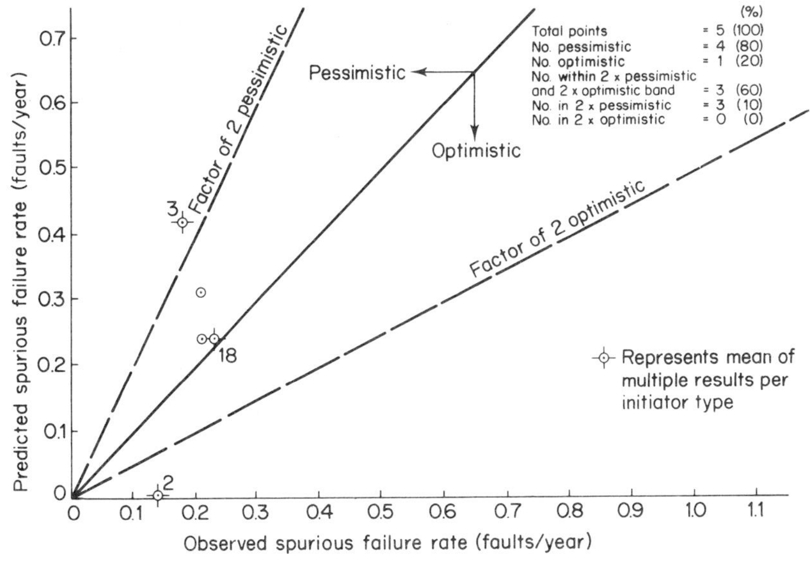

Fig. 4.6 Plot of the mean observed failure rates from twenty-three items of field data on five protective channel equipments against predicted values (spurious faults)

A breakdown of the total faults recorded in column 3 into fail-safe, fail-danger and neutral effect faults was also provided by the operators.

Figures 4.6 and 4.7 show plots of the predicted fault rates with the corresponding observed values for spurious and dangerous failures for all equipments, i.e. a total of twenty-three observed data values. However, only five points are plotted since each point represents the mean value of several similar equipments for which independent data was collected. This will be seen by reference to Table 4.11 which lists the population of each equipment. It will be seen that all points lie within the 1 : 3 ratio of predicted to observed results. For the total fault rate the predictions are optimistic, i.e. more observed faults occurred than were predicted. For the fail-danger results predictions were again slightly optimistic, 60 per cent being a factor of approximately 1.5 and 40 per cent (two equipments) approximately 2.2. These could be because although all observed faults may be recorded they may not all be what may be termed 'random'. Some observed faults could be design and/or operating mode faults, e.g. overstressed by temperatures, pressures, environment etc., and others may be due to some common influence affecting more than one item. This would tend to increase the observed number over the predicted value. Investigations of other field data have shown similar results.

The graphs represented by Figs 4.6 and 4.7 plot the ratio, observed failure-rate/predicted failure-rate versus observed failure-rate. A best-fit line through the points demonstrates a definite correlation, i.e. the prediction

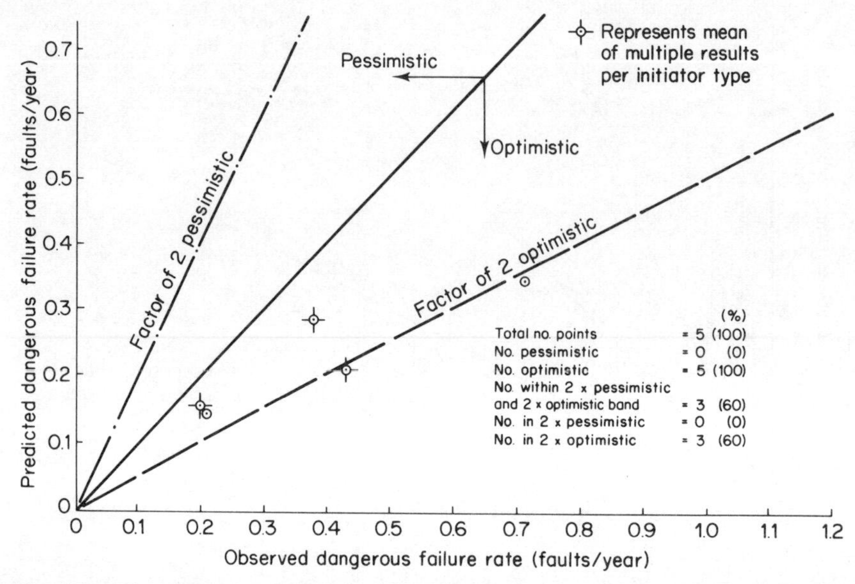

Fig. 4.7 Plot of the mean observed failure rates from twenty-three items of field data on five protective channel equipments against predicted values (dangerous faults)

method over-estimates low failure-rate equipment but underestimates high failure-rates.

It could be argued that errors associated with prediction should be pessimistic, particularly for safety-related systems, and since equipment used in such systems of necessity require to be of high reliability, i.e. low failure-rates, then an over-estimate of low failure-rate equipment is in the right direction. For equipment with high failure-rates the need for predictions is not so great since failures would manifest themselves and hence a quick record of operating failures would be available. Thus the underestimates of prediction are quickly recognized by comparison with early field data.

4.4.2 Diverse equipments and systems

Various exercises to correlate the prediction of different types of equipments and systems have been undertaken or, in some cases, have been more comprehensive and detailed than the one quoted for the protective system trip initiators (Green and Bourne[49]), and consisted of an investigation of about fifty different system elements which were examined and ratio, r, of the value of the observed failure-rate to the corresponding predicted failure-rate has been obtained for each system so studied. It was found that the distribution of r due to various causes of variation was approximately lognormal. Also it was found that the median value of r is 0.76, that the chance of r being within a

Table 4.11 Summary of failure data—all modes

Invent. No.	Item population	Total failures	Break-down failures	Danger failures	Operating time: Item Years	Calendar time	Breakdown failure-rate	Danger failure-rate	Total failure-rate
1	3	7	5	0	13.5	15.45	0.324 (002)	0	0.453 (001)
2	15	80	50	22	62.4	72.15	0.693 (005)	0.305 (006)	1.109 (004)
3	24	22	22	0	102.88	118.5	0.186 (009)	0	0.186 (008)
4	18	15	15	0	75.906	87.588	0.171 (013)	0	0.171 (012)
5	18	15	15	0	75.906	87.588	0.171 (015)	0	0.171 (014)
6	24	21	21	0	102.88	118.5	0.177 (017)	0	0.177 (016)
7	39	215	107	44	165.28	190.67	0.561 (019)	0.231 (020)	1.128 (018)
8	6	19	3	13	27	30.9	0.097 (024)	0.421 (025)	0.615 (023)
9	6	0	0	0	27	30.9	0	0	0 (027)
10	12	19	2	3	54	61.8	0.032 (029)	0.049 (030)	0.307 (028)
11	3	3	7	1	13.5	15.45	0.065 (033)	0.065 (034)	0.194 (032)
12	3	1	0	0	13.5	15.45	0	0	0.065 (036)
13	3	1	1	0	13.5	15.45	0.065 (038)	0	0.065 (037)
14	15	66	8	20	67.5	77.25	0.104 (040)	0.259 (041)	0.854 (039)
15	15	30	4	14	62.4	72.15	0.055 (045)	0.194 (046)	0.415 (044)
16	3	8	1	4	13.5	15.45	0.065 (049)	0.259 (050)	0.518 (048)
17	3	5	0	2	13.5	15.45	0	0.129 (053)	0.324 (052)
18	3	9	0	6	4.422	4.5	0	1.33 (056)	2.0 (055)
19	18	88	9	55	81	92.7	0.097 (059)	0.593 (060)	0.949 (058)
20	51	8	0	2	215.32	246.64	0	0.008 (064)	0.032 (063)
21	3	0	0	0	13.45	15.45	0	0	0 (066)
22	3	1	1	0	13.5	15.45	0.065 (068)	0	0.065 (067)
23	3	1	1	0	13.5	15.45	0.065 (070)	0	0.065 (069)

1 Operating time and calendar time are measured in item years.
2 Failure rates are measured in failures per unit per year.
3 All failure rates are based on calendar time and *not* operating time.
4 Data reference numbers are shown after the relevant failure-rate in parentheses, e.g. (001) etc.

factor of 2 of the median value is 70 per cent and that the chance of the ratio being within a factor of 4 is 96 per cent.

On recent evidence, in looking at systems and elements which have been analysed at the design or production stages, and the data on reliability performance subsequently quoted during a period of operation, then reasonable agreement is shown. It may be deduced that if the techniques are carefully applied then the normal values yielded for the failure-rates of the system studied can reasonably be expected to be within a factor of 2 either way of the actual figure. This is typically illustrated in Fig. 4.8 which shows the predicted values against the observed values (in faults per year) resulting from an investigation covering over 130 items of diverse equipments of an electronic, pneumatic or mechanical nature, also control, instrumentation, electrical and mechanical systems. Figure 4.9 is a plot of the same data which shows the ratio of the observed to predicted failure-rates against the proportionate frequency of such ratio being less than or equal to the specified ratio value. This

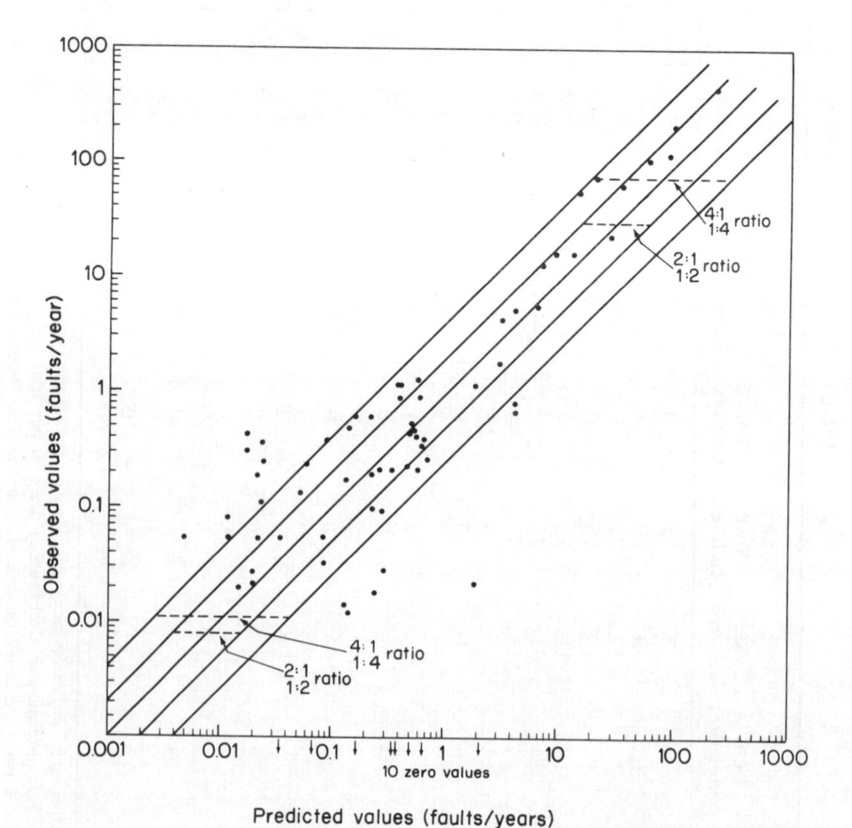

Fig. 4.8 Predicted values against the observed values (in faults per year) resulting from an investigation covering over 130 items of diverse equipment and systems

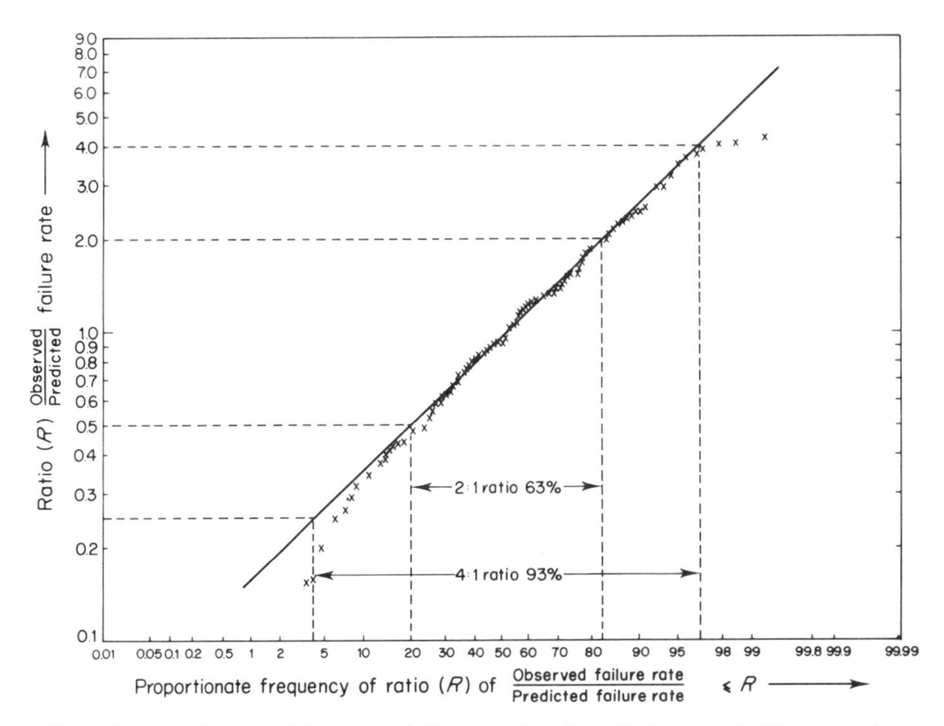

Fig. 4.9 Plot of ratio of (observed failure rate/predicted failure rate) (R) versus the cumulative proportion of values of the ratios $(R) \leq R$ for 130 items of engineering equipments and systems

plot indicates that 63 per cent of the items analysed have a ratio of observed to predicted failure-rate within a factor of 2 : 1 and 93 per cent within a ratio of 4 : 1, and is further commented upon by Snaith[120].

4.5 Limitations of Conventional Estimating Techniques

The estimation of the various reliability techniques really commences from the time of the original design concept and it is the designer who can initiate the preparation of the item inventory. This can be done in conjunction with the reliability assessor as part of the overall safety assessment of the protective equipment and system.

However, this is not often the case and other people later in the process may not have the insight of the original designer in deducing correctly the modes of failures and their effects. It is not sufficient to have some abstract statistical data but it requires to be 'married' properly with the appropriate 'materialistic' aspects of the equipment and system.

Furthermore, in the original design specification and its assessment the various performance characteristics such as response, accuracy and reliability require to be defined and quantified. These characteristics need to be in a form such that later in the overall reliability programme the estimating of the characteristics cannot be left for a statistician just to do the best. This approach has from experience produced data but it is not possible to correlate them in any particular way because they represent different hierarchical levels than those required to prove the original design intent. In the previous section on correlation this is illustrated by the characteristic of failure-rate where the failure-rate prediction must be of the same form as that being monitored and analysed in practice.

The techniques described for the monitoring and analysis of the various reliability characteristics tend to require relatively large samples where classical statistical analysis is undertaken. There is basically, for example, available the method of the Poisson assumption and making a one-side confidence statement on the basis of zero, one or more failures. Of course this can lead into various discussions on the validity of the assumptions. In general there does not appear to be among the classical statistical techniques, methods which can be readily used in a 'dynamic' sense for the analysis of the slow build-up of information, i.e. in sample testing or field experience.

With highly reliable equipment of the type found in protective systems and the overall system itself tend to produce only small samples of failures of real safety significance. This is as would be expected by the original design intent for a system and equipment of high reliability. Hence, in practice, it is found there is a combination of methods at play which extends from 'hard' scientific method to 'soft' subjective judgement which, when combined with the statistical data, enables probabilistic statements to be made. This interplay is multi-disciplinary involving engineering, statistics, reliability technology, manufacturing and plant operation. This study programme has therefore included looking at the real world situation on different nuclear reactor installations taking the protective system as the example for study. The build-up of the actual chronological information is considered and the techniques which appear to be experimentally feasible have been applied and the results presented. This investigation is considered later in the chapter.

4.6 The Bayesian Approach

4.6.1 Bayes Theorem

As indicated earlier the frequency information may be generated slowly and although it is actual fact it is often not sufficient to give the means of analysis of the events of interest. In engineering the engineer uses feel or subjective judgement and develops a belief in the work being undertaken. Hence the present study programme investigates the results of taking belief into account in a probabilistic way as well as the more factual use of the actual frequency of events devoid of any subjectivity. Since it is proposed to use Bayes Theorem which was given in an essay in 1763 (see Crellin[121]), it is discussed in

Appendix 6 and may be quoted in the following form:

$$P(A/B) = P(A)\frac{P(B/A)}{P(B)} \tag{4.2}$$

where $P(A)$ = 'Prior' probability which is what is known before additional evidence becomes available.

$P(B/A)$ = 'Likelihood' which is how probable is the new evidence, assuming that hypothesis A is true.

$P(B)$ = The probability of the evidence B alone.

$P(A/B)$ = The 'posterior' probability of A now that evidence B has been added to the existing knowledge.

It is clear that Bayes Theorem gives a format which is convenient for continually updating probability estimates on the basis of gaining new knowledge. Obviously the process may be continued if further new knowledge C becomes available. The new 'prior' is $P(A/B)$, the new 'posterior' is $P(A/BC)$ and so on. The probability $P(B)$ of the evidence of B alone may not be easy to handle but it can be written:

$$P(B) = \sum_{A_i} P(B/A_i)P(A_i) \tag{4.3}$$

when A is a discrete random variable. If A is a continuous variable then:

$$P(B) = \int_A p(B/A)\,p(A)\,\mathrm{d}A \tag{4.4}$$

Hence Bayes Theorem is sometimes written in the form for the discrete cases as:

$$P(A/B) = P(A)\frac{P(B/A)}{\sum_{A_i} P(B/A_i)P(A_i)} \tag{4.5}$$

For the continuous case it is written with an integral sign.

4.6.2 The application of the Bayesian approach

Let a statistic y be the random times to say $t_1, t_2, \ldots t_i$ of a population having a failure pdf of $f(t, \alpha)$ where α is some shape factor. In the classical statistical inference approach α is assumed to be a predetermined constant. Hence from prior experience the belief in the value of α may be represented by a pdf designated as $g(\alpha)$.

Consequently there exists a conditional pdf $g'(\alpha \mid y)$ which is the posterior and redefines $g(\alpha)$ in the light of the experiment y. Hence from equation 4.2 this can be written:

$$g'(\alpha \mid y) = \frac{g(\alpha)f(y \mid \alpha)}{f(y)} \tag{4.6}$$

Since $f(y)$ is simply a normalization constant for a fixed y this gives:

$$g'(\alpha \mid y) = g(\alpha) \cdot f(y \mid \alpha) \tag{4.7}$$

where $f(y \mid \alpha)$ is the likelihood function and modifies the prior $g(\alpha)$ to the posterior $g'(\alpha \mid y)$.

In the experiment described the likelihood function L is based on a set of N independent observations from a total set having a probability distribution represented by $f(t_i, \alpha)$, hence

$$L = \prod_{i=1}^{N} f(t, \alpha) \tag{4.8}$$

where α is a chosen constant.

The basic problem is to find the one value of α which gives a best fit to the observations. This will occur when L reaches a maximum as α is varied, i.e. $\partial L/\partial\alpha = 0$. When L reaches a maximum, its logarithms will also, so for mathematical convenience, $\log L$ is used for the optimization process and

$$\log L = \mathscr{L} = \sum_{i=1}^{N} \log f(t_i, \alpha) \tag{4.9}$$

Unless otherwise stated, all logarithms in this book are to the base 'e'.

Equation 4.9 bears some resemblance to the expression for entropy used in information theory inasmuch as each additional failure time contributes an extra 'bit' of information $\ln f(t_i, \alpha)$.

This process may be illustrated for the exponential distribution by assuming for the experiment that

$$f(t, \alpha) = \alpha\, e^{-\alpha t} \tag{4.10}$$

then

$$\mathscr{L} = \sum_{i=1}^{N} \log \alpha - \alpha t \tag{4.11}$$

$$= N \log \alpha - \alpha \sum_{i=1}^{N} t_i \tag{4.12}$$

$$\frac{\partial \mathscr{L}}{\partial \alpha} = \frac{N}{\alpha} - \sum_{i=1}^{N} t_i$$

$$= 0 \text{ for best fit} \tag{4.13}$$

which gives

$$\frac{1}{\alpha_1} = \frac{1}{N}\sum_{i=1}^{N} t_i = \bar{t}(\tau) \tag{4.14}$$

where $\bar{t}(\tau)$ is the average t over the time interval τ and α_1 is the optimum value of α.

This derived value of α would give the best fit for the observed data in the absence of any prior information. However, it is assumed that from experience there is prior information which would suggest that α would have a different value from that in the foregoing derivation and this, say, is nominally α_0 with a distribution given by

$$g(\alpha) = \frac{1}{\alpha_0} e^{-\alpha/\alpha_0} \tag{4.15}$$

Equation 4.15 represents the belief in an expected α_0 for the variable α.

It is, therefore, now possible to proceed to modify the likelihood function on the foregoing observations by the prior belief expressed by $g(\alpha)$. This gives

$$\begin{aligned} L &= \prod_{i=1}^{N} f(t_i/\alpha) \cdot g(\alpha) \\ &= \prod_{i-1}^{N} (\alpha e^{-\alpha t}) \left(\frac{1}{\alpha_0} e^{-\alpha/\alpha_0} \right) \end{aligned} \tag{4.16}$$

Taking logarithms

$$\begin{aligned} \log L = \mathscr{L} &= \sum_{i=1}^{N} \left(\log \alpha - \alpha t_i - \log \alpha_0 - \frac{\alpha}{\alpha_0} \right) \\ &= N \log \alpha - N \log \alpha_0 - \frac{N\alpha}{\alpha_0} - \alpha \sum_{i=1}^{N} t_i \end{aligned} \tag{4.17}$$

$$\frac{\partial \mathscr{L}}{\partial \alpha} = \frac{N}{\alpha} - \frac{N}{\alpha_0} - \sum_{i=1}^{N} t_i \tag{4.18}$$

$$= 0 \text{ for best fit} \tag{4.18}$$

which gives

$$\frac{1}{\alpha_2} = \frac{1}{\alpha_0} + \frac{1}{N} \sum_{i=1}^{N} t_i$$

where α_2 is the best fit value of α with the prior information. This shows that the optimum value of α deduced from observations alone has been modified by the extra term $1\alpha_0$. Furthermore, it will be noted that at $t = 0$ (i.e. $\Sigma t_i = 0$) then $\alpha = \alpha_0$ as the optimum value which is of course the prior information only.

4.6.3 Practical considerations in the application of the Bayesian approach

In the present study the Bayesian approach is investigated as a tool for the analysis of information derived from the actual operation of plant systems and this work is described later. However, it is pertinent at this stage to point out that there is also an area of difficulty in the application of any technique and in the Bayesian application it tends to be in the development of assigning the prior distribution. Furthermore, this involves consideration of other estimates for instance the objective of the Bayesian measurement, the decision parameters which are chosen and the available actions. In addition, it is necessary to consider a complete posterior as well as just the prior. It is considered that for the present work that a reasonable assumption is that an appropriate index of reliability is that of failure-rate and that cyclic dependent modes could be considered as initially not requiring specifically to be modelled.

A review of previous experience in the practical application of Bayesian techniques gave some definite pointers for the present research work which were worthy of further investigation. Crellin[121], stated that in the aerospace field an effective posterior had been employed based on the use of the Gamma distribution for the prior on the failure-rate.

In the original application it was assigned that the prior mean equalled the predicted failure-rate (θ_0) and the assumption was made that the appropriate decisions could be obtained by consideration of the posterior mean. Extensive pre-posterior analysis was undertaken which considered the adequacy of the posterior mean to the decision for various imagined data situations and it was determined that the prior coefficient of variation (m^2) should be such that $0.75 < m < 1.75$. It was decided to use $m = 0.75$ when no failures were present in the data ($\kappa = 0$) and $m = 1.75$ when failures were present ($\kappa > 0$). The posterior mean was calculated with m as determined by the failure circumstances. This permitted the 'tracking' of a single convenient decision parameter for each of a large number of components and easily monitors the trend of the posterior mean.

It is reported by Crellin that some subsequent studies of ground-based electronics showed observed rates for generic items to be distributed as gamma distributions with the associated coefficient of variation for each falling within the range of $0.75 < m < 1.75$. Furthermore, the plot which was carried out by Green and Bourne[49] (Fig. 13.4), shows a cumulative plot of the ratio $r = \theta_{obs}/\theta_{pred}$ (observed failure-rate/predicted failure-rate) to be distributed as a lognormal with $m = 0.75$. This plot had been obtained mainly from ground-based electronics and since the lognormal differs little from the gamma, it would appear to support such an assignment process.

The investigation of the correlation between the observed failure-rate and predicted failure-rate discussed previously in this book shows in Fig. 4.9 a ratio r for a much wider range of engineering equipments and systems. However, similar lognormal distribution characteristics are shown for the ratio r, although the values of the parameters are slightly different than for the equipment involving mainly electronics. This strengthens further the assignment and investigation of a gamma prior. Waller *et al.*[122] have investigated in detail the time-to-failure modelled by an exponential distribution with failure rate θ and the Bayesian analyses of this assumed model. This leads to a family of gamma distributions which provide conjugate prior models for θ. The information has been specifically presented in graphical form for the engineer for selected percentiles. However, it was found necessary to extend the work from this reference so as to make it applicable to the present study, due to the high values of reliability under investigation. An important contribution which is described by Waller *et al.*[122] in their Appendix C is a theorem on the asymptotic behaviour of gamma percentiles which is as follows:

If, as $\alpha \to O^+$, $\kappa(\alpha)$ is defined by

$$\frac{1}{\Gamma(\alpha)} \int_0^{\kappa(\alpha)} e^{-x} x^{\alpha-1} \, dx = p \tag{4.19}$$

and if $0 < p < 1$ then

$$\kappa(\alpha) \sim p^{1/\alpha}$$

where $f(\alpha) \sim g(\alpha)$ means

$$\lim \frac{f(\alpha)}{g(\alpha)} = 1.$$

The proof of this theorem has been abstracted and is given in Appendix 7.

The precision on some incomplete gamma subroutines of calculation deteriorates as α approaches zero. The precise calculation of the necessary gamma percentiles for small values of α becomes difficult. To address this problem the foregoing theorem of equation 4.19 can be used. Let θ denote the pth percentile of gamma prior with parameters α and β.

Then

$$p = \frac{1}{\beta^{\alpha}\Gamma(\alpha)} \int_0^{\theta_p} \theta^{\alpha-1} e^{-\theta/\beta} \, d\theta = \frac{1}{\Gamma(\alpha)} \int_0^{\theta_p/\beta} x^{\alpha-1} e^{-x} \, dx \tag{4.20}$$

It may be noted that

$$h(\theta) = \begin{cases} \dfrac{\theta^{\alpha-1} e^{-\theta/\beta}}{\beta^{\alpha}\Gamma(\alpha)}; \theta > 0 \\ 0 \text{ otherwise} \end{cases} \tag{4.21}$$

therefore

$$\begin{aligned} p &= \int_0^{\theta_p} h(\theta) \, d\theta \\ &= \frac{1}{\beta^{\alpha}\Gamma(\alpha)} \int_0^{\theta_p} \theta^{\alpha-1} e^{-\theta/\beta} \, d\theta \end{aligned} \tag{4.22}$$

from which

$$p = \frac{1}{\Gamma(\alpha)} \int_0^{\theta_p/\beta} x^{\alpha-1} e^{-x} \, dx \tag{4.23}$$

From the theorem of equation 4.19,

$$\theta_p/\beta \sim p^{1/\alpha}$$

Thus $\theta_p \sim \beta p^{1/\alpha}$ for small values α and this result can be used to extend the table in Waller *et al.*[122], their Appendix A to smaller values of α.

The Bayesian procedure which has been discussed in this section deals with statistical knowledge. Engineering experience which improves the existing knowledge can only be included by entering into the prior and the appropriate definition of the particular statistical process. Questions of common cause failure potential have not been discussed and necessarily modelled in the foregoing process and common-mode failure is commented upon later in this book. References on the general applicability and rigidity of the Bayesian approach are given in Appendix 6. However, the foregoing Bayesian process has been used as part of the general experimentation in the analysis of operating plant systems and equipment and where possible the results have been compared with classical methods.

4.7 Analysis of Operating Plant Systems

The need for reliability control and management has already been discussed in the previous chapters. It has always been an important factor in achieving the safe operation of nuclear reactors. During the past two decades there has been an extension of the earlier engineering qualitative approach which tended to be somewhat deterministic to the quantified probabilistic approach. Hence it appeared profitable to consider basically two types of plant, first one which essentially used the earlier engineering approach but had the benefits of later thinking and secondly plant which used both the probabilistic approach from its initial concept as well as the earlier engineering qualitative approach.

As mentioned in Chapter 1, Section 1.2.3 the UK nuclear-power programme produced the world's first nuclear power station at Calder Hall which was formally opened on 17 October, 1956. The plant at Calder consists of four gas-cooled, graphite-moderated reactors and in parallel with this development at Chapelcross a further nuclear power station was constructed and completed later with four reactors similar to those at Calder.

Both the Calder and Chapelcross plants are still operational and the original design/operation and assessment were based on the earlier engineering approach. Therefore, the opportunity has been taken to illustrate retrospectively some of the ideas for reliability and implementation which have been part of the present study.

The second type of plant which utilized the probabilistic approach in its original design/operation and assessment is represented by the Steam Generating Heavy Water Reactor (SGHWR) which was formally opened in February 1968. Since this date it has been operational and generating electricity. In both cases the protective system and the shutdown system equipment have been taken as examples for considering some methods of viewing the prior assessments and actual experience. At the same time the opportunity has also been taken to consider any sample testing information applicable to the systems and their equipment. Furthermore it can be seen if any patterns or trends emerge to indicate techniques which may be applicable to view and assess reliability growth. For this purpose the information has been considered purely chronologically as it would be generated by the plant itself or any trials which may constitute sample testing.

Other kinds of nuclear plant systems operating in France and in the USA are considered in Appendices 8 and 9, pages 234 and 252, respectively.

4.8 Gas Cooled Type Reactor

Figure 4.10 shows a general view of two of the four Calder reactors and the overall design and construction features are given in various references (e.g. *The Engineer*[123] and Stretch[124]). A simplified diagram of the reactor is shown in Fig. 4.11. The commissioning of the Calder Hall first Reactor is described by Stretch[125].

The main safety features of this particular type of reactor tend to revolve round its relatively slow response to a wide range of fault conditions. Furthermore the behaviour under most fault conditions is to stabilize safely. An important feature is the large graphite heat sink in the core which markedly affects the time constants for any transient thereby delaying the time to melt down for an appreciable period of time. In the context of considering protective systems this is important as it makes emergency shutdown actions and

Fig. 4.10 General view of Calder Hall Power Station showing two of the four reactors

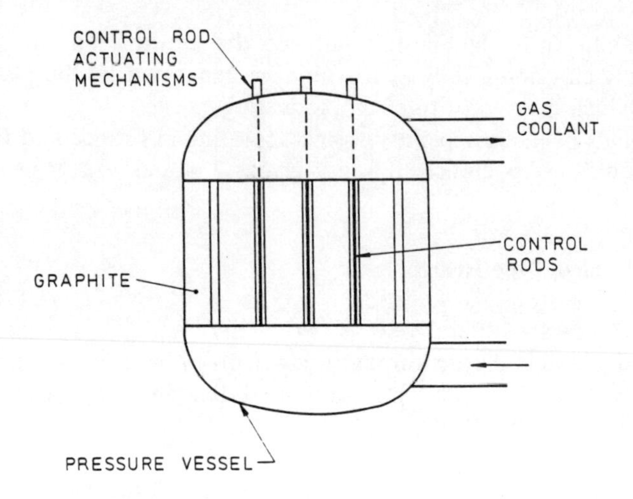

Fig. 4.11 Simplified diagram of a gas-cooled type reactor

other mitigating procedures realistic. Hence because of the slow fault-response time the control rods for emergency shutdown can act purely under gravity. Furthermore there are a relatively large number of control rods, i.e. forty-eight in number which helps in giving increased assurance of shutdown action. In effect, this is an example of the 'domino effect' mentioned in Chapter 1, Section 1.3 whereby the accidental condition is not allowed to develop by stopping a possible sequential action.

4.8.1 Control rod actuating equipment and system

The forty-eight vertical control rods move vertically by means of actuating mechanisms and by absorbing neutrons the control rods can be used to control the rate at which nuclear fission takes place. It may be noted that the actuating mechanisms combine the functions of control and shut-off. Vertical operation of the rods permits the use of a gravity fall for shut-off. Controlled braking is used so that a high launching speed can be attained with a slow touchdown. The complete equipment and system is described by Green[126] and Ghalib and Bowen[127].

Each control rod has an operational travel of 21 ft and is suspended by a low-cobalt, stainless-steel, flexible cable, which is edge-wound between the side cheeks of a drum. The actuating mechanism is shown in Fig. 4.12; a simplified functional diagram is in Fig. 4.13 and a cross-section in Fig. 4.14.

The shut-off performance of each rod is initial acceleration not less than 2 ft/sec^2, maximum speed 4 ft/sec, a travel of 18 ft 6 ins in a time not greater than 5 sec and touchdown speed not greater than 6 ins/sec.

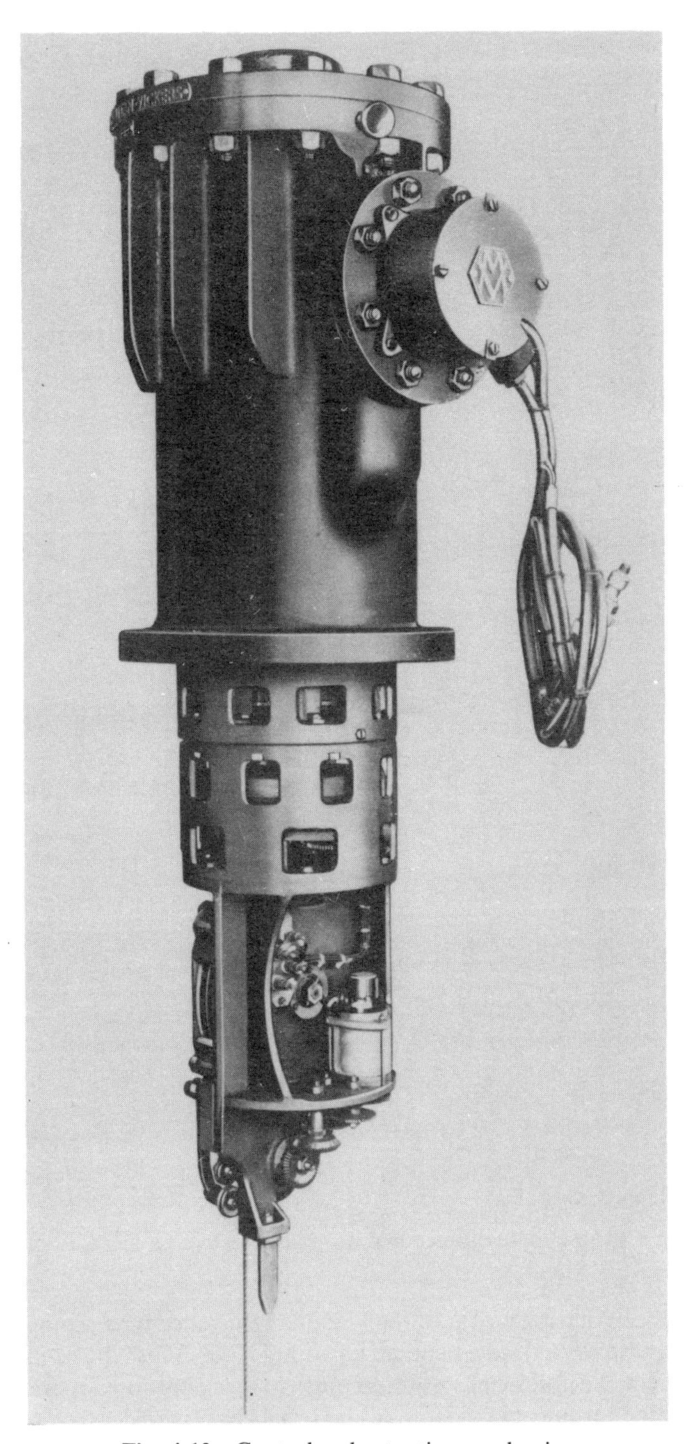

Fig. 4.12 Control-rod actuating mechanism

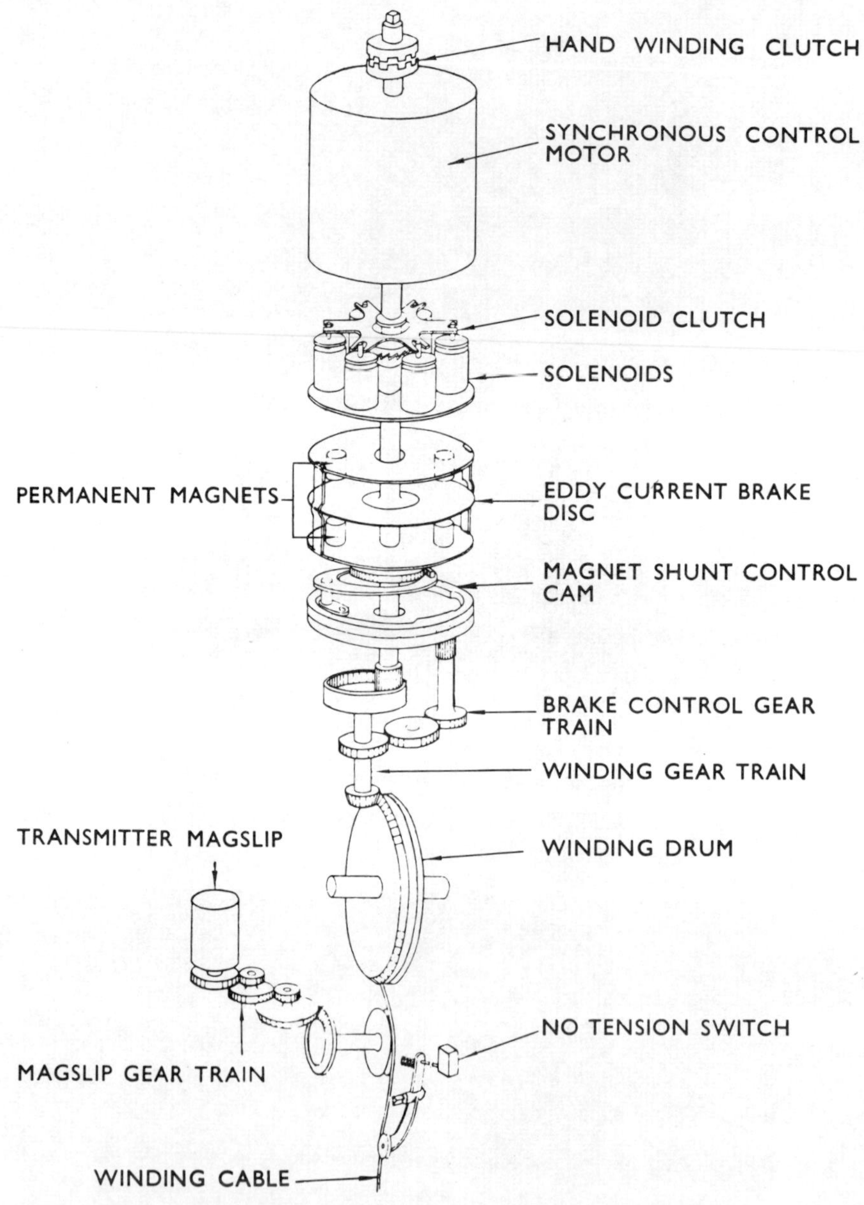

Fig. 4.13 Simplified functional diagram of actuating mechanism

A synchronous driving motor of the variable reluctance type is employed. It has a three-phase wound stator and an unwound rotor. This type of motor was chosen to meet the following requirements: (a) synchronous operation, (b) positive holding torque at standstill, (c) no brushgear (in order to give greater reliability), (d) high torque-to-volume ratio, (e) ability to work in an ambient

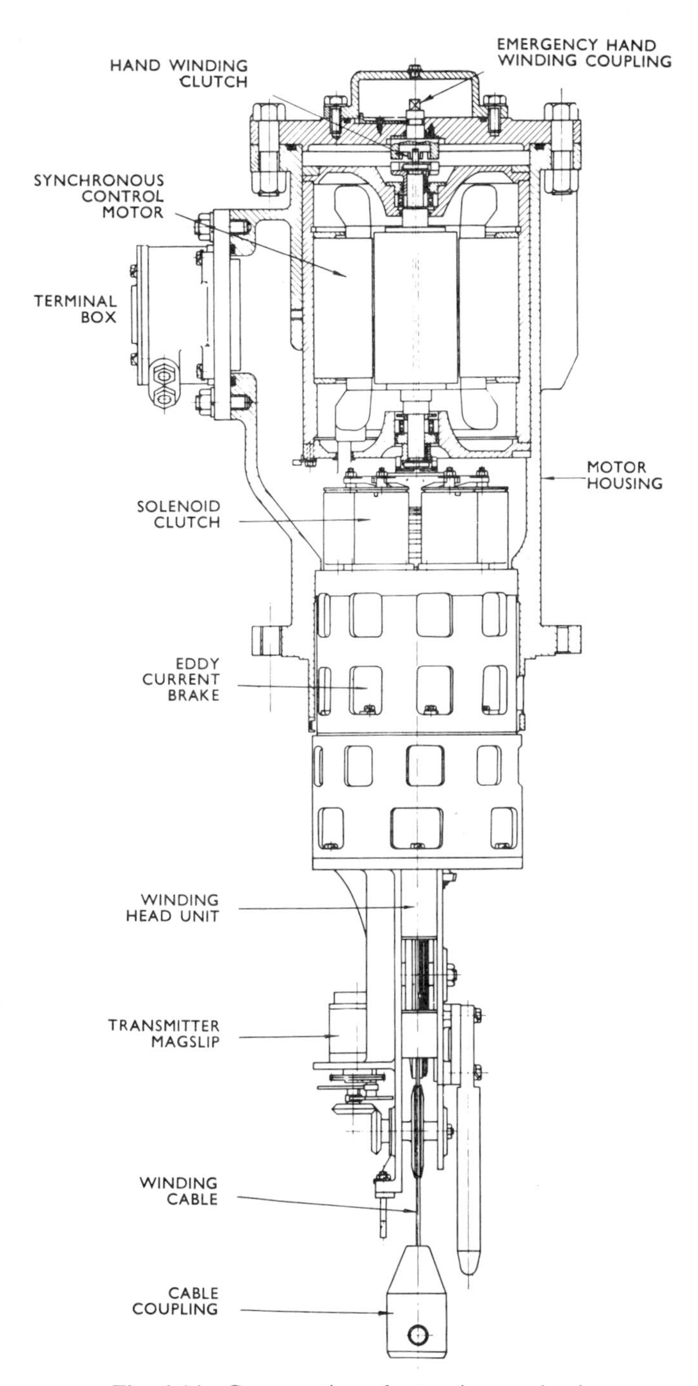

Fig. 4.14 Cross-section of actuating mechanism

temperature up to 100 °C, and (f) low-speed operation, so that a small gear ratio could be employed to permit back driving by the rod during shut-off.

A solenoid clutch is fitted between the motor shaft and the remainder of the mechanism. Its purpose is to disengage the rotor of the motor from the system, thereby reducing the inertia to be accelerated during shut-off. The clutch is operated by six solenoids energized from the same supply as the motor. Should the rotor of the motor fail to be disengaged when the supply is opened then a shut-off action will still result due to the de-energized motor being back driven.

If there is a loss of drive, an emergency winding handle can be attached which mechanically engages the clutch. The eddy-current brake provides a controlled fall of the rod. It consists of two sets of permanent magnets with alternatively opposed north and south poles. An eddy-current, copper-clad steel disc rotates in the air gap between the magnets. The braking torque on the disc is controlled by varying the air gap magnetic flux with a mechanical shunt operated from a cam. The correct rate of fall under shut-off conditions is programmed by the profile of the cam.

The drive from the motor is via single-stage spur gearing and right-angled bevel gears giving a total reduction of 20 : 1.

Linear measurement of the rod position by a transmitter magslip (synchro) driven by a gear train from a cable-driven measuring wheel. A lever-operated switch with a follower wheel on the winding cable detects 'no-tension'. This switch is used in conjunction with a second switch operated by the magnet shunt control cam to detect 'overrun or lost rod' conditions. The second switch is also used to check that the cam is correctly set.

The original lubrication of the mechanism was by a dry lubricant involving a molybdenum disulphide (MoS_2). This is due to the absence of oxygen and water vapour under operating conditions and the use of normal oils was considered undesirable. However, after several years of plant operation a suitable grease for use in the reactor became available and all mechanisms were then grease lubricated.

The overall system provides for the operation of the control-rod actuating mechanisms into two groups, (a) 'coarse'—up to sixty actuating mechanisms, and (b) 'fine'—up to four actuating mechanisms. Coarse and fine supplies to any of the actuating mechanisms can be selected by means of changeover switches. The coarse supply is generated by one of two identical frequency convertors (Fig. 4.15), the other frequency convertor set being used as a standby. Fine control supplies are provided by two sine-potentiometers. Further description of the overall system may be found in the references already given.

However, it should be noted that an emergency shutdown of the reactor is initiated from the emergency push button on the reactor control desk or from one of the shutdown circuits. All the line contactors and 415 V circuit breakers are opened, thus removing the control supply from the main bus-bars and thereby de-energizing the actuating mechanisms so that the rods give shut-off

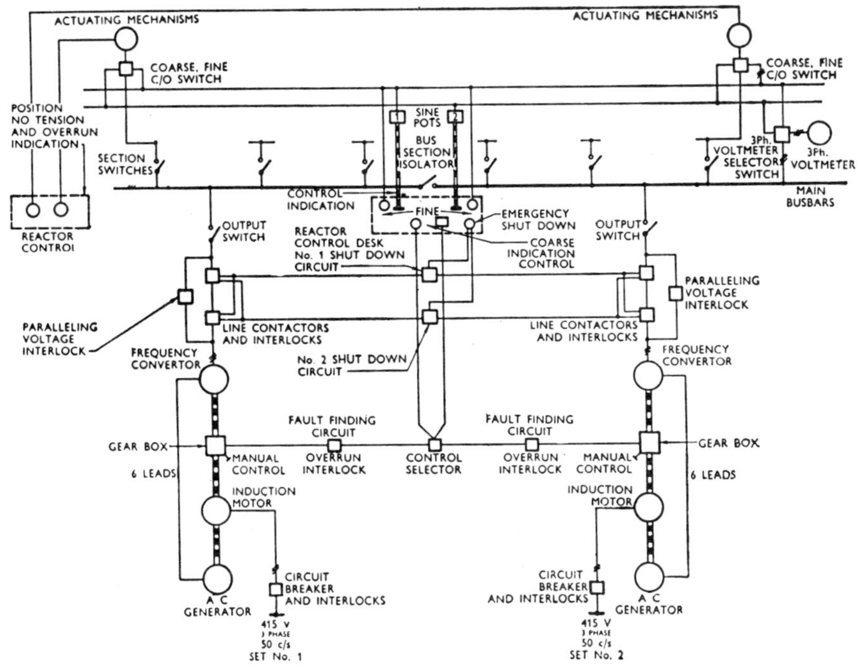

Fig. 4.15 Simplified schematic diagram of control-rod system

action. At the same time the sine-potentiometer supply is tripped so that the fine rods also give a shut-off action.

4.8.2 Theoretical failure-rate prediction of control-rod actuating mechanism

A theoretical failure-rate prediction has been undertaken independently using data from sources other than this particular Gas Cooled type reactor. This analysis has been undertaken as an experiment to see if a prior prediction could be derived to give the basis for creating artificially what would have been an initial assessment at the design stage. The intention is to use this estimate to retrospectively consider the growth programme which actually took place and to experiment with various methods of updating the prior estimate. The techniques used have already been described in earlier chapters by which the modes of failure are analysed and their effects studied. Failure rates are then assigned on a component part basis according to the stressing and environmental conditions by assigning 'K' factors to weight the failure rates. This analysis is concerned with failure of any type during actual operation in the reactor has been defined as 'failure of the mechanism to meet is operational specification'. The actuating mechanisms have been assumed to operate mechanically on average about 10^{-3} of the total reactor operating time.

Table 4.12 Theoretical estimate of actuating failure rate in service

Component/Sub-assembly	Estimated failure-rate/10^6 hr	Published failure-rate/10^6 hr	Source	K factor
Category D1				
ECB rotor shaft/bearings	0.02	2.2	Earles	10^{-2}
ECB/WD gearing and coupling	0.03 (0.02) S1	5.0	Earles	10^{-2}
Synchronous motor	0.01	0.8	Earles	10^{-2}
Wiring and terminals to motor	0.07	6.5	AHSB(S)R117	10^{-2}
Hand winding mechanism	0.01	—	—	—
Solenoid coils	0.05	—	—	—
Wiring and terminals to solenoids	0.05	4.8	AHSB(S)R117	10^{-2}
Solenoid plungers and spider arms	0.02	—	—	—
Clutch drive assembly	0.01	—	—	—
Clutch spring	0.01	0.01	Earles	1
Clutch plates	0.10	—	—	—
Journal bearings (2) and thrust bearing (1)	0.01	1.0	Earles	10^{-2}
ECB assembly	0.05 (0.02) S1	—	—	—
Brake actuating mechanism	0.05 (0.02) S1	—	—	—
Brake-cam gear train	0.01	0.9	Earles	10^{-2}
Winding drum	0.01	1.0	Earles	10^{-2}
Control-rod cable	0.02	—	—	—
Sub-total	0.49 (0.05) S1			
Category D2U				
Fixed guide pulley, jockey pulley, etc.	0.01	1.0	Earles	10^{-2}
Magslip gear train	0.01	0.9	Earles	10^{-2}
Magslip	0.05	—	—	—
Wiring and terminals to magslip	0.05	8.0	AHSB(S)R117	10^{-2}
Slack wire microswitch	0.06	2.0	AHSB(S)R117	3×10^{-2}
Wiring and terminals	0.03	3.5	AHSB(S)R117	10^{-2}
Overrun microswitch	0.06	2.0	AHSB(S)R117	3×10^{-2}
Wiring and terminals	0.03	3.5	AHSB(S)R117	10^{-2}
Sub-total	0.30			
Category S1				
Items associated with category D1 above	0.05	—	—	—
Category S3				
Pressure seals and gaskets	0.02	—	—	—

Total Category D = 0.79/10^6 hr
Total category S = 0.07/10^6 hr
Total Failure Rate in Service = 0.86/10^6 hr
Or 0.0075 faults/mechanism/reactor operating year
Or 0.35 faults/reactor operating year

Table 4.12 gives a breakdown of the mechanism into components or sub-assemblies with failure rates on a reactor operating hours basis. Failures during service are also categorized according to the effect each failure will have on reactor operation, and coded as follows:

FIRST	Character—safety aspect
D	Dangerous or tending to reduce reactor safety
S	Safe or not tending to reduce reactor safety
SECOND	Character—operational area affected
1	Affecting control-rod movement
2	Affecting operator control
3	Affecting areas other than reactor control

In addition the letter U is added as a third character to those faults in the dangerous category which may not be revealed when the mechanism is operated.

In Table 4.12, given a summary of the results obtained it may be noted that the source references for data have been taken from early references and not from the latest sources which may have included data subsequently derived from this Gas Cooled type reactor. As an example, report AHSB(S)R117 was published in 1966 and Earles is data from USA sources earlier than 1966. The results give a total failure rate of $0.86/10^6$ operating hours per mechanism. For a reactor using forty-eight control-rod actuating mechanisms this is equivalent to one fault per three reactor operating years. Almost half of the total faults are estimated to occur in the control, instrumentation and warning circuits.

4.8.3 Analysis of control-rod actuating mechanism drop failure modes

An estimate has been made of those failure modes which would cause a control rod actuating mechanism to give a control-rod drop outside specification or to completely fail to drop. The analysis of these various items was undertaken under the headings given in Table 4.13 where the types of operation are designated:

(A) Fails to release
(B) Fails to complete drop
(C) Outside drop time specification

The overall results obtained are discussed in Section 4.8.4.

4.8.4 Comments on results

If faults concerned with the shut-off function of the mechanism are considered, then failure can be defined as not meeting the drop specification. This could be brought about by any kind of excess friction in the moving parts or mis-aligned parts, or in the eddy-current brake, or by non-disengagement or partial disengagement of the clutch. Indication or lack of operator control faults are not relevant to this particular analysis.

Table 4.13 Component failure-rate analysis table headings

Component or sub-assembly		Operating conditions				Failure data (Rate = faults/10^6 hr)						
Description	Number	Temp. °C Approx.	Loading	Movement Damage	Wear	Mode	Basic Rate	Source	K Factor	Field rate	Estimated rate	Effect of failure category

With reference to the shut-off function the total fault rate was estimated to be 0.12 faults/10^6 hr, about one-seventh of the overall fault rate.

The assessment shows for failure to release the rod a rate of 0.004 faults/10^6 hr is indicated and for failure to complete the rod drop the rate is 0.008 faults/10^6 hr which gives a total 'dangerous' failure rate for rod drop to be 0.012 faults/10^6 hr (or 1.2×10^{-8} faults/hr).

The above assessment takes a much simplified view from which the drop failure probability can be derived. For a more detailed analysis it would be necessary to isolate failures which might be revealed and then evaluate the probability of such failures which might be corrected. It would be necessary to evaluate the damage to transit back into the reactor which might cause drop failure and the chance it would not be corrected. It would be necessary to evaluate the setting up of the mechanism after overhaul and the possible consequences of operator errors on drop probability.

The analysis has attempted to put all this together simply by giving an approximate estimate of the possible outcome of all these routes to failure in terms of revealed or unrevealed faults during service.

The difficulty with the really dangerous faults of failure to release or complete the drop is that we are trying to synthesize a rare event from component events which may be even rarer. There are strong indications from this analysis that the failures in these mechanisms are almost entirely related to human activities and that to synthesise rates from data using 'K' factors may not be the most appropriate approach if it is required to understand the processes which lead to failure; which is obviously a much more direct method of finding the ways of improving reliability.

4.8.5 Life testing of control rod actuating mechanism

After tests had been completed on individual components extensive life tests were carried out on two actuating mechanisms, which were run in an atmosphere of pure dry CO_2 with not more than 0.01 per cent by volume O_2 impurity at a pressure of 100 psi. Figure 4.16 shows the actuating mechanisms seated on the test pressure vessels, lagged to simulate the thermal effects of the biological shield of the reactor. Heating was incorporated equivalent to the heat flow up a charge tube. The actuating mechanisms were run for successive tests at the highest rod speed of 50 in/min and reversed automatically at the top and bottom of the rod travel. The measurement of the rod travel from the top to the bottom was 608 cm. Periodically the rod was dropped with a gravity fall by de-energizing the synchronous driving motor or on some occasions the solenoid clutch.

During the tests the following main points were studied: wearing properties of MoS_2 treated gears and bearings, temperature rises, effects of thermal cycling, and the functional operation of components. Special measuring techniques were developed to obtain this information without opening up the actuating mechanism.

Fig. 4.16 Actuating mechanisms on test pressure vessels

Table 4.14 Summary of the first series of tests carried out on two actuating mechanisms

Stage	Test details	Results
Mechanism V		
1	Original design except eddy-current brake cam profile	After 3853 cycles of operation the 3:1 ratio internal gearwheel and pinion failed completely. Appreciable wear on crown wheel and pinion. Transmitter magslip stop sheared
2	3:1 main drive internal gearwheel and pinion replaced to original dimensions but pinion manufactured in different material and dipped for dry lubrication	After 2029 cycles of operation the 3:1 ratio main-drive gearwheel and pinion showed little signs of wear
3	Internal gearwheel pinion from Stage 2 dipped for dry lubrication. Standard brush-type magslip transmitter	After 2091 cycles of operation 3:1 ratio main-drive pinion showed signs of appreciable wear. Transmitter magslip satisfactory
Mechanism W		
1	Original design except transmitter which was standard brush type	After 2053 cycles of operation, 3:1 ratio internal gearwheel showed signs of appreciable wear. Transmitter magslip satisfactory

The description of these tests is given in Green[126 and 128]. For the first series of tests two mechanisms were run continuously for several tests at their highest speed. After 3853 complete cycles of operation the first-stage pinion in the mechanism failed due to the wearing away of each tooth. Modifications to the pinion and internal wheel were suggested to give improved life. Successive tests each of approximately 2000 cycles of operation incorporating these modifications gave improved life.

At no time when the mechanisms were de-energized did they fail to perform a shut-off action dropping with a gravity fall. Table 4.14 gives a summary of the first series of tests carried out on the two mechanisms. Figure 4.17

Table 4.15 Number of rod drops during the first series of tests

Stage	Number of rod drops
Mechanism V	
1	24
2	16
3	27
Mechanism W	
1	17

Table 4.16 Number of rod drops during the second series of tests

Stage	Number of rod drops
Mechanism W reconditioned	
1	94

shows a chart which is typical for a test stage giving information on the acceleration measurements, thermal cycling etc., against cycles of operation and calendar events.

Although the life tests were primarily intended as tests to destruction on a continuous-running basis, the friction of the mechanism was measured by de-energizing the driving motor as discussed so that this information has been used to derive from the tests the number of rod drops which took place. Table 4.15 gives this information for the various stages of the tests, corresponding with Table 4.14.

The actuating mechanism *W* was reconditioned to the normal production standard after its previous run of 2053 operations and given an accelerated life test as in the previous series of tests. The main objective was to see how many raising and lowering operations the mechanism could complete without any attention before it failed to lift the weight of the control rod. The mechanism was operated at the highest rod speed of 50 in/min and failed after 8376 complete operations. It was found that the primary cause of failure was the wear on the solenoid clutch but the internal gear and pinion had nearly reached the end of their lives. The number of rod drops under gravity has been derived as ninety-four for these tests and quoted in Table 4.16.

4.8.6 Analysis of life testing results on control actuating mechanisms

The values of the control-rod drop times when correlated against the completed control-rod up and down movements at that time from Fig. 4.17 for instance indicated linearity as the 'best-fit' correlations for the majority of the test stages. Basically there were five test stages, four on mechanism V and the fifth on mechanism W.

Of these the first, second and fifth test stages were basically important from the point of view of improvements to the mechanism, design changes and the final proving in the fifth test stage and for these stages the correlation coefficient squared (r^2) values were determined.

The range of 'test-fits' examined were

$$y = a + bx \qquad \text{(linear)} \tag{4.24}$$

$$y = ax^b \qquad \text{(power law)} \tag{4.25}$$

$$y = a + b \log x \qquad \text{(logarithmic)} \tag{4.26}$$

$$y = ae^{bx} \tag{4.27}$$

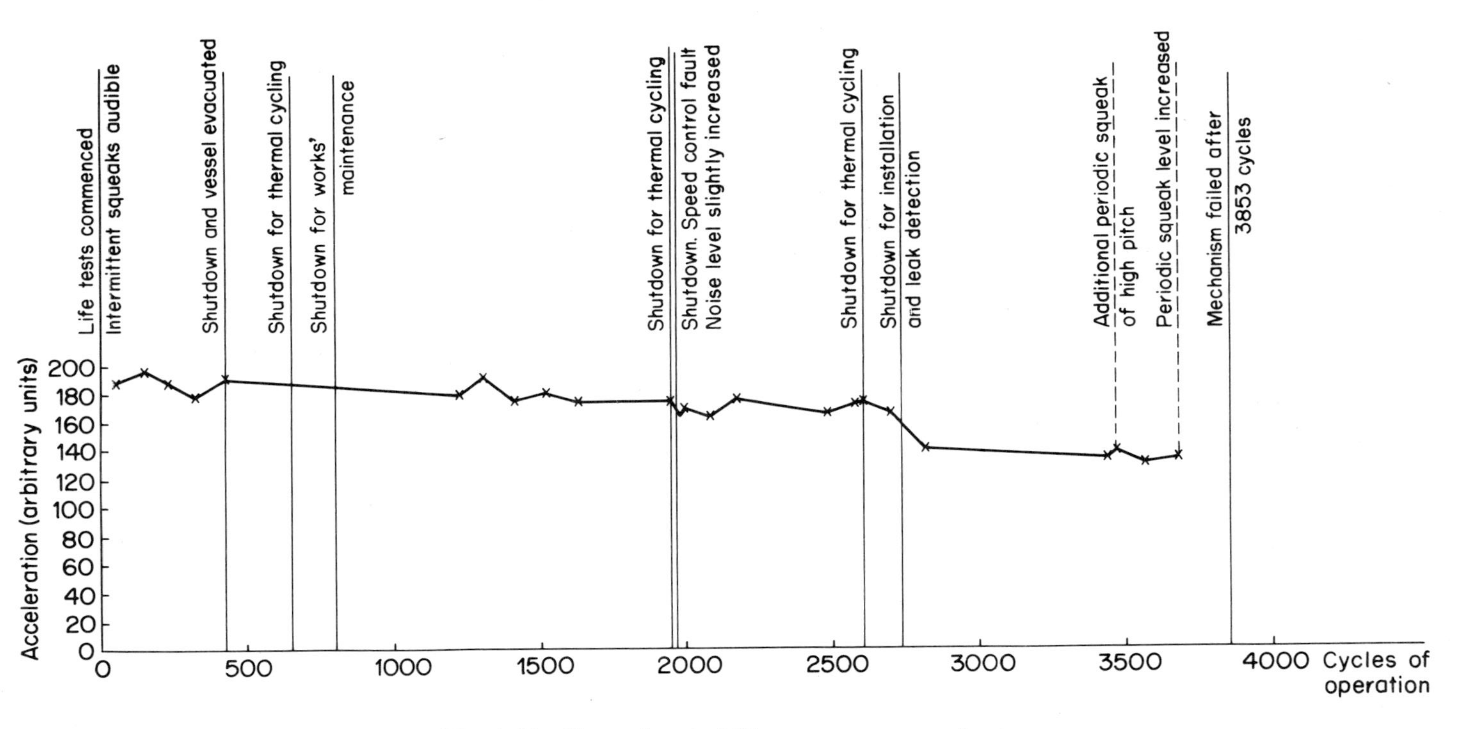

Fig. 4.17 Chart of typical life test stage on mechanism

where y = the point value of drop-time
x = the point value of the associated life-test cycles at the time of the drop.

Thus the data for all the test stages received fairly detailed examination using adequate exploratory test-fit models.

The first test stage data were adopted as typical of all the tests carried out, and although not given in detail here, these fuller analyses supported the use of the data first test stage as a typical basis for further work. These data enter into the classical and Bayesian statistical analyses described in this chapter.

The first test stage examination indicated for its drop-time against completed cycles a correlation with a linear best-fit of

$$y = 196.5 - 0.015x \tag{4.28}$$

$$\text{with } r^2 = 0.79$$

A collation of the data for the cycles between test-drops in the first test stage and the associated values of the drop-time is shown in Tables 4.17 and 4.18. It is evident that Poisson characteristics were generally observed which were considered important in the allocation of gamma model forms for the Bayesian prior, pre-posterior and general analysis.

In Table 4.18 is illustrated the majority distribution, being Poisson, with reference to the cycle period which has been arranged in order of reducing regression coefficients. It is obvious that a marked difference exists between the '100 cycle' and '200 cycle' regression coefficients, and those for the remainder which are for 1000, 900, 400 and 700 cycles. This enabled the decision to be made to use the 100-cycle or 200-cycle data because of the

Table 4.17 Distribution of cycle times between rod drops (First test stage)

Basic cycle period	Best pdf distribution found			Coeff. of regression
	Poisson	Power	Exponential	
100	Yes			0.946
200	Yes			0.803
300		Yes		0.050
400	Yes			0.070
500		Yes		0.094
600			Yes	0.0253
700	Yes			0.1867
800		Yes		0.1595
900	Yes			0.2051
1000	Yes			0.3893
Totals	6	3	1	

Table 4.18 Majority distribution (Poisson) with reference to cycle period; and associated regression coefficients (First test stage)

Basic cycle period	Coeff. of regression
100	0.946
200	0.803
1000	0.3893
900	0.2051
400	0.0500
700	0.0253

highly Poissonian features. It may be noted that 100-cycle data from the first test stage with an $r^2 = 0.946$ (see Table 4.18) were chosen for use in all subsequent analyses described.

The 'event rate' for the first test stage, taken as the ratio of drop-times on test and the number of cycles completed was twenty-three events in 3600 cycles, with the number of cycles being rounded to the lowest completed 100-cycle period. Hence the Poisson mean-rate parameter used was 23/3600 giving 0.639 events per 100 cycles. The variation of the event drop-times compared with the best regressed drop-times per cycles line was found to be linearly reducing and had a distributional form which was highly Gaussian. The value for the test-drop timing fit was therefore a Gaussian distribution about the mean line giving an r^2 value of 0.973.

From the description of the life-tests given in the previous section it was indicated that for the first test stage it was a test to destruction and as such was an accelerated test at 100 : 1 control rod speed compared with that under evential reactor working conditions. Hence it is pertinent to consider the relationship of such tests with reference to actual reactor working conditions and the various points are considered as follows.

(a) An assessment of the control-rod mechanism using current reliability technology experience indicated a failure-rate for the mechanism with reference to its shut-off function to be 0.12 fault per 10^6 h. Of these, 90 per cent were estimated to be of nuisance value only in reactor use but about 10 per cent could be considered of a 'serious malfunction' type. Depending upon conditions at the time of the particular serious malfunction fault, it was possible to have the highest category of fail-to-danger type of fault taking place. Hence the serious malfunction or fail-to-danger type of fault was estimated to occur at a rate of 0.012 fault per 10^6 hr. In the first test stage actual destruction took place which not only involved serious malfunction but at the actual time of occurrence may have been displaying fail-to-danger characteristics. Therefore from a safety point of view the 10 per cent serious malfunction failure rate needed no further adjustment on these grounds.

(b) The life tests for the five test stages were accelerated to run at 100 times normal reactor rod speed. Taking into account the variation of the rod-speed control and the frequency convertor cycles which are available on the actual reactor for both coarse and very fine adjustments of control-rod position it was estimated that the control-rod movement time during operation would be about 10^{-3} of that of normal reactor operational time.

Therefore for 1 year (8760 hours) then 1000 frequency convertor cycles would be necessary for a complete control-rod cycle which is a full up and down travel, the electrical frequency during the control-rod mechanisms being approximately 3 cycles per minute and from the 10^{-3} time factor already given the following must be derived:

$$\text{Total reactor hr/year} = 8760 \text{ hr}$$

$$10^{-3} \text{ reactors hr/year} = 8.76 \text{ hr}$$

$$= 526 \text{ min}$$

time for 1000 cycles (i.e. a = 1000/3 min complete up and down) movement in reactor

From the foregoing it is seen that 8.76 h/year corresponds with 1.58 cycles of movement on the accelerated life tests. Therefore the 3853 cycles of test stage no. 1 which was deliberately run to destruction without any remedial attention represents some 2400 reactor years and the total five test stages which were trouble free gives 18 370 cycles of movement which represents 11 650 reactor years. It may be noted that the '90 per cent trouble free' of the overall failure given by the theoretical assessment represents some 13 700 trouble-free reactor years. Hence there is a good correspondence between the two trouble-free times of running.

4.8.6.1 Confidence Intervals in Data (Classical and Bayesian)

(i) *Classical reliability estimation.* Green and Bourne[44] (p. 342) indicates that if the times between failures of an equipment are shown to be exponentially distributed with a mean rate of θ, then it can be shown that the associated γ probability confidence interval is given by:

$$\frac{\bar{\theta}\chi_1^2}{2n} \leqslant \theta \leqslant \frac{\bar{\theta}\chi_2^2}{2n} \tag{4.29}$$

where $\chi_1^2 = \chi^2(\nu = 2n; p = \frac{1}{2}(100 + \gamma)\%)$

$\chi_2^2 = \chi^2(\nu = 2n; p = \frac{1}{2}(100 - \gamma)\%)$

n = sample size

The percentage points of the chi squared (χ^2) distribution are given in standard tables (e.g. Murdoch and Barnes[130]).

It was shown in the previous section that the assessed failure-rate defined as fail-to-danger or serious malfunction was 1.2 faults/10^8 hr. The 'sample' size indicated by this assessed failure rate would be one fault per 8.3×10^8 hr.

For the purposes of illustration which is used later in applying field failure-rate confidence bands using a 93 per cent confidence band, a '4 : 1/1 : 4 either way' concept has been chosen which also corresponds with the correlation given in Fig. 4.9. Hence equation 4.29 becomes

$$\frac{[1.2 \times 10^{-8}][\chi^2_{0.97}(2 \times 1)]}{(2 \times 1)} \leqslant \theta \leqslant \frac{[1.2 \times 10^{-8}][\chi^2_{0.04}(2 \times 1)]}{(2 \times 1)} \tag{4.30}$$

which gives

$$\frac{[1.2 \times 10^{-8}][0.0609]}{2} \leqslant \theta \leqslant \frac{[1.2 \times 10^{-8}][6.4337]}{2}$$

or

$$0.0365 \times 10^{-8} \leqslant \theta \leqslant 3.86 \times 10^{-8} \tag{4.31}$$

This may be compared with the assessed estimate of 1.2×10^{-8} faults as the failure-rate.

Applying these figures to determine the reliability over a period of, say, 10 years for a 93 per cent confidence band and calculating the reliability from the exponential $e^{-\theta t}$ gives:

$$0.99664 \leqslant R(t \mid \theta) \leqslant 0.999968 \tag{4.32}$$

(ii) *Bayesian reliability estimation.* In applying the Bayesian analysis to the present data set it may be assumed that the failure-rate θ is not constant and can vary. For the present case it has been shown that a Poisson series of events exists and the value of θ would depend on two variational descriptors, α and β, giving rise to a 'gamma prior' model for θ described by:

$$f(\theta \mid \alpha, \beta) = \begin{cases} \dfrac{\theta^{\alpha-1} e^{-\theta/\beta}}{\beta^{\alpha}\Gamma(\alpha)} & \text{for } \theta, \alpha \text{ and } \beta > 0 \\ 0, \text{ otherwise} \end{cases} \tag{4.33}$$

For this model the reliability estimates would be derived from:

$$e^{-\left\{\frac{\beta t[\chi^2_{\gamma_u}(2n + 2\alpha)]}{2\beta \sum t_i + 2}\right\}} \leqslant R(t \mid \alpha, \beta) \leqslant e^{-\left\{\frac{\beta t[\chi^2_{\gamma_l}(2n + 2\alpha)]}{2\beta \sum t_i + 2}\right\}} \tag{4.34}$$

where α and β are the two variational descriptors to be determined
t would equal the time period for which a reliability would be required to be found
n would represent the 'n-lot' of the sample data
γ_u and γ_l would represent upper and lower percentiles of the reliability confidence sought

Σt_i would be the time-summation over which the 'n-lot' of the sample occurred.

Data and the procedures taken from Waller *et al.*[122] have been used to establish the parameters α and β. These have also been discussed in a previous section, particularly for the gamma percentile, p, which is given in Appendices C1 and C2 of Waller *et al.*,[122] and is defined by:

$$P = \frac{1}{\Gamma(\alpha)} \int_0^{\kappa(\alpha)} e^{-x} x^{\alpha-1} \, dx \tag{4.35}$$

and if $0 < p < 1$, then $\kappa(\alpha) \simeq P^{1/\alpha}$.

The proof for this asymptotic theorem is given in the appendices quoted. In particular for a test α,

$$P_{\text{upper}} = \frac{1}{\Gamma(\alpha)} \int_0^{P^{(1/\alpha)}_{\text{upper}}} e^{-x} x^{\alpha-1} \, dx \tag{4.36}$$

and

$$P_{\text{lower}} = \frac{1}{\Gamma(\alpha)} \int_0^{P^{(1/\alpha)}_{\text{lower}}} e^{-x} x^{\alpha-1} \, dx \tag{4.37}$$

where the integral is described as that of the incomplete gamma function. The confidence band, p, if of course (P_{upper}-P_{lower}).

It will be noted that if $P_{\text{upper}}/P_{\text{lower}}$ is determined, the $\Gamma(\alpha)$ terms cancel out, P_{upper} and P_{lower} would notionally be known and fixed and the problem resolves using for example an iterative procedure to finding the single value, α.

In Waller *et al.*[122] it shows that $\beta \sim p^{-(1/\alpha)}$ so that if $\beta p^{1/\alpha} \sim$ unity for the three parameters then this is an indication that the correct values of α and β have been determined. Hence the determination of α and β for the percentiles such as 0.97 and 0.04 are treated according to the following procedure:

$$p = \text{confidence band} = 0.93$$

$$= P_{\text{upper}} - P_{\text{lower}}$$

and the ratio $P_{\text{upper}}/P_{\text{lower}}$ can be found from equations 4.36 and 4.37.

Hence for this case with

$$P_{\text{upper}} = 0.97$$

and

$$P_{\text{lower}} = 0.04$$

then

$$\frac{P_{\text{upper}}}{P_{\text{lower}}} = 24.25$$

from which $\alpha = 0.788$, and $\beta = 1.096$.

Applying these factors to equation 4.34 gives the reliability for 10 years

(876 000 hr) as follows:

Bayes gives $0.9951 \leqslant R(t \mid \alpha, \beta) \leqslant 0.9998$

and from equation 4.32

Classical gives $0.996644 \leqslant R(t \mid \alpha, \beta) \leqslant 0.999968$

Therefore, comparing these statements for the ninety-three per cent confidence band calculated the surest limits, (lower values of R in both extremes) are given by the Bayes results. The confidence limits are very slightly wider for the Bayesian approach compared with the Classical but bearing in mind the very high reliabilities involved the engineering significance is trivial.

4.8.6.2 Reliability Growth During the Accelerated Life Tests with 'Duane' Interpretation

The five test stages on the prototypes of the control-rod actuating mechanisms were such that after the test to destruction of the first test stage then support testing was undertaken in the second test stage after the obvious malfunctions observed in the first test stage had been corrected and improved.

In the fifth test stage a long duration test of the final prototype design which represented production standards was undertaken. The intermediate third and fourth test stages were carried out to consider minor modifications. However, the tests did not actually exhibit a failure to drop the control rod but the main evidence appeared to be a reduction in acceleration in the tested drop-time through the life of any test stage.

Hence for the main test stages, the reduction in acceleration in dropping against test cycles was analysed and found to be linear with the main data as shown in Table 4.19.

The main feature of interest and the most sought after improvement lies in maintaining a high acceleration which is reflected in the negative gradient reduction which took place progressively through the test stages. Summarizing from Table 4.19 these test gradients are shown in Table 4.20.

Table 4.19 Analysis of acceleration

Test stage no.	Linearity formula	Test cycles	Equivalent reactor years
1	$y = 196.52 - 0.015x$	3675	2434
2	$y = 188.03 - 0.0073x$	2030	1282
5	$y = 176.52 - 0.0046x$	3840	5268

where y = acceleration (in arbitrary units)
x = number of lift test cycles completed.

Table 4.20 Acceleration test gradient constants

Test stage no.	Test gradient reduction constant per cycle	Equivalent reactor years
1	0.015	2434
2	0.0073	1282
5	0.0046	5268

Table 4.21 Analysis for 'Duane type' plot

Test stages aggregated (1)	Equivalent reactor years (2)	$\frac{\Sigma \text{ Equivalent reactor years}}{\Sigma \text{ Test-gradient reductions}}$ (3)	Log of (2)	Log of (3)
1	2434	2434/0.0015	3.3863	5.2102
1 and 2	3716	3716/0.0223	3.5701	5.2218
1, 2 and 5	8984	8984/0.0269	3.9535	5.5237

The test gradient reduction constant given in Table 4.20 may be considered as a goodness of performance parameter and the lowering of these constants for the successive test stages is an indication of the betterment of reliability.

A 'Duane' procedure has been applied which would be to summate any parameter related to equipment malperformance, together with the summation of equivalent time, cycle of tests, equivalent reactor years to produce a typical 'Duane type' plot (Table 4.21). Section 5.2.1, page 180, further discusses the Duane approach.

The 'Duane' procedure has been used to give the plot shown in Fig. 4.18 and the reliability growth constant of the data, namely the slope α, is equal to 0.5585. This general growth reliability process is discussed later in chapter 5 but it is sufficient to say that from the positive slope of the plot arising out of the test stages that there is a positive growth in reliability.

4.8.7 Field data considerations on control rod actuating mechanisms

In considering data from a safety point of view the definition of failure would imply a failure of the control-rod drop under gravity when required and no such failure was known to have taken place. Such a drop is considered as a discrete event in time. In general terms this is concerned with the cumulative distribution of times to failure and this will be derived by means of some test at some instant in time. In practice, however, there may be a gradual deterioration but the information which is available is whether at the time of test the actuating mechanism was in a working state (i.e. meeting its specification), or in a non-working state.

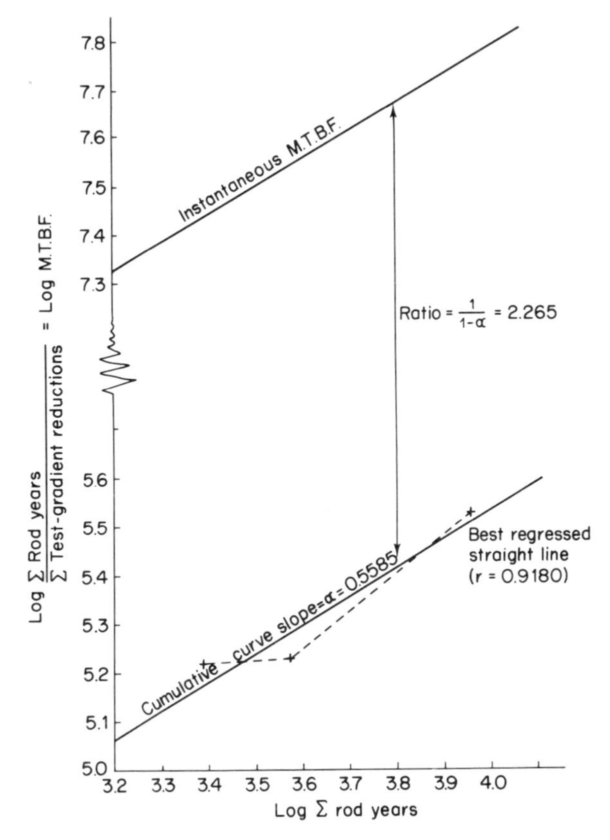

Fig. 4.18 Duane-type plot for control-rod actuating mechanism life test

Subject to certain assumptions that any shutdown with the actuating mechanisms remaining untouched, *in situ* in the reactor does not affect them and that maintenance of the actuating mechanisms restores them to 'mint' condition then the reactor running time is the essential time of interest. It is required to identify a number of particular times that a number of actuating mechanisms were known to be in a working state and a number in a state which has been defined as 'failed'. From this information the probability that an actuating mechanism will be in the working state after the elapse of these times can be estimated.

The observed failures from specific tests may be typically analysed in the categories given in Table 4.22.

If it can be assumed that the failures follow an exponential distribution (i.e. are random at a constant rate), then this may be used for analysis purposes where:

$$P = \exp(-\theta t) \tag{4.38}$$

Table 4.22 Types of test failures for actuating mechanisms

Type of test failure
Visual inspection
Drop test
Torque test
Drop test and visual
Torque test and visual
Torque test and drop test
Clutch plate failures
Electrical and instrumentation cables

and θ estimated by fitting a weighted regression line through the origin to logarithmic form of equation 4.38, i.e.

$$\log_e P = -\theta t \tag{4.39}$$

The fits may possibly indicate that the exponential distribution is not truly applicable but there are advantages in using the exponential distribution because of its simplicity. Hence the upper confidence limit for the estimate of θ may be found at a relatively high level and it may be sufficient if an exponential distribution is to be utilized for safety and reliability analysis to use this estimate. The upper confidence point is assumed to be sufficiently pessimistic to allow for errors in sampling and poorness of fit. Clearly, the acceptance of such an assumption would be dependent upon the sensitivity of the particular reliability analysis being undertaken to the distribution and value of parameter taken.

If it is assumed that Weibull distribution can be taken for the failure distribution with the actuating mechanisms subject to failure from the start of life, then the location parameter can be made zero. The probability of success P may be written as:

$$P = \exp\left(-\frac{t}{\tau}\right)^m \tag{4.40}$$

where τ = characteristic time
m = a shape or wear-out parameter
t = time

It can be shown (Keller[56]), that the

$$\text{MTTF} = \tau\left(\frac{1}{m}\right)! \tag{4.41}$$

The values of n and τ may be obtained by a least square method and the mean time to first failure calculated.

However, for the purposes of this study this mean time to first failure could be interpreted as a mean time between failures because of the assumption of perfect repair at overall. This would mean that on an actuating mechanism which had been failed '*n*' times is treated in the manner as that obtained for one failure on '*n*' separate actuating mechanisms.

4.8.8 Control rod drops as discrete events

An approach to the rod-drop analysis is to consider on each occasion a rod is dropped that it is a single independent trial in a binomial scheme. This implies assuming that a rod will fail to drop with the same probability on each occasion and is independent of the particular rod, length of time in the reactor and the number of previous successful drops. The confidence levels for a binomial parameter are derived in the usual way as discussed by Green and Bourne[49]. An interesting assumption that could be made which appears to be true, is there is no evidence from the data on the MoS_2 lubricated and grease lubricated actuating mechanisms that the two populations behave differently. Therefore it is possible to consider the total number of separate drops if required.

This analysis should be used with caution and for this reason it is useful also to draw on other sources of information. However, if a growth programme is being followed then the analysis could be valuable in updating a prior to give a posterior subject to the assumptions made.

4.8.9 Actuating mechanisms in service failures for a Gas Cooled type reactor

Various changes took place during the earlier years as already indicated in a previous section. During a subsequent period a sample of the types of failures which are understood to have occurred are shown in Table 4.23. It should be noted that these data are for in-service reactor operating conditions and do not include faults found for example in the workshop. During this period all the actuating mechanisms were grease lubricated and the general procedures for maintenance and operation were then well established. For a reactor (48 actuating mechanisms) it was estimated that this gave a total in-service rate of one fault per 2.5 reactor operating years.

Table 4.23 Types of faults in service for control rod actuating mechanisms

Fault	Effect of fault
Motor cable (short circuit)	Rod dropped
Motor terminal (open circuit)	Rod dropped
Magslip (sticking)	Lack of control indication
Slipping clutch	Spurious rod movement

4.8.10 Control rod actuating mechanisms workshop drop tests

It is not readily possible to measure the detailed dropping characteristics of the actuating mechanisms *in situ* in the reactor so the workshop test data have been considered for any trends. The drop test undertaken in the workshop involves a tachogenerator driven by the control actuating mechanism cable which produces a voltage proportional to the cable or rod speed and a time velocity curve is recorded. The basic test specification limit for the velocity after 1 sec is defined, based on the tachogenerator output.

Test rod results on the actuating mechanisms which are collected may be re-examined. The measurements which were made for the dropping of a rod may be considered for example purely for the velocity after 1 sec. The reason for selecting this test was there appeared to be a similarity between the measurement of initial acceleration in the life tests described in section 4.8.5 and the workshop velocity test.

Statistical tests made of the velocity after 1 sec information indicated that sufficiently accurate comparisons could be made if a normal curve was fitted in every case and the mean and standard deviation calculated. This is an interesting result because it gives a simple approach for general modelling purposes. Variations could be detected depending upon various factors such as the length of time operating in the reactor but the corresponding velocity changes within the context of the specification testing were not significant.

4.8.11 Protective system for a Gas Cooled type reactor

In the present study it has been found that naturally extending from the so-called 'deterministic' methods of engineering the methods later applied to this reactor tended to use more quantified probabilistic approaches. This is seen in the main objective originally laid down as being 'to limit the fuel-can temperature in the event of any credible accident' being extended for example in one particular case as the 'steady-state temperature distribution shall be such that in the event of one specific type of reactor fault, the sum of the probabilities of can temperatures reaching a particular temperature shall not exceed 0.1 with 90 per cent confidence for the whole reactor'. The idea being expressed with the confidence statement allowed for errors in the model and was also a function of the number of thermocouples used in normalizing the model and the number of thermocouples used for measurement purposes in estimating the standard deviation.

It would be considered useful to describe in general terms the overall automatic protective system. A point to be noted is that there have been a number of modifications to the system which has been a characteristic observed in all the growth programmes of development and operation. Hence a brief description will be given relating to the system in its early operation as it was at this time that a quantified reliability prediction was undertaken for the protective system.

The protective channels of the system at this time were as shown in Table

Table 4.24 Protective channels for reactor automatic protective system

No. of channels	Description
3	Fuel-element can temperature
2	Rate-of-change of pressure
3	Shutdown amplifiers
2 groups out of 3	Out-of-balance flow
2	Blower OCB tripped
2	Excess pressure
2	Reactivity
1	Low-power recorder
2	Counters
1	Emergency shut push-button for manual operation by operator

4.24 and the general arrangement and configuration are as shown in Fig. 4.19 in simplified schematic form. From Fig. 4.19 it can be seen that when the reactor is operating normally all the series contacts are closed and the relays energized. If for any reason the series circuit is broken due to a reactor-fault condition, for example abnormally high fuel-element temperature can be measured, then the relays in the 110 V AC 'safety circuits' would be de-energized and trip all the supplies to the control-rod actuating mechanisms. The control rods would then be allowed to fall under gravity into the reactor core to shut down the reactor.

Furthermore, it may be noted that basically the system is designed to fail to safety and that the emergency shutdown circuits are duplicated and that any emergency shutdown device operates in both circuits. Either circuit will shut down the reactor.

Using quantified reliability techniques of prediction in the early years the probability of the protective system failing on demand was estimated as not to be worse than 10^{-4}, making various assumptions in connection with the response of the various protective channels and distributions of failure testing and periodicity. This estimate of the probability of failure of the protective system has been used as a prior prediction in the present study.

In addition the degree of belief in these predictions has been taken from the correlation exercises described in this chapter. Thereby the study has proceeded along the lines of sampling the information which can be derived within the bounds of reasonable effort in the analysis and associated tasks. Table 4.25 gives events which have been used for this study on a reactor of the Gas Cooled type for a sample year of operation.

The general objective in this analysis has been to select appropriate techniques for updating prior information and to experimentally investigate their application in deriving posterior information. This has been analysed assuming that information has been derived in its true chronological perspective, i.e. as seen at the particular time in question. By this means an outline of the

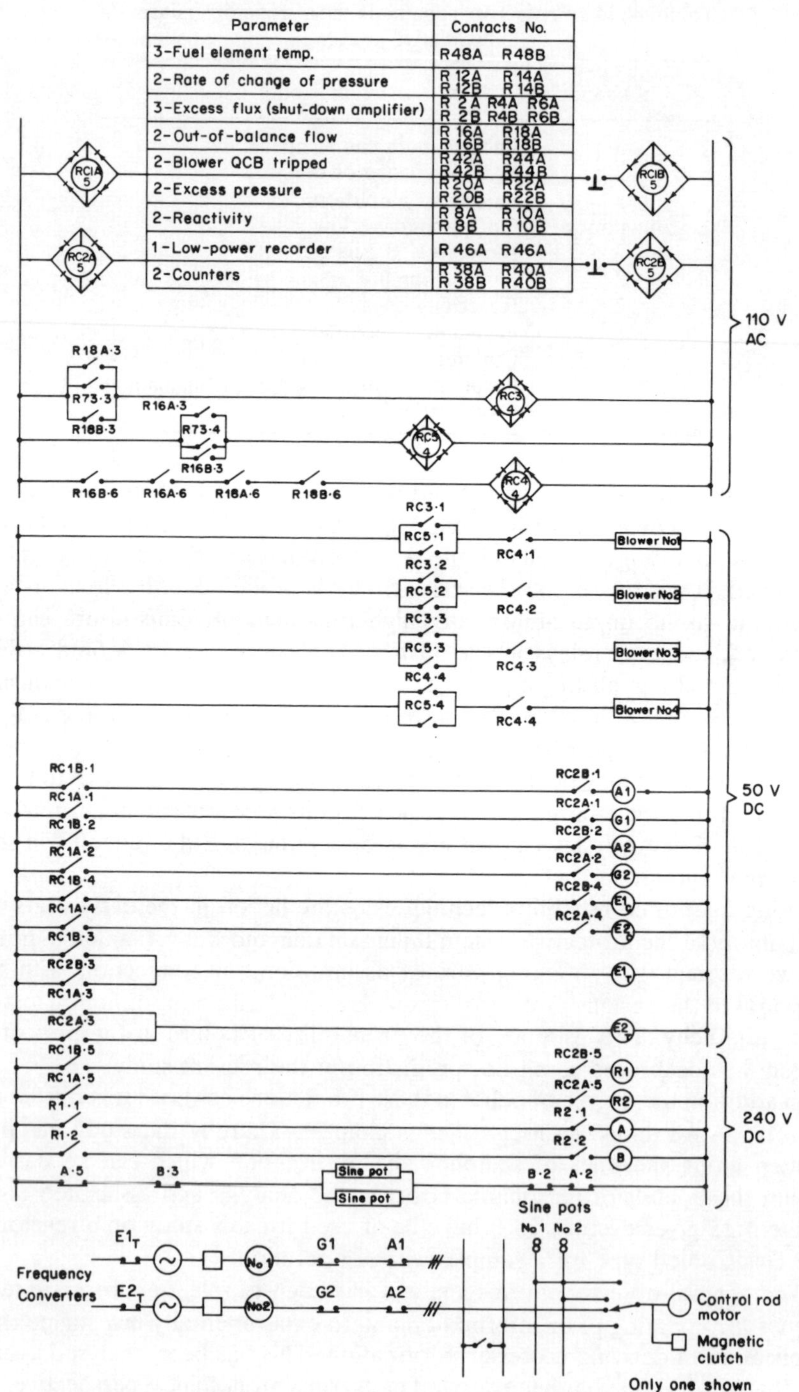

Fig. 4.19 Automatic protective system simplified schematic diagram

Table 4.25 Events incurring use of control-rod mechanisms and automatic protective systems: sample year of operation

Number of events	Reason for demand on control-rod mechanisms	Automatic protective system motivation
2	Manual shutdown	
3	Reactor trip	Protective channel

Note: No failures of either control-rod mechanisms or automatic protective system.

characteristics of the growth programme has been derived together with techniques which have been developed for their particular applicability.

4.8.12 Data calculations for Gas Cooled type reactor protective system

1. (a) System prediction of the automatic protective system (failure probability) = 1×10^{-4} on demand (per reactor per year).
   (b) Data for the automatic protective system (average)
      (i) three trips/year, on demand, automatic
      (ii) five trips/year, on demand, including manually-initiated.
   (c) With respect to this Gas Cooled type reactor, only two sets of data can be used for comparison purposes, these being:
      (i) the 1×10^{-4} initial assessment of failure of the system on demand
      (ii) The 20 years in-life data for a single reactor of the sixty trips (automatic) in 20 years and, the 100 (automatic and manual) in the same 20 years.

NB: No 'accelerated life-testing' of a reactor automatic protective system can feasibly be carried out, so only the two groups of data can be used as a check on the initial assessment, together, with passage of time, the year-by-year aggregation of APS usage.

2. (a) Experience of this Gas Cooled type reactor is assumed to indicate that on average, activation of the automatic protective system occurs 'automatically' about three times per year. If 'automatic' and 'manual' initiated trips are added together, the average is some five actuations of the automatic protective system per year (*per reactor* protective system).
   (b) Hence in the *worst case* (i.e. five automatic protective system actuations per year) and a probability given by the assessor that there is a probability of the order of 1×10^{-4} failure of the system on demand, the serious APS system failure rate is, on average, as high as $(5)(1 \times 10^{-4})$ failures/year. That is the failure of the APS to operate automatically and correctly becomes an 'ordinary' failure-rate of the system, of $f = 5 \times 10^{-4}$ faults per year (8760 hr). This corresponds to an equivalent *single* failure event (unity fault) in $(8760)\,[1/(5 \times 10^{-4})]$ h, i.e. one fault in some 17.5×10^{6} h.

(c) From earlier calculations in this work a '4 : 1/1 : 4' tolerance zone has been used giving parameters as follows:

probability band = 93%
upper probability = 0.97 (= P_{upper})
lower probability = 0.04 (= P_{lower})

and for 'Bayesian' work related to the above figures see section 4.8.6.1 which derived the following:

$\alpha = 0.77967$
$\beta = 1.09755$

(d) 'Classical estimations' can be found for the failure-rate bounds, also the system reliability range(s). These are as follows from the theory used earlier in this chapter:

(i) Classical confidence limits related to the point estimate of failure-rate

$$\frac{\bar{\theta}\chi_1^2}{2n} \leqslant \bar{\theta} \leqslant \frac{\bar{\theta}\chi_2^2}{2n} \tag{4.42}$$

where $\bar{\theta}$ = estimated system failure-rate
χ_1^2 = χ^2 function where ν = degrees of freedom of 'n' faults, with an upper probability of P_{upper}, see equation 4.29
χ_2^2 = χ^2 function where ν = degrees of freedom of 'n' faults, but with the lower probability, P_{lower} used.
n = number of faults: this as stated above = one fault (related to 17.4×10^6 hr).

(ii) Hence $\frac{\bar{\theta}\chi_1^2}{2n} \leqslant \bar{\theta} \leqslant \frac{\bar{\theta}\chi_2^2}{2n}$ becomes

$$\frac{\overline{(5.7143 \times 10^{-8})(\chi_1^2)}}{2n} \leqslant \bar{\theta} \leqslant \frac{\overline{(5.7143 \times 10^{-8})}}{2n}\chi_2^2 \tag{4.43}$$

NB: 5.7143×10^{-8} is the 'failure-rate' at a maximum of five events per year, corresponding to one fault (failure on demand of the APS) in some 17.5×10^6 hr.

$$n = 1$$

$$\chi_1^2 = \chi_{0.97}^2 2(1) = 0.060944$$

$$\chi_2^2 = \chi_{0.04}^2 2(1) = 6.43374$$

Hence the failure-rate bounds are indicated by the inequality

$$\frac{(5.7143 \times 10^{-8})(0.060944)}{2} \leqslant \bar{\theta} \leqslant \frac{(5.7143 \times 10^{-8})(6.4374)}{2} \tag{4.44}$$

$$17.41\ldots \times 10^{-9} \leqslant \bar{\theta} \leqslant 183.82 \times 10^{-9}$$

(related to an estimated θ, $\bar{\theta}$, of 5.7143×10^{-8} failures per hour).

(iii) The 'classical', say 5 years, reliability bounds would be given by

$$e^{-(\theta_{\text{upper}})t} \leqslant R(t \mid \theta) \leqslant e^{-(\theta_{\text{lower}})t} \quad (4.45)$$

where $t = 5$ years, i.e. 4.38×10^4 hr.
Hence

$$e^{-(183.82\times 10^{-9})(4.38\times 10^4)} < R(t \mid \theta) < e^{-(17.41\times 10^{-9})(4.38\times 10^4)}$$

i.e. $$e^{-(8.05\times 10^{-3})} \leqslant R(t \mid \theta) \leqslant e^{-(0.76256\times 10^{-3})}$$

i.e. $$0.9920 \leqslant R(t \mid \theta) \leqslant 0.9992 \quad (4.46)$$

(e) (i) 'Bayesian' estimators can also be evaluated for the failure-rate range (based on the estimated $\bar{\theta}$), and the corresponding reliability range (again for 5 years) for

$$p(\theta \mid \alpha, \beta) = \left\{ \frac{\theta^{\alpha-1} e^{-\theta/\beta}}{\beta^{\alpha}\Gamma(\alpha)} \text{ for } \begin{matrix}\theta \\ \alpha \\ \beta\end{matrix} \right\} \text{all} > 0 \quad (4.47)$$

$$0 \text{ otherwise}$$

the usual Poissonian based model with Γ posterior conjugate model, as described in this chapter, applies and from the earlier calculations

$$\alpha = 0.77967$$

$$\beta = 1.099755$$

$$P_{\text{upper}} = 0.97$$

$$P_{\text{lower}} = 0.04$$

$$P_{\text{band}} = 0.93$$

'Failure-rate' $\simeq$ one fault in 17.5×10^6 hr.

(ii) 'Failure-rate' range is given by the inequality

$$\frac{(\beta)(\chi^2_{P_{\text{upper}}} 2(n+\alpha))}{(2)(17.5 \times 10^6) + 2} \leqslant \theta \leqslant \frac{(\beta)(\chi^2_{P_{\text{lower}}} 2(n+\alpha))}{(2)(17.5 \times 10^6) + 2} \quad (4.48)$$

In the above n = unity, (the single failure in 17.5×10^6 hr)

$$\chi^2_{0.97} 2(1+\alpha) = \chi^2_{0.97} 2(1 + 0.77967) = \chi^2_{0.97} 3.55934$$
$$= 5.97886\ldots \times 10^{-8}$$

Similarly

$$\chi^2_{0.04} 3.55934 = 0.4706$$

Hence 'failure-rate' range by $P_{\text{band}} = 0.93$ is

$$\frac{(1.09755)(5.97886 \times 10^{-8})}{(2)(1.09755)(17.5 \times 10^{6}) + 2} \leqslant \theta$$

$$\leqslant \frac{(1.09755)(0.4706)}{(2)(1.09755)(17.5 \times 10^{6}) + 2}$$

i.e. $1.708 \ldots \times 10^{-15} \leqslant \theta \leqslant 1.3446 \times 10^{-8}$

(iii) Bayesian reliability range limits are derived from

$$e^{\left\{\frac{\beta t[\chi^2_{0.97} 2(n + \alpha)]}{(2\beta)(\sum t_i) + 2}\right\}} \leqslant R(t \mid \alpha, \beta) \leqslant e^{-\left\{\frac{\beta t[\chi^2_{0.04} 2(n + \alpha)]}{(2\beta)(\sum t_i) + 2}\right\}} \tag{4.49}$$

Only t in the above has not, so far, been defined. For comparison with the earlier classical methods,

$$t = 5 \text{ years} = 4.38 \times 10^{4} \text{ hr.}$$

However, care must now be taken in that the earlier classical model was constant in its terms whereas in the above expression one is combining α and β derived from a *prior* assessment and with t (= 5 years) data experience.

As no faults occurred in this time

$$\alpha \rightarrow (\alpha + 0) = 0.77967$$

But
$$\beta \rightarrow \left(\frac{1}{\beta} + 4.38 \times 10^{4}\right)^{-1}$$

i.e.
$$\beta = \left(\frac{1}{1.09755} + 4.38 \times 10^{4}\right)^{-1} = 2.2831 \times 10^{-5}$$

i.e.
$$\sum t_i \rightarrow 17.5 \times 10^{6} \text{ hr} + 5 \text{ years}$$
$$= 17.5 \times 10^{6} \text{ hr} + 4.38 \times 10^{4} \text{ hr}$$
$$= 17.5438 \times 10^{6} \text{ hr}$$

The Bayesian reliability range thus becomes (for 5 years)

$$e^{-\left\{\frac{(2.2831 \times 10^{-5})(4.38 \times 10^{4})[\chi^2_{0.04} 2(1 + 0.77967)]}{(2)(2.2831 \times 10^{-5})(17.5438 \times 10^{6}) + 2}\right\}}$$

$$< R(t \mid \alpha, \beta)$$

$$< e^{-\left\{\frac{(2.2831 \times 10^{-5})(4.38 \times 10^{4})[\chi^2_{0.97} 2(1 + 0.77967)]}{(2)(2.2831 \times 10^{-5})(17.5438 \times 10^{6}) + 2}\right\}}$$

i.e. $0.999442 \leqslant R(t/\alpha, \beta) \leqslant 1.000000$ (4.50)

(f) *Summary*
Initial assessment

'fail-danger' of the APS = 1×10^{-4}
average actuation of the APS = 5 times/year

Failure-rate Ranges
(a) Classical $1.741\ldots \times 10^{-8} \leq \theta \leq 18.382 \times 10^{-8}$
(b) Bayesian $1.708\ldots \times 10^{-15} \leq \theta \leq 1.3446 \times 10^{-8}$

Reliability Ranges
(a) Classical $0.991981\ldots \leq R(t/\chi, \beta) \leq 0.999238$
(b) Bayesian $0.999442\ldots \leq R(t/\alpha, \beta) \leq$ approximate unity

(g) As a final comment, it must be remembered that normally, for a system, three stages are passed through until normal system operation. These are
(i) Initial engineering assessment
(ii) Accelerated component/system life testing
(iii) Subsequent derived life-experience
It is not feasible to carry out (ii) on any realistic APS system, such that the results in this section only reflect the combination of (i) and (iii). Section 4.9 (SGHWR) has data in which (i), (ii) and (iii) are all present and can reflect their individual influences.

4.9 The Automatic Protective System for the Winfrith Steam Generating Heavy-Water Reactor (SGHWR)

The SGHWR is a pressure-tube direct-cycle reactor, as shown diagrammatically in Fig. 4.20, in which light-water coolant is boiled in the pressure tubes which contain the fuel elements. The electrical power output is designed to be 100 MW (electrical). There is part moderation by the coolant but the main moderation is by the heavy-water moderator which is contained in the calandria. The fuel is uranium dioxide pellets in zircaloy cladding. This reactor was fully commissioned at the end of 1967 and a general account of it is given by Moore and Holmes[131] with a description of the engineering design in Bradley *et al.*[132]

The protective system of the SGHWR has been chosen for study as, unlike the Gas Cooled type protective system of the previous section, it was designed using the conventional engineering techniques and these were extended to include full probabilistic analysis in the early stages of design. Essentially the protective system is in two parts—an 'X' and a 'Y' trip system as shown in Fig. 4.21. Both these systems can cause liquid shutdown tubes to operate, drain the moderator and have other functions. The liquid shutdown system consists of twelve loops each one of which is similar to that shown schematically in Fig. 4.22. The important additional function of the 'X' trip system is the operation of emergency cooling-water valves but this emergency

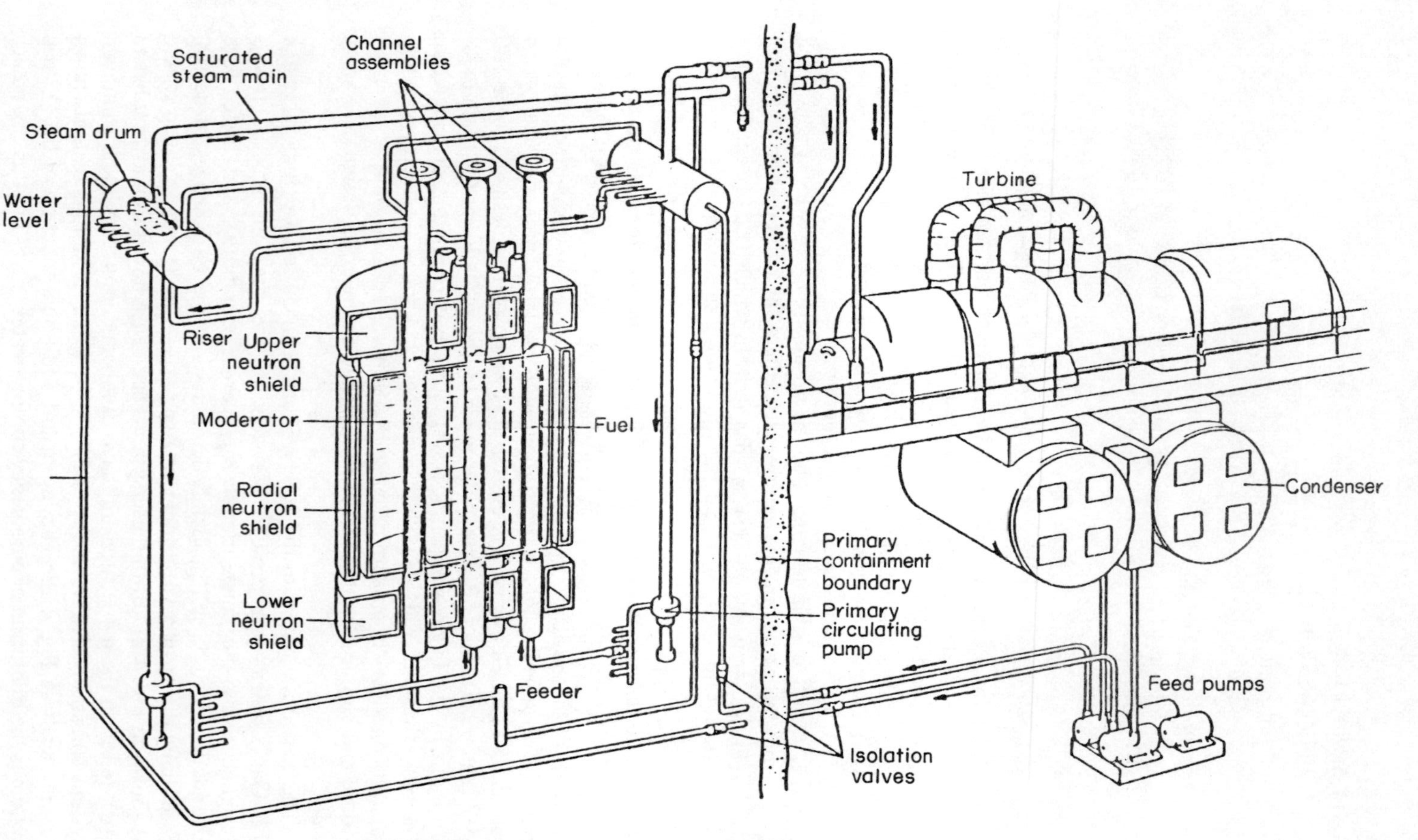

Fig. 4.20 Schematic diagram of SGHW reactor and turbine

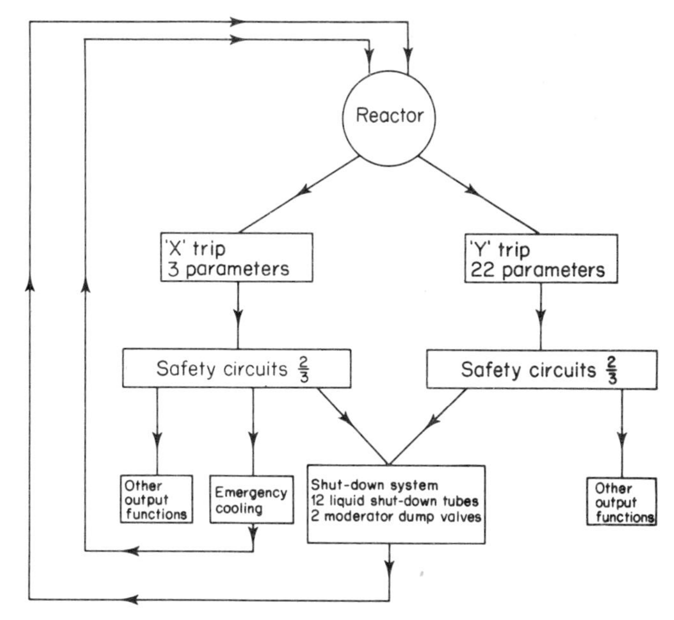

Fig. 4.21 Schematic diagram of SGHWR protective system

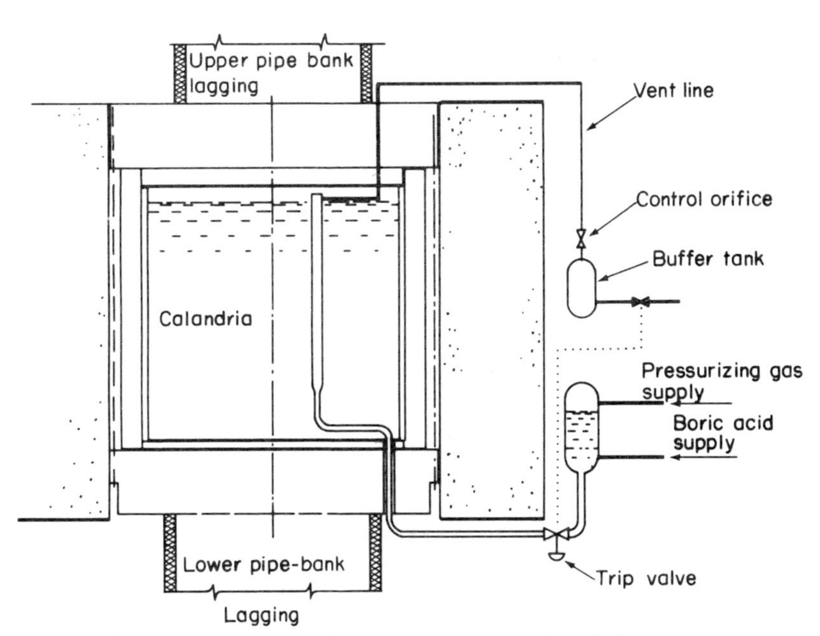

Fig. 4.22 Steam-generating heavy-water reactor liquid shutdown system

Table 4.26 Protective circuit parameters for SGHWR

'X' trip parameters

1. Primary containment rate of change of pressure.
2. Primary containment pressure high.
3. Primary containment top hot-box rate of change of pressure.
4. Primary containment bottom hot-box rate of change of pressure and CO_2 vault rate of change of pressure.
5. Extra low drum level.

'Y' trip parameters

6. Negative rate of change of pressure across main circulators.
7. Negative rate of change of pressure across natural uranium loop circulators (not now in use).
8. Steam drum pressure high.
9. Under voltage trip (N_1 and N_2 circulators) (replaced by phase sensitive circulator current trips).
10. High differential pressure across circulator pairs.
11. Negative rate of change of main circuit pressure.
12. Steam-drum water level high.
13. Under voltage trip (S_1 and S_2 circulators) (replaced by phase sensitive circulator current trips).
14. Superheat steam temperature high (superheat not provided).
15. Steam activity high.
16. Condenser vacuum low.
17. Negative rate of change of pressure across cluster loop circulators.
18. Neutron flux high.
19. Period low.
20. Pulse rate high.
21. Steam-drum water level low.
22. Cluster loop pumps electric supply failure.
23. Two-element loop pumps electric supply failure.
24. Cluster loop steam-drum pressure high.
25. Cluster loop steam-drum water level low.
26. Two-element loop steam-drum pressure high.
27. Two-element loop steam-drum water level low.

cooling system is not part of the study undertaken. The parameters for the 'X' trip system are those which are intended to sense most rapidly a burst circuit within the primary containment. The 'Y' trip parameters are mainly those which detect unsafe plant conditions caused, for example, by loss of coolant flow, excess neutron flux, etc. Table 4.26 gives a list of the protective circuit parameters. A more general account of the protection and control of the SGHWR is given by Wray *et al.*[133]

4.9.1 Liquid shutdown system for SGHWR

As in the Gas Cooled type reactor where the control-rod actuating mechanisms have been investigated in a previous section as a means for rapidly inserting negative reactivity by the use of control rods so the equivalent system on SGHWR has been studied. Under reactor fault conditions boron poison is forced into shutdown tubes which have already been mentioned in order to rapidly shut down the reactor. It may be noted that there is a second independent method of shutting down by draining the moderating liquid but this is a more lengthy process. The liquid shutdown system operates on the basis of the insertion of lithium borate solution at a pre-determined speed into twelve separate shutdown tubes. Each tube is self-contained in that it is provided with its own trip valve, pressurized poison storage vessel (poison head tank), and buffer tank. Common supplies are provided for poison, pressurizing gas, rinsing water and purge helium gas.

The shutdown system is considered to be primed when the main trip valves are closed and magnetically latched and connected into the safety circuit with the poison at a pre-determined level in the poison head tank and an over-pressure of helium gas, again at a pre-determined value. The shutdown tubes are continuously purged with low-pressure helium to remove any moisture which may be present and to ensure the correct conditions of dry helium in the shutdown tubes when a trip occurs.

In this study only the performance reliability of the liquid shutdown system has been considered from the original theoretical predictions for system reliability characteristics to its ultimate operating history in the actual plant. Here again the main objective has been to consider trends and to experiment with techniques of analysis which may be pertinent in the 'reliability growth' process.

4.9.2 Theoretical reliability prediction of liquid shutdown system

The reliability assessment undertaken at the design stage was intended to see what boundary value for the probability of failure on demand should be used as a prior estimate for the purposes of safety evaluation. Use was made of methods previously described and also given by Bourne[58]. This detailed analysis has been re-examined. The probabilities of failure were deduced for an individual shutdown tube and for the complete system of twelve shutdown tubes.

Various items of the system were given an engineering appraisal and their modes of failures and effects investigated and categorized in the manner already described in previous sections. Typically the main shutdown valve was analysed in this manner. Figure 4.23 shows a photograph of the actual valve on a test bed and Fig. 4.24 shows the valve assembly in simplified diagrammatic form.

However, prior to any testing information the various modes of failure were considered along the following lines. Component faults which caused the valve to operate when not required to do so were considered to be fail-safe, as were neutral faults which had no effect on performance. Component faults which prevented the 'front' valve from being fully opened or the 'rear' valve being fully closed, after a shutdown signal had been received were

Fig. 4.23 SGHWR main shutdown valve assembly

considered to be fail-dangerous. Fail-danger faults were divided into separate categories, revealed (r) and unrevealed (u).

By this means with associated failure-rates the overall failure-rate was estimated together with the total failure-rates associated with the categories fail-safe, fail-danger unrevealed and fail-danger revealed, as shown in Table 4.27 for the complete shutdown valve. Similarly, the various items were analysed for each sub-system failing to meet a given requirement on demand was assessed as given in Table 4.28. Assuming a 12-week test interval the probability of failure of a single sub-system (loop) was predicted to be 3.1×10^{-2}.

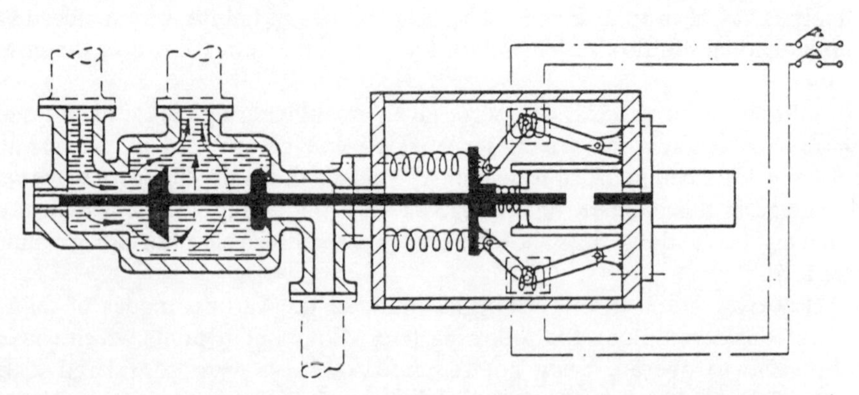

Fig. 4.24 Simplified diagram of the liquid shutdown valve—SGHWR. A view of the value in the 'shutdown' position (shutdown tube connected to the poison head tank)

Table 4.27 Assessment of failure rates of the SGHW main shutdown valve

	Failure-rate (faults/year)			
		Dangerous		
	Safe	u	r	Overall Total
3-way valve Valve mechanism	0.310	0.085	0.067	0.462

Table 4.28 Probability of failure of liquid shutdown system from theoretical analysis

System requirements for successful operation	Approximate probability of system failure on demand (Depends on test interval)
12-out-of-12 sub systems	2×10^{-1}
11-out-of-12 sub-systems	2×10^{-2}
10-out-of-12 sub-systems	2×10^{-3}
9-out-of-12 sub-systems	9×10^{-5}
etc.	

4.9.3 Sample test information on shutdown sub-systems

The results of sample testing were reappraised to deduce the systems reliability performance making assumptions concerning the statistical distribution of system faults in space or time. The theory of the technique used is given by Green and Bourne[49] under the headings of confidence limits and sample testing. It will be seen from the discussion that the accuracy of the reliability estimate improves as the tested sample increases in size and as the general pattern of failure is better known. Where the sample sizes are relatively small and no further information is available, it can be generally assumed that failures can take place randomly with a mean constant rate over the time-scale or number of tests of interest. Using this type of assumption, estimates can be made at certain confidence levels of the chances of failure or mean failure-rates using the Poisson sampling distribution, which is also described by Green and Bourne[49]. Hence it may be derived, for example, that if one failure has taken place in 100 events, this leads to an estimate of the mean chance of failure of 0.01 per event, or to an estimate of not more than 0.02 per event with 60 per cent confidence, or not more than 0.05 per event with 96 per cent confidence. In this manner the estimates of the chance of failure have been made from the results of tests.

Certain tests were carried out on prototype shutdown valves in the development of the liquid shutdown system. Characteristically, as it had been the trend indicated previously, this led to the discovery of certain faults which were subsequently corrected and the modified valves were subjected to further tests. Furthermore, the prototype valves were also tested in conjunction with a Loop Rig which simulated the complete liquid shutdown system as if it were installed in the reactor. At the completion of the development programme fourteen production shutdown valves were manufactured, twelve were installed in the SGHWR with two valves kept as spares. Subsequent to the installation, further tests were carried out on the actual reactor system. It is this general history which is being re-examined to see what relevant data can be derived.

4.9.4 Prototype main shutdown valves

On number one prototype valve 363 test firings appeared to have been carried out in the pressurized range of 350–450 psi. Various failures took place in the development programme and it would seem that several different types of faults occurred.

From general experience of reliability analysis it would be expected that the fault rate is probably not constant, being higher at the beginning of the programme than at the end. Hence the test programme was divided into a number of intervals and an examination made of the estimation of the chance of failure in each interval in order to derive the general trend. The results of this analysis are given in curve A of Fig. 4.25 as also discussed by Green[109]. In this figure the probability of failure per firing with 95 per cent confidence is plotted against the number of firings from the results obtained with the Number One prototype valve. It will be observed that the initial probability of failure over the first few firings is quite high. At 363 firings which was the extent of the test programme on the Number One valve the curve actually finishes and at this point shows a probability failure of approximately 0.02. Hence, assuming that not more than one other fault would have occurred if the test firings had been extended to 1000 firings the curve has been extrapolated, as shown dotted. This indicates a trend towards a probability failure per firing of 0.01 based on the possible results from a programme of 1000 firings.

A more optimistic approach would be to assume that the true test information should be taken from approximately the last 100 firings when most of the 'teething' problems had been solved. Using the same approach with the information as before leads to curve B in Fig. 4.25. It will now be observed that the evidence for the curve now ceases at 100 firings and a failure probability of approximately 0.03 has been demonstrated with 95 per cent confidence at this point. Further extrapolation up to 1000 firings, is shown dotted on the curve and assumes that not more than one fault would have occurred in this total number of firings. Thereby it is seen that the final probability figure is about 0.005 per firing. Hence, it would be considered a reasonable assump-

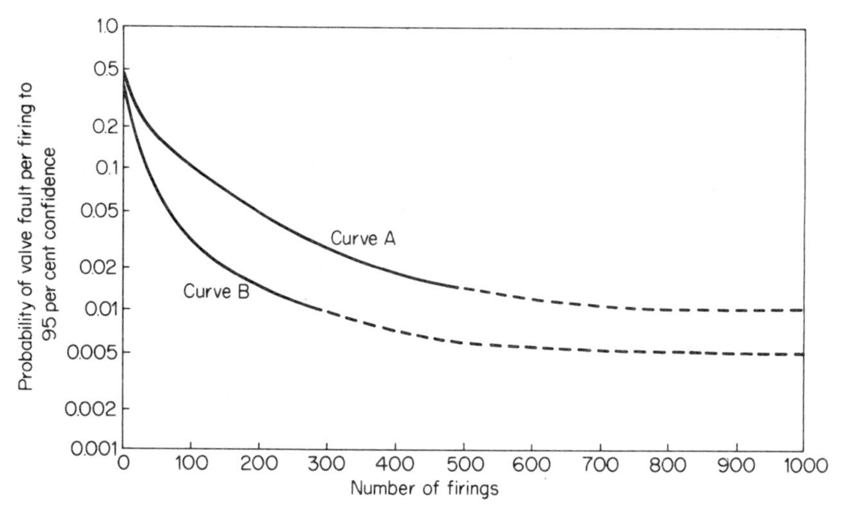

Fig. 4.25 Probabilities of valve faults per firing

tion to make that the prototype tests on the valve Number One lead to an estimate of a fault occurring per firing which lies in the probability range 0.005 to 0.01.

In the pressurized range of 350–450 psi 109 firings were carried out on number two prototype valve. No significant faults developed on this valve during the test programme. Taken on their own, these successful firings led to a probability of failure estimate of 0.028 with 95 per cent confidence. It would be seen, as would be expected, that this point lies on the curve B of Fig. 4.25.

Approximately 100 successful firings on valve Number Two, together with the last 100 successful firings on valve Number One, can give an estimate at the 200 firings point which comes to 0.015 chance of failure per firing with 95 per cent confidence. This point also as expected lies on curve B of Fig. 4.25.

4.9.5 Loop rig tests on production valves

Fourteen production shutdown valves which had been manufactured and subsequently tested on the loop test rig were each subjected to about fifteen successful firings on the loop test rig. Assuming samples in type and number can be added cumulatively, this gives a further sample of 210 firings without fault. Hence, adding these to the previous 200 successful firings on the two prototype valves, the estimate of probability per firing is 0.0073 at 95 per cent confidence, which may be compared with curve B of Fig. 4.25. On the other hand, if all prototype firings are included, together with prototype faults, the total number of pressurized firings is 682, and this leads to a failure probability of 0.013 with 95 per cent confidence, which may be compared with curve A of Fig. 4.25.

4.9.6 Installation on site

The pressurized 792 firings on site brings the total of all pressurized firings to 1474 or to 1211 taking into account only the 100 successful firings of Number One prototype valve. Additionally, two valves had developed a partial leakage fault on the back seat during the site tests. Depending on whether or not the initial prototype faults are included so the total number of fault occurrences on the main valve ranges between two and six. Failure probability for the main valve on the combined data amount to 0.0052 at 95 per cent confidence or 0.01 at 95 per cent confidence if the prototype faults are included. It is seen that these estimates are almost exactly what would be expected from the trend shown in Fig. 4.25.

The site tests also provided some additional information on other parts of the overall system as well as the main shutdown valve. The auxiliary valve, which developed one fault during site tests, had also been subjected to 474 total firings. This yields an estimated failure probability per firing for the auxiliary valve of 0.0032 with 95 per cent confidence. On site the complete liquid shutdown system had been subjected to a total of 792 pressurized firings without any reported defects. This gives an estimate of the failure probability per firing for the complete system, excluding the main shutdown valve assembly, of 0.004 with 95 per cent confidence.

4.9.7 General test results

Estimates of the failure probability per firing for each SGHWR liquid shutdown system as derived from test results on prototype valves, production valves and complete systems and calculated at the 95 per cent confidence level are:

Complete system (excluding main valve assembly)	0.004
Auxiliary valve	0.003
Main shutdown valve	0.01

The calculated values make the assumption that all the faults that developed would have led to an unacceptable degradation in system performance. In addition these results have been obtained from a series of tests and firings which took place at a fairly rapid rate compared with the ultimate rate under normal reactor operating conditions. On this basis, a safety assessment value for the failure of each complete shutdown loop system on demand is indicated as being in the range 0.01 to 0.02.

4.9.8 Liquid shutdown comparison between prediction and actual failures over 6 years

The data available from operational experience on the plant over a period of 6 years have been examined. In this examination it has been assumed that a dangerous failure had occurred if on receipt of a shutdown signal any loop failed to operate according to the particular specification involved.

It is of interest to note that only one failure was of the common-mode type which took place early in the life of the reactor and can be classified as a commissioning failure. This was due to the failure of the original material used in the back seats of the main valves which prevented the loops from operating correctly but did not prevent the loops from firing. There were other early life failures which can be discounted although they may be expected in any 'reliability growth' programme. These included rapid deterioration of valve-seat rubbers which were readily remedied by changing the material and high mortality of micro-switches which were replaced by types of a more robust nature.

However, failures which occurred during the actual operation of the reactor were classified as safe, neutral or partial failures and dangerous failures. Under the failure criteria generally defined it is the dangerous failures which are of particular concern from a safety assessment point of view which was the basis of the original prediction.

There were four dangerous faults which occurred in the individual loops which were all associated with failure to retain the pressure after firing. However, with reference to the inbuilt redundancy in having twelve loops installed these failures did not constitute a hazard to the reactor in the general sense.

Under the failure criteria assumed these faults were analysed as random unrevealed failures in the dangerous category preventing a loop from operating correctly. Therefore, with twelve loops operating for 6 years giving 72 loop years of operation this gives a mean failure rate of 0.056 faults per year. From the prediction already described which was based on a 12-week test interval it will be seen that the probability of failure was predicted as 3.1×10^{-2} for each loop. Hence, the predicted failure-rate is approximately 0.12 faults per year which is around twice the mean failure-rate resulting from actual operating experience. Referring to the correlation analysis described in this chapter for over 130 systems it will be seen that this is conformity with the expected accuracy the methods employed. Furthermore, it is shown in the analysis that both modes of failures which actually occurred which were leakage past the back seat and leakage across the seat of the helium valve were found on examination to be among the predicted failures. As a demonstration reactor it facilitated proof testing so that the actual test interval was less than the assumed 12 weeks which could be beneficial.

4.9.9 SGHW protective system classical and Bayesian analysis

(a) The previous discussion in section 4.8 and calculations related to the automatic protective system (APS) of a Gas Cooled type nuclear reactor are referred to. The final comments in that section indicated that normally, for a 'system' (i.e. not necessarily a complete plant) three stages preferably were involved from the initial conception up to the final stage of a routinely working plant. These were:

(i) An 'Initial Engineering Assessment'—which for high-integrity plant under modern conditions involved a quantitative evaluation in terms of the probability (very low for high hazard plant) of serious malfunction occurring on the plant.
(ii) Accelerated life-testing of important sections of the plant or vital components of the plant, in order to prove the capability of these units to perform their duty to specified limits.
(iii) A monitoring process, or the accumulation of subsequent plant in-service performance data, in order to corroborate the continuing successful operation of the plant itself.

(b) The previous Gas Cooled-type reactor analyses in section 4.8 stated that in using classical or Bayesian methods for the prediction/monitoring, etc., for plant management, in the case of reactor APS, Stage (a)(ii) above was not feasible to carry out, and that the data used in section 4.8 could only synthesize the failure-rate/reliability evaluations required.
(c) With respect to the important SGHWR shutdown system (APS), a vital part of this system, namely the liquid-boron injection valves, could be subjected to accelerated life-testing which corresponds with the previous stage (a)(ii). Therefore the present part of the study embraces data inclusion from all three stages in order to determine features, validities and the acceptability of the vital sub-units.

It may be noted that apart from information later detailed in this section, the classical and Bayesian methods of section 4.8 are used identically as previously in that section. Below are given the data arising from the full three stages required of plant/units, etc. Tedious repetition of parts of the section 4.8 analysis and detail are not duplicated.
(d) Following from the above discussion, the data now used are as follows:

Stage (i) The initial design condition set for 'failure-on-demand' was probabilistically aimed to be no greater than 1 failure in 10^5 demanded shutdown events.

Stage (ii) For consistency in comparison, probability conditions (e.g. upper limits, lower limits and probability bands, etc.), will be the same as in the earlier section 4.8. It will be recalled that the 'upper probability' where used was 0.97; the 'lower probability' was taken as 0.04, and the 'confidence band' was thus 0.93. Again it will be recalled that these led to values of α and β in section 4.8 of 0.77967 and 1.09755 respectively.

Relating to Stage (ii), the relevant data are that some 1774 firings of the important liquid shutdown valve were all 'successful'. That is there were zero 'failures-on-demand' in these firings. Quoting data from the later stage (Stage (a)(iii)), some 32 000 hr of operation indicated that 71 'on-demand' actuations of the valve system occurred, and from the system design where in 1 'on-demand' actuation operated a number of valves (10 has been taken as a reference figure for use in the later calculations), the 71 'demands' in 32 000 hr with successful operation of 10 valves at all times, constituted $[(71 \times 10)/(32\,000)]$ (8760) valve actuation demands per year, i.e. 195 actuations per year. Compared to the 710 valve actuations in 32 000 hr, the Stage (ii) accelerated life-tests represent zero failures of any valve in an equivalent normal in-service life of 66 434 hr.

Stage (iii) Seven-hundred-and-ten successful valve firings in 32 000 hr without a single failure is equivalent to zero failures in 22.72×10^6 hr.

Summarising the foregoing gives the following results:

1. In-service valve 'operations on demand' (revealed) from Stage (iii) data equals 195 operations per year. When combined with the 1×10^{-5} probability of a failure occurring, this implies a failure rate (mean) of 1 failure per 22.26×10^6 hr.
2. Test firings (all successful) yield a data value (for the Stage (ii) test firings) of zero failures in 66 434 hr. Bayesian use of all/any available relevant data, can 'add' data such as this into the gamma conjugate model implied in this section, by means of modifying the α and β values.
3. In-service life (the 32 000 hr already referred to) represents zero failures in 22.26×10^6 hr.

(e) Modification of the α and β values for the various stages

(i) Previous work indicated $\alpha = 0.77967$, $\beta = 1.09755$. These values stem entirely from the probability/belief conditions related to equipment of this type. *Specifically, α and β are not controlled initially by the '1 in 10^5 demands' considerations*. This condition merely modifies prior and posterior gamma-model evaluations.

(ii) Following from (i) above, a *single failure in 10^5* actuations is the first modification to the values of α and β. This condition is equivalent to an assumption of 1 failure in 45×10^6 hr.

(iii) The life-testing 'Stage (ii)' data combines *zero faults* in 66 343 hr into the gamma-model.

(iv) The in-service 'Stage (iii)' data combines *zero faults* in $(32\,000 \times 10)$ hr to the modification of α and β.

(f) Modifications to α for n failure events merely changes α to $(\alpha + n)$.

Modifications to β for n failure events (the 'n' events above) in τ hr changes β as below:

$$\beta \rightarrow \left(\frac{1}{\beta} + \tau\right)^{-1}$$

A set of n and τ values can successively modify α and β in the manner indicated.

(g) The α and β modifications are given as follows and are used in later calculations:

(i) *Initially*

$$\left.\begin{array}{l}\alpha = 0.77967\\ \beta = 1.09755\end{array}\right\}$$ for the initial and known, 0.04, 0.93, 0.97, probability conditions.

The '1 failure in 10^5 actuations' condition with the number of valves and valve operations required is equivalent to 1 failure in 45×10^6 hr $= 222.22 \times 10^{-6}$ faults/hr.

Thus $\alpha \rightarrow (0.77967 + 1) = 1.77967$

$$\beta \rightarrow \left(\frac{1}{1.09755} + 45 \times 10^6\right)^{-1} = 222.22 \times 10^{-6}$$

(ii) *The accelerated life-test data*

$$\alpha \rightarrow (0.77967 + 1 + 0) = 1.77967$$

$$\beta \rightarrow \left(\frac{1}{1.09755} + 45 \times 10^6 + 66434\right)^{-1} = 221.89 \times 10^{-6}$$

(iii) *Plant in-service data*

$$\alpha \rightarrow (0.77967 + 1 + 0 + 0) = 1.77967$$

$$\beta \rightarrow \left(\frac{1}{1.09755} + 45 \times 10^6 + 66434 + 320\,000\right)^{-1}$$

i.e. $\beta = 220.33 \times 10^{-6}$

(h) (i) From (g)(i) to (iii) the failure-rates have been calculated in classical and Bayesian form and from these reliability evaluations have been made. The reliability evaluations are related to a specific time period; this has been taken as 5 years, this having a near equivalence to 32 000 actual in-service hours for the system.

(ii) *Reliability Evaluations*

Case (i) Using the original α and β values combined with the serious malfunction of valve in 10^5 demands for shutdown on the system.
Classical $0.1012 \leqslant R(t \mid \theta) \leqslant 0.9999997092$ (perhaps better expressed as $1 - (2.908 \times 10^{-7})$
Bayes $0.9977102 \leqslant R(t \mid \alpha, \beta) \leqslant$ effectively unity.

Case (ii) Adding the accelerated life-testing results to the data.
Classical $0.1014 \leqslant R(t \mid \theta) \leqslant 1 - (2.904 \times 10^{-7})$
Bayes $0.9997714 \leqslant R(t \mid \alpha, \beta) \leqslant$ effectively unity.

Case (iii) Adding the in-service 32 000 hr service data.
Classical $0.1033 \leqslant R(t \mid \theta) \leqslant 1 - (2.886 \times 10^{-7})$
Bayes $0.9997730 \leqslant R(t \mid \alpha, \beta) \leqslant$ effectively unity.

(j) The results given in (h) above have not been stated to an unnecessary degree of accuracy. They do indicate the difficulty, with engineering systems of extremely high integrity, of placing a useful judgement on reliability figures obtained. Additionally, they do show the tendency for positive reliability improvement, and appropriate forms of malfunction could modify, through the total plant-life, the evaluated reliability figures and therefore serve as a means of plant-safety monitoring.

4.10 SGHWR Operational Experience Over 5 Years and Some Aspects of Unavailability

In the predictive assessments of the reliability and availability of the Winfrith SGHWR which have already been discussed in a previous section, the work was largely based on a negative exponential distribution type of model where the density function is

$$p(t) = \theta e^{-\theta t} \tag{4.51}$$

and θ = mean failure rate.

Hence the probability of one (or more) failures by the time is found by integrating equation 4.51 which gives

$$P(t) = 1 - e^{-\theta t} \tag{4.52}$$

If r is the repair rate then probability of one (or more) repairs by time t is given by

$$P_r(t) = 1 - e^{-rt} \tag{4.53}$$

The mean unavailability of the particular system U may be defined as

$$U = \frac{\theta/r}{1 + (\theta/r)} \tag{4.54}$$

It can be shown by Keller[134] that equation 4.54 is true for all failure and repair distributions, provided θ and r are calculated as the reciprocals of the mean values, where θ is the total mean rate and assuming an exponential mode is given by

$$\theta = \sum_{k=1}^{n} \theta_k \tag{4.55}$$

and similarly r is the total mean repair rate for the system.

From equation 4.54 the mean repair rate for the system may be derived as

$$r = \theta(1 - U)/u \tag{4.56}$$

Table 4.29 gives the overall predicted failure rates and repair rates applying to equations 4.55 and 4.56. The predicted repair rates are taken from a plant

Table 4.29 Predicted mean failure and repair rates for liquid shutdown and automatic protective systems

Subsystem	Overall predicted fault rate/hr	Overall predicted repair rate/hr
Liquid shutdown	8.4×10^{-5}	5.3×10^{-2}
Automatic protection	6.2×10^{-4}	1.7×10^{-1}

availability analysis described in the UKAEA report[135] and were directed at making a general study rather than one carried out in depth.

From the times between failures, assuming the exponential distribution model, the probability of exactly n failures occurring by time T is given by the Poisson distribution:

$$p_n(T) = \frac{(\theta T)^n}{n!} e^{-\theta T} \tag{4.57}$$

For the investigation undertaken in the UKAEA report[135] the total operational time was 43 000 hr and the analysis has been arranged to make a comparison between the predicted mean number of faults, the predicted range at some confidence level and the observed number of faults for the liquid shutdown and protective systems. The results are given in Table 4.30.

In the case of the repair time and resultant outage time, the assumption has been made that the time of repair can be modelled by the exponential distribution and therefore the probability of completing m repairs in a total time t is given by the gamma distribution

$$g_m(t) = \frac{r^m t^{m-1}}{(m-1)!} e^{-rt} \tag{4.58}$$

The observed number of faults, m, observed in the operational time of 43 000 hr, has been taken and compared with the predicted mean outage time t, the predicted range of outage time at some confidence level and the observed outage time for the liquid shutdown and protective systems. The results are shown in Table 4.31 for the upper and lower 95 per cent confidence levels.

Table 4.30 Comparison of predicted and observed number of faults for liquid shutdown and protective systems

Subsystem	Predicted number of faults		Observed faults	Comment
	95% range	Most probable		
Liquid shutdown	0.25–6.6	3	3	Good agreement
Automatic protection	18–35	26	9	Prediction pessimistic

Table 4.31 Comparison of predicted and observed outage time for observed number of faults for liquid shutdown and automatic protective systems

Subsystem	Predicted outage time (hr) 95% range	Predicted outage time (hr) Most probable	Observed time (hr)	Comment
Liquid shutdown	16–120	38	38	Good agreement
Automatic protection	30–85	47	277	Prediction optimistic

In order to present the foregoing information in graphical form showing the nature of the Poisson distribution function as an envelope rather than in histogram form, the cumulative Poisson, the gamma distribution function and the cumulative gamma for the liquid shutdown and protective systems have been calculated. These have been done from equations 4.57 and 4.58 and cumulative forms from equations 4.59 and 4.60 which are as follows:

$$\text{Cumulative Poisson } P \text{ (up to } n \text{ in } T) = \sum_{k=0}^{n} \frac{1}{K!} (\theta T)^n e^{-\theta T} \qquad (4.59)$$

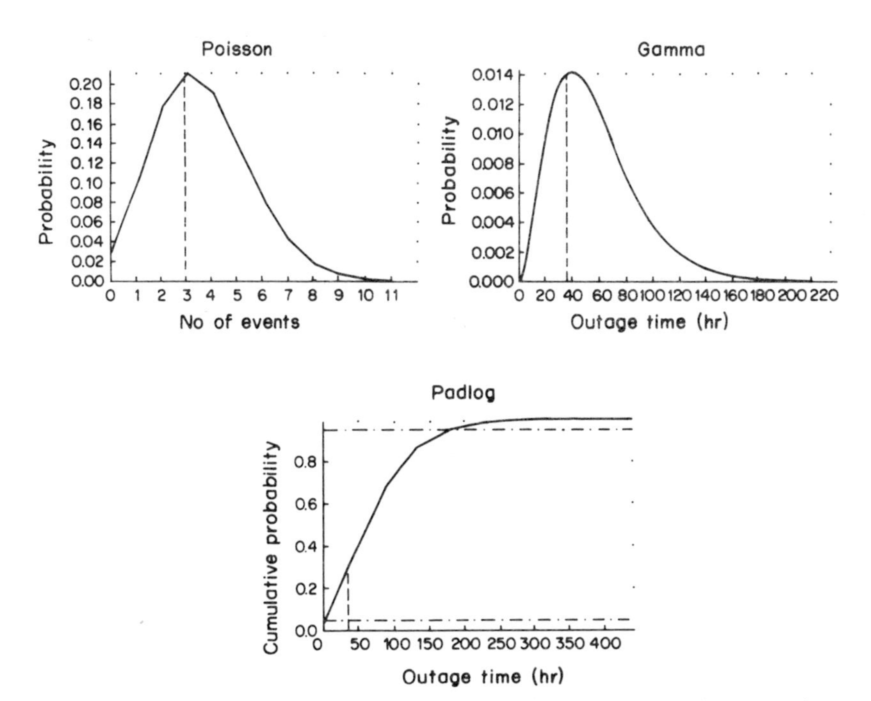

Fig. 4.26 SGHWR comparison of predicted and observed performance in graphical form for the liquid shutdown system (exponential model)

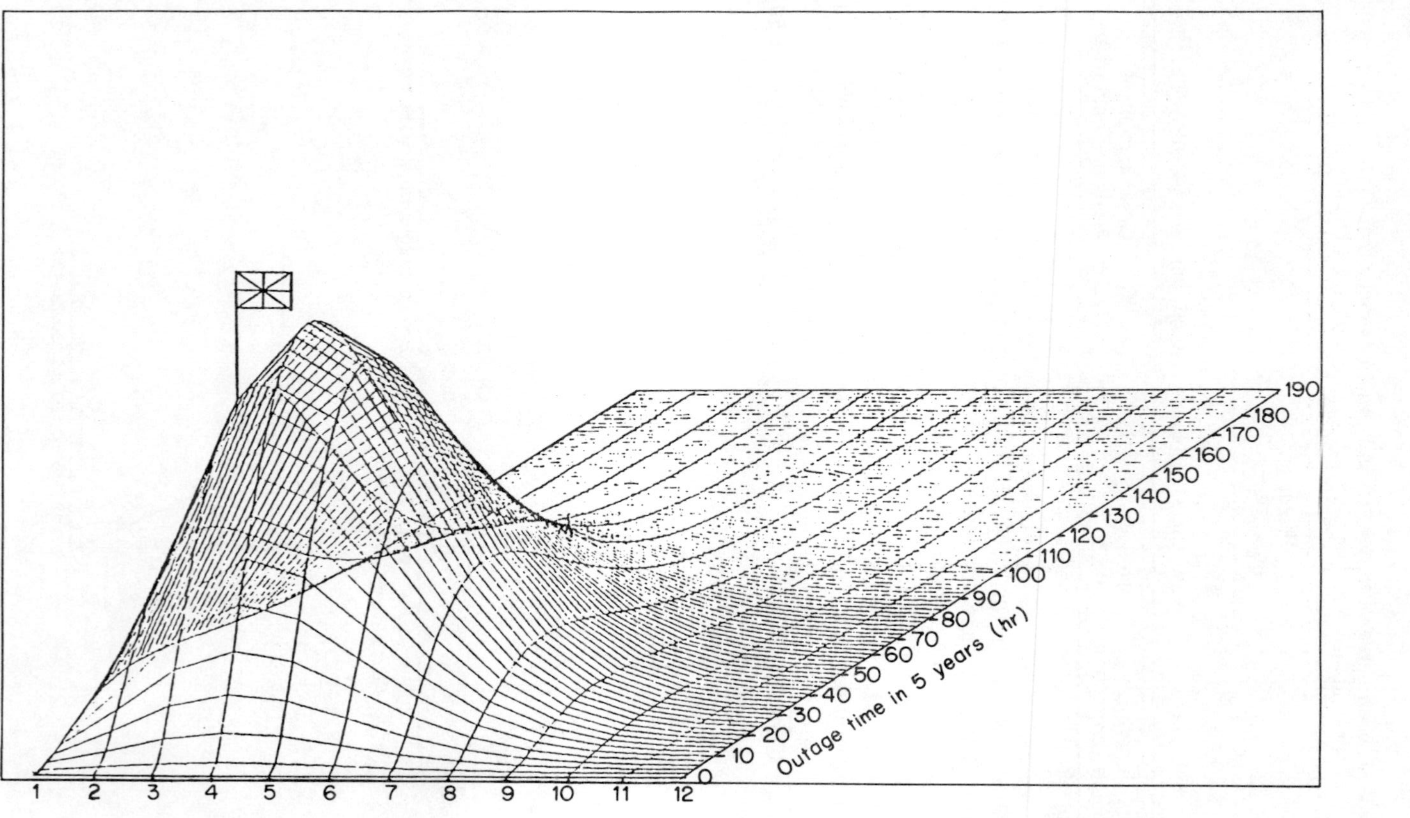

Fig. 4.27 Pictorial representation of the number of events in 5 years for the liquid shutdown system

$$\text{Cumulative Gamma } G(t \text{ for } n) = \frac{\gamma(n, rt)}{(n - 1)!} \tag{4.60}$$

The upper and lower 95 per cent levels are shown in the cumulative graphs and the observed performance is shown as a dotted line.

Figure 4.26 shows the comparison of the predicted and observed performance for the characteristics of fault rate and repair rate for the liquid shutdown system. The results are also shown for the unavailability of the system. For this purpose a computer code called PADS (Plant Availability Distribution Synthesis) was used. The code has the strategy of generating a cumulative repair distribution from the component failure and repair characteristics and this uses a Monte Carlo technique to obtain the cumulative distribution of outage time for the particular system of components.

A further pictorial representation of the combined probabilities of the number of events and the outage time distribution is shown in Fig. 4.27 for the liquid shutdown system. The height of the 'hill' is a scaled representation of the probability of the predicted number of events and associated outage time. More precisely this may be expressed as joint probability density distribution n in n and outage time, viz:

$$\text{Height} \propto \frac{(\text{Probability of } n \text{ faults}) \times (\text{Probability of } t \text{ outage time})}{\text{Most likely combination of } n \text{ and } t}$$

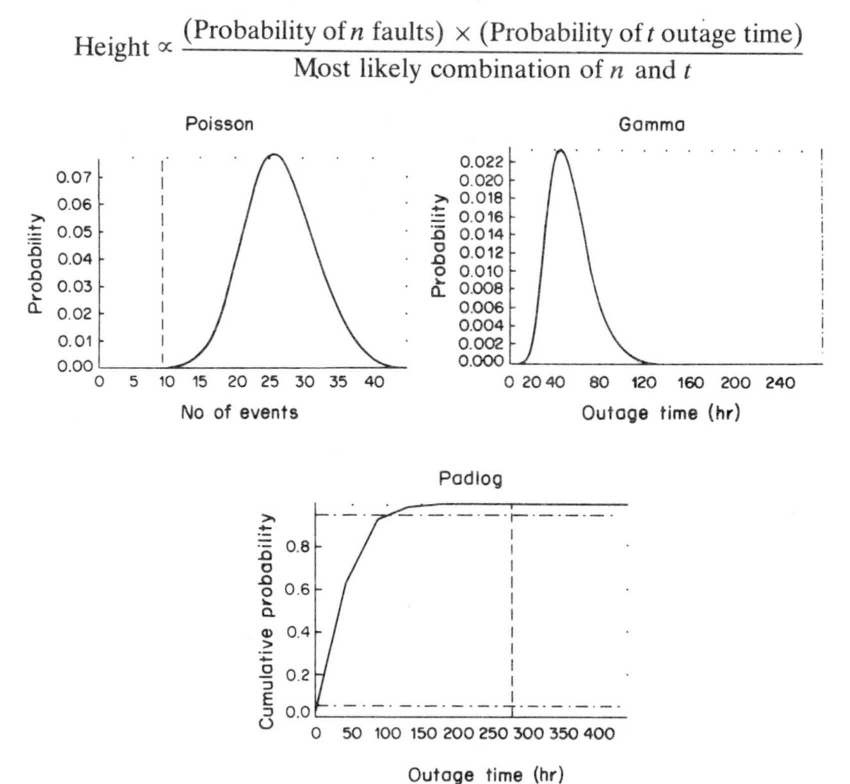

Fig. 4.28 Comparison of the predicted and observed performance in graphical form for the protective system (exponential model)

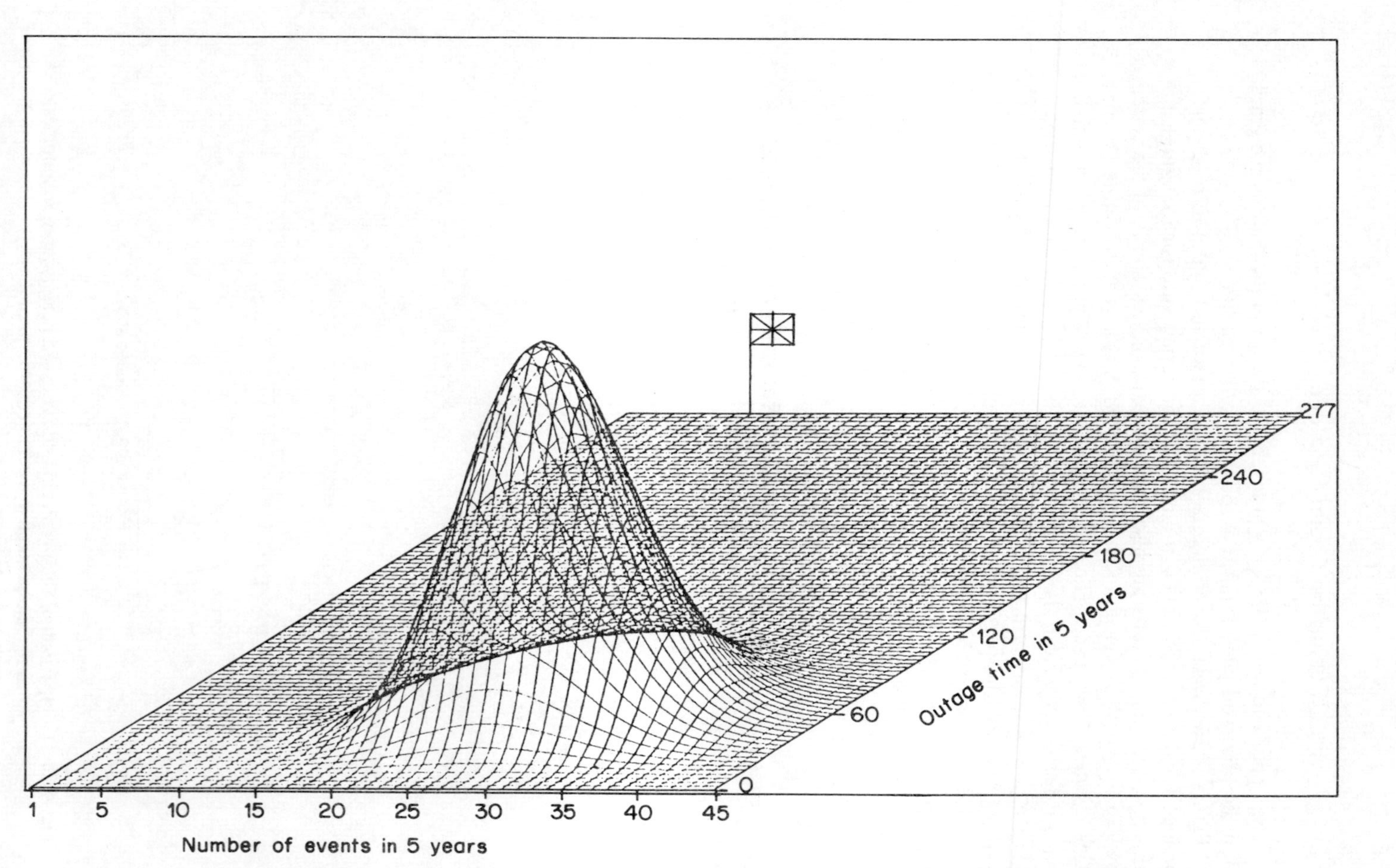

Fig. 4.29 Pictorial representation of the number of events in 5 years for the automatic system

The other axes being the integer number of faults and outage time as a continuous variable. The ‘flag’ represents the number of faults and outage time which actually occurred in the liquid shutdown system over a 5-year period and have already been indicated as dotted lines in Fig. 4.26. It will be seen that there is good agreement for the characteristics depicted.

Similarly for the protective system Fig. 4.28 shows the comparison of the predicted and observed performance in graphical form and Fig. 4.29 gives a pictorial representation of the number of events in 5 years. It will be noted that the mean failure rate is pessimistic and the mean repair time is optimistic.

From similar studies of such systems there appears to be a general trend for the number of faults predicted to be greater than the actual for the predicted repair time to be generally less than that observed. For the particular automatic protective system under discussion the prediction of the failure-rates was carried out in depth from a safety point of view. Whereas for the repair rates the predictions were not carried out in such depth. It is indicated from the present study and analysis that the repair modelling requires to have a similar in-depth treatment but this would be more significant in considering the availability of the output of the plant rather than its safety. As the systems become of greater complexity then in the safety case the analysis tends to be deeper and more searching than in the study of the available case which tends to have greater complexity.

CHAPTER 5

A Total Approach to Reliability Growth and Control

5.1 Reliability Growth and Control

The study described so far has commenced with a need from a safety point to protect plant which may involve high-risk situations which if allowed to go uncontrolled may lead to severe consequences. In the case of the nuclear reactors for example, a criterion can be developed, as discussed in chapter 1, to lay down a quantified probabilistic statement as a requirement for the automatic protective system which is called upon to operate under reactor fault conditions. Hence the first phase in any controlled programme of design and operation of a protective system is to establish the performance requirement with a properly specified reliability criterion as indicated in Fig. 5.1. Such a criterion will involve considerations which may include factors at a higher hierarchical level than the particular protective system or even the plant itself and may require to take into account humanitarian and environmental factors as discussed in chapter 1.

In considering the various factors involved it should be borne in mind that the generation time for the development of a particular generic type of plant may be over a period of many years. As an example of this growth it may be noted that the original programme of nuclear power (HMSO[4]), in 1955 specified 'that in order to obtain breeding of plutonium with a large positive gain factor, a fast reactor will be required'. It appears it was appreciated that the small core of a fast reactor could present difficult technical problems and that the solution of these would take time. The development of such a reactor and other types usually follows through four main stages and has been discussed by Palmer and Platt[136] and later in 1970 by Bainbridge[137]. These stages are in general terms as follows.

(a) Critical facilities and zero power uncooled.
(b) Experimental power reactor is built so that engineering problems can be studied on a scale which is realistic.
(c) Prototype reactor is built and at this stage economic operation must be sort after and considered.
(d) Full-scale power reactor.

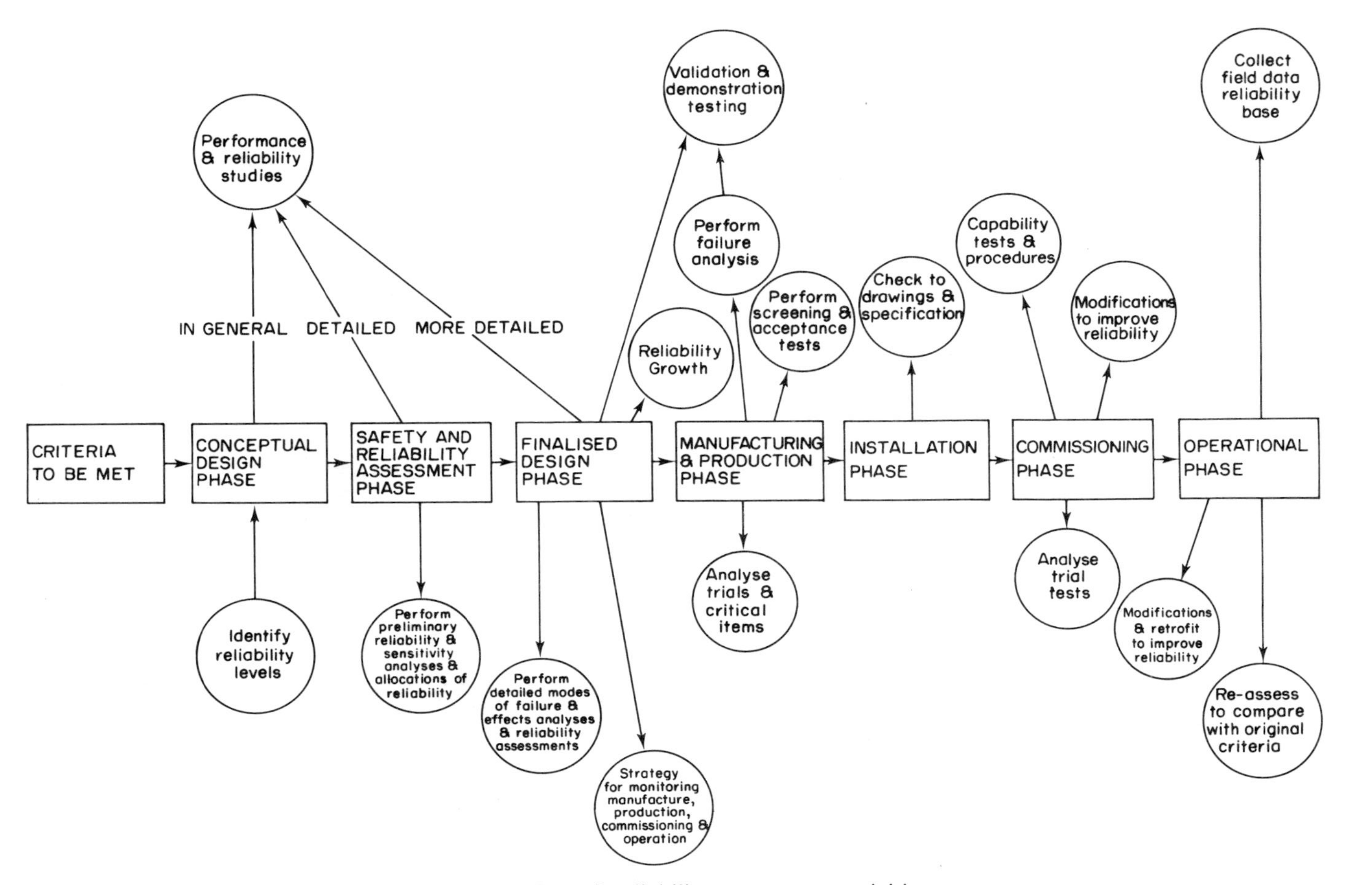

Fig. 5.1 Overall reliability programme activities

It is indicative of the long generation time for some generic plant development that in 1970 Bainbridge[137] was commenting on the prototype fast reactor which at this date was not yet operational but formed part of the fast-reactor breeder programme in the UK culminating in a full-scale power reactor at some future date.

Whether it be a nuclear or non-nuclear plant some form of growth will evolve at the higher hierarchical level which gives the direction and aims by which the automatic protective system requires to be designed and made operational on the plant. It is the aim for the high-integrity type of protective system which has been discussed at length in this book to operate on demand when required to do so. Hence the question of reliability is an outstanding characteristic for the protective system.

However, even after the growth of the generic type of plant and the decision having been made for a specific design to be undertaken then the design and ultimate operation of the automatic protective system may be over a period of time in excess of say 5 years. In the case of very complex and costly plants this period can be even extended and there is a need for projection at every phase to ensure that the system reliability is meeting the requirement for the plant in such a way that the safety is adequate, Hensley[138] has commented on some aspects of this subject. This leads to some form of reliability management and control and the preceding chapters have investigated the various factors involved both from the point of view of projection techniques and the results of operational experience. It is from these studies that the basic elements of reliability control are now considered from a total point of view.

The conceptual design phase requires a general and broad evaluation not only as part of a progressive design review procedure but in conjunction with an independent safety and reliability assessment as described in chapter 2. The strategy for the demonstration of the adequacy of the safety and reliability should be generally defined and the target allocations made for the different reliability levels of the system. It has been already shown in the present investigation that reliability during the whole life cycle of the system and plant can be measured and there are controllable parameters. The strategy adopted should carefully consider the reliability characteristics to be specified. In the investigations carried out the characteristic of failure rate as a parameter has been considered for example with particular proof-test intervals which can be specified during the design process, measured during test in many cases and observed and controlled during field operation.

It is considered important from the studies carried out that a 'matrix' for the complete protective system requires to be prepared by commencing with an inventory of the system items giving the design performance parameter estimates. At the conceptual design stage the assessor is particularly interested in the sensitivity of the overall system performance with reference to changes in the reliability parameters which are going to be 'tracked' through the whole cycle from design to ultimate operation. Furthermore, the

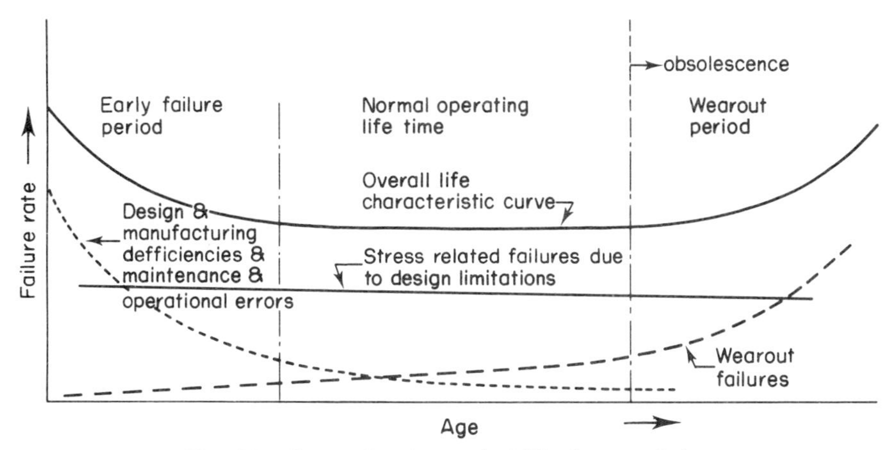

Fig. 5.2 Curve showing typical life characteristics

safety and reliability analyses are normally concerned with an envelope or boundary approach which has been discussed in previous sections.

From the studies carried out so far it is evident that the various items investigated have shown the 'bath-tub' curve characteristic as indicated in Fig. 5.2. Furthermore it is evident that the various studies which have been undertaken in this book indicate a 'reliability growth' process.

However, it may be noted that the reliability analyst undertaking the assessment may challenge the design with a series of 'what if' questions. In practice it will be found that there are many more questions than there are answers. Therefore from a safety assessment point of view, the empty 'schema' or 'matrix' to be used in building up to a prior estimate for the system requires to have an inbuilt memory for these questions. Although some of the questions may have no answers there may be no doubt cast on future quantified prior characteristic predictions. On the other hand it may be that doubts could exist which can only be resolved later in the reliability programme and will at the design and assessment stage require careful consideration of the nature of the 'statistical experiment' which will take place as the design naturally evolves to the ultimate phase of operation. Changes and modifications must be catered for in this strategy and the limits of the human memory recognized. Clearly this returns to the point made in an earlier chapter on having a proper filing system and various computer codes and storage facilities are available for this purpose. Table 5.1 shows part of a print-out development for recording and filling in the matrix. This print-out is for a temperature protective channel involving a thermocouple and temperature amplifier.

Each entry in Table 5.1 is showing the progress of the assessment so that at any stage there is available a compilation of the present state of the assessment. The questions which cannot be answered together with relevant drawing information, performance characteristics, await the assessment result of

Table 5.1 Example of part of a print-out in the assessment at the design stage phase

ITEM TYPE	REF.OR.DWG.NO.	PERFORMANCE RANGE	ACCUR	RESP	F.RATE	COMMENT.	DATE	INIT.	LOCATION
001 CHR./AL. T/C	W21196FO14	0-1000	3.5	3.0	.01	SATISFACTORY	27JAN78	AEG	FUEL EL.5
002 T/C CONNECT.	NOT KNOWN	0-500	5.0		.01	QUESTIONABLE	14FEB78	AEG	REACTOR DOME
003 COMP.CABLE	W21053/029	0-50	.5				7MAR78	AEG	PILE CAP
004 T.T.A.	TS 757-2273	200-600	7.5	1.0	.42	INADEQUATE	19APR78	AEG	SAFETY CUBIC.
005 T.T.A.MODS.	H.P.5.1.2.1.					NEW F.R.DATA	7JUN78	AEG	N/A
006 COMP.CABLE	W21053/029	0-50	.5		.01	SATISFACTORY	11JUL78	AEG	PILE CAP
007 CHR./AL. T/C	W21196F014	0-1000	3.5	3.0	.01	SATISFACTORY	28JUL78	AEG	FUEL EL.5
008 T/C CONNECT.	NOT KNOWN	0-500	5.0		.01	QUESTIONABLE	19AUG78	AEG	REACTOR DOME
009 COMP.CABLE	W21053/029	0-50	.5				27SEP78	AEG	PILE CAP
010 T.T.A.	TS 757-2273	200-600	7.5	1.0	.42	INADEQUATE	3OCT78	AEG	SAFETY CUBIC.
011 T.T.A.MODS	H.P.5.1.2.1					NEW F.R.DATA	31OCT78	AEG	N/A
012 COMP.CABLE	W21053/029	0-50	.5		.01	SATISFACTORY	20NOV78	AEG	PILE CAP

the designer's information. In practice a highly categorized matrix type of recording system can lead to inflexibility because the modifications etc., tend to be presented to the assessor as they occur and in general it is in a random fashion. In other words the assessment must proceed naturally in phase with the evolution of the design. If a basic 'matrix' can be created at the early design phase it obviously leads to a much easier and better method of information control.

5.2 Reliability Growth

The experience gained from the investigations carried out by the author are in conformity with general experience of not always being able to get it right first time. Hence the priors which have been discussed in earlier chapters have also been compared with posterior information and it is clear that there needs to be a process of growth and improvement until some target or criterion is met. Hence the question of whether the prior is such that the reliability programme enables it to be achieved is certainly salient as the manufacturing and production phase of the system and its equipment is entered. In newly designed military equipment it was quoted by Selby and Miller[139] that investigation showed that the mean time between failures (MTBF) achieved was often no more than 8 per cent to 20 per cent of its ultimate potential level. It was also shown by Selby and Miller[139] that a properly organized 'test and fix' programme can be of great assistance in closing this reliability gap. Obviously this argument of the extent of the reliability gap goes on through all the various phases following manufacture and production. However, quite often it is the first time that the 'paper-work system' is converted into some form of 'material or hardware system' and from an engineering point of view it is required to make the most use of any trials or tests carried out.

One of the features of this transformation from paper to materialistic form is to reveal hidden weaknesses and deficiencies in the equipment for example the systematic type of failure. This has already been seen in the analyses carried out in previous sections as for instance revealed in the valves in the liquid shutdown system investigated and discussed in chapter 4. The reliability growth process may be illustrated as shown in the simplified block diagram of Fig. 5.3. Broadly, the failures which are of the systematic type will require careful examination and those which are found significant will lead to some rectification process in conjunction with the designer and other interested parties such as the reliability assessor. The other failures which may be classed as random will require a restoration process so that the equipment may be repaired for further trials. This is a fairly standard procedure which is described by BS4200[140], particularly for electronic equipment but similar principles hold for other types of equipment.

These types of trials and tests can be used to obtain initial reliability information but the conditions of the tests require to be carefully studied. The tests themselves may not be under the same environmental conditions, it is often

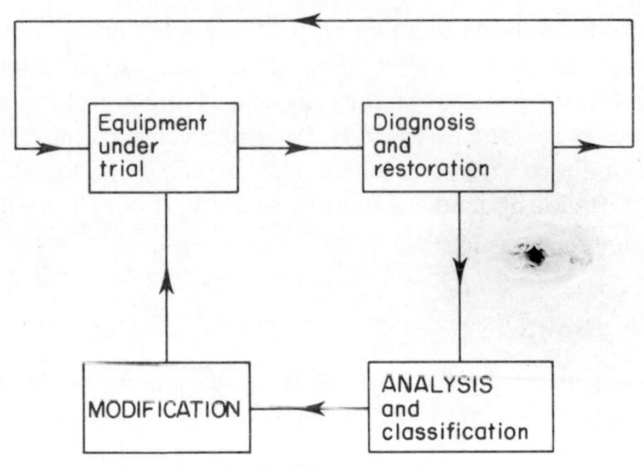

Fig. 5.3 Reliability growth process

not easy to set up correctly the true conditions. In the case of life testing as already illustrated in the previous chapter, this may represent accelerated testing particularly where the equipment is of very high reliability and catastrophic failure information is required. Various techniques of analysis exist for estimating the reliability characteristic of interest such as failure rate and some of the techniques have already been illustrated. Typically two periods of testing time may be selected, one at the start of the test and the other at the termination of the test, selecting periods with approximately equal numbers of failures. Thereby giving the initial cumulative number of failures F_1 in the increment of time Δt, with the initial failure rate θ_1 being given by:

$$\theta_1 = \frac{\Delta F_1}{\Delta t_1} \tag{5.1}$$

and similarly for the final failure rate θ_2:

$$\theta_2 = \frac{\Delta F_2}{\Delta t_2} \tag{5.2}$$

The improvement factor or ratio of growth can now be 'tracked' from:

$$\text{Ratio of growth in failure rate} = \frac{\theta_2}{\theta_1} \tag{5.3}$$

and confidence limits obtained in the conventional manner from a set of Poisson tables.

However, other techniques using a Bayesian approach may be applicable which have been discussed in previous chapters.

5.2.1 The Duane approach

Empirical models can be derived and it is part of the general reliability technology to search for such models. Crow[141] gives a version of the Duane model which may be stated as:

$$F = F_0 t^{\beta} \tag{5.4}$$

where F = the cumulative number of failures at test time t
F_0 and β are constants.

From equation 5.4

$$\frac{dF}{dt} = F_0 \beta t^{\beta - 1} \tag{5.5}$$

$$= \theta$$

where θ = the instantaneous failure rate.

Putting $\alpha = 1 - \beta$ then equation 5.5 becomes

$$\theta = \theta_0 t^{-\alpha} \tag{5.6}$$

where $\theta_0 = F_0(1 - \alpha)$ = constant.

This representation is of the original form given by Duane. By analysis and experiment in the testing situation the constants θ_0 and α may be derived for the model. The basis is then created for the prediction and planning future programmes of testing.

Usually equation 5.6 is represented graphically on log-log scales by representing equation 5.6 in the form:

$$m = m_0 t^{\alpha} \tag{5.7}$$

where m = mtbf
$m_0 = 1/\theta_0$ = a constant.

Figure 5.4 shows such a plot giving a straight line with slope α. The horizontal scale starts at some arbitrary time, say 100 hr, in other words it is a suppressed zero or otherwise at time zero this would give an infinite failure rate or zero mtbf.

The instantaneous value of the mtbf is shown by the straight line for 'm' and the cumulative mtbf (m_c) is shown as a dotted line and gives the value t/F at any particular time.

From equations 5.4 and 5.5 by division gives

$$\frac{F}{\theta} = \frac{F_0 t^{\beta}}{F_0 \beta t^{\beta - 1}} \tag{5.8}$$

$$\frac{F}{\theta} = \frac{t}{\beta} \tag{5.9}$$

Substituting $F = t/m_c$ and $\theta = 1/m$ gives

$$m_c = m\beta = m(1 - \alpha) \tag{5.10}$$

It will be seen that m_c is proportional to m as a line parallel to that of m in

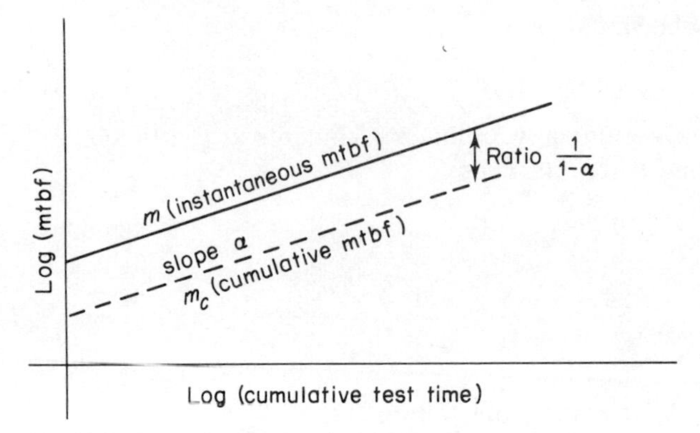

Fig. 5.4 Duane plot showing instantaneous and cumulative mtbf

Fig. 5.4 with a constant ratio $(1 - \alpha)$. This enables m_c to be derived from the failure and test-time data and the value of m can be derived directly at any time t. Furthermore, the value of m can be read off the graph at some early point and the growth ratio can then be readily calculated for the particular intervening period of test. These methods are described generally in the literature for example by Mead[142].

One of the main reasons for describing this Duane model and plot is that there are indications of its applicability to control-rod actuating mechanisms, for example as studied in chapter 4. Figure 5.5 showing control-rod system reliability growth trends has been reproduced from Brandt *et al.*[143] and shows

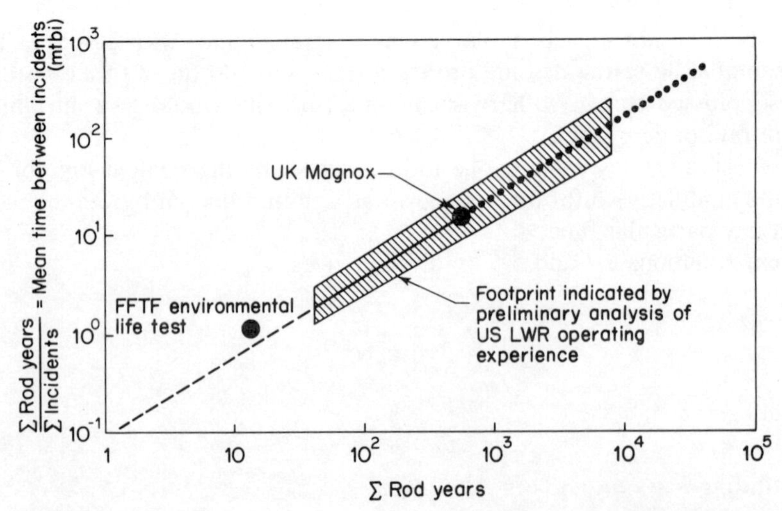

Fig. 5.5 Control-rod system reliability growth trends. (Reproduced by permission of A. M. Smith and G. L. Crellin from Brandt *et al.*, 1978, *ENS/ANS Tropical Meeting on Nuclear Power Reactor Safety, Brussels*)

all the characteristics of a Duane plot. Furthermore, the point marked 'UK Magnox' is for the type of control-rod actuating mechanisms already discussed and analysed in chapter 4. The following interesting suggestion was made by Brandt *et al.*[143]: 'This suggests that control rod mechanisms of various designs form a characteristic product family—and that LBR control rod mechanisms may well follow the same trend' (p. 8). This information is obviously useful in helping in decision-making as it predicts the level of failure frequency which can be expected and assists in the planning of future tests.

5.3 Early-life Plant Defects and Breakdowns

Analyses which have been carried out particularly in the early-life phases of a plant show that a small percentage of the defects give rise to a very high percentage of the work load on the plant. This is particularly true at commissioning for instance and Fig. 5.6 illustrates the character of the work involved in rectifying defects. This has been investigated in the literature for example by Marsh and Ferguson[144] who discuss this in more detail and can be used to assist in planning where to apply the effort as it is not normally possible to deploy the work load equally across the whole plant. Such investigations can lead to Pareto type distribution for the service times where for example a queuing system is involved as discussed by Harris[145].

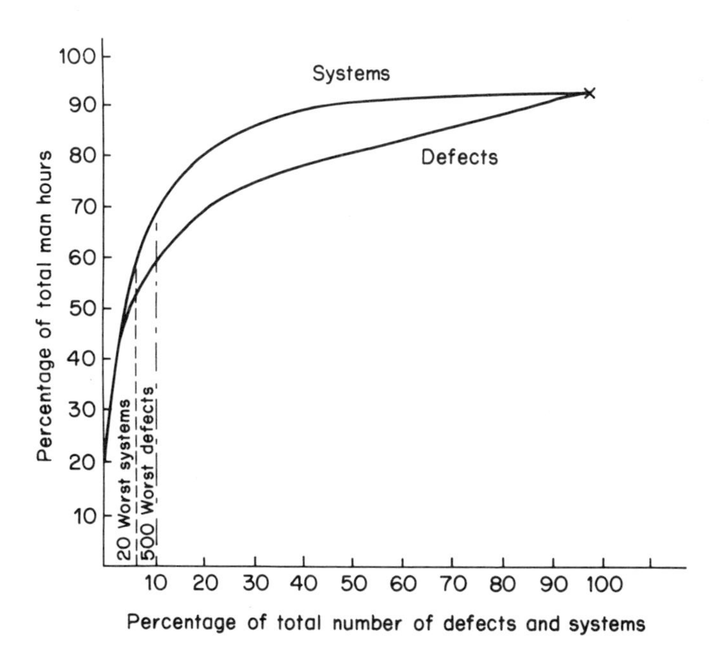

Fig. 5.6 Plot showing the percentage of man hours spent on repair and maintenance versus the percentage of defects and systems

Hence if the design work and modification activity which is inherent in the reliability growth process were not confined to those items which occur within, say, the worst 15 per cent then the result may be a large effort being involved in correcting a relatively large number of small sources of troubles. On the other hand it is also necessary to consider the safety significance of the sources of troubles. Investigations by Moore[146] have also shown a significant correlation between what appear to be nuisance defects and major safety defects. As it can distort the distribution of effort originally intended in the programme careful control is necessary to ensure that defects and failures of major significance receive the appropriate attention on the timescale incorporated in the overall safety strategy.

5.4 The Overall Reliability Programme

The overall cycle from the original design concept through to the ultimate

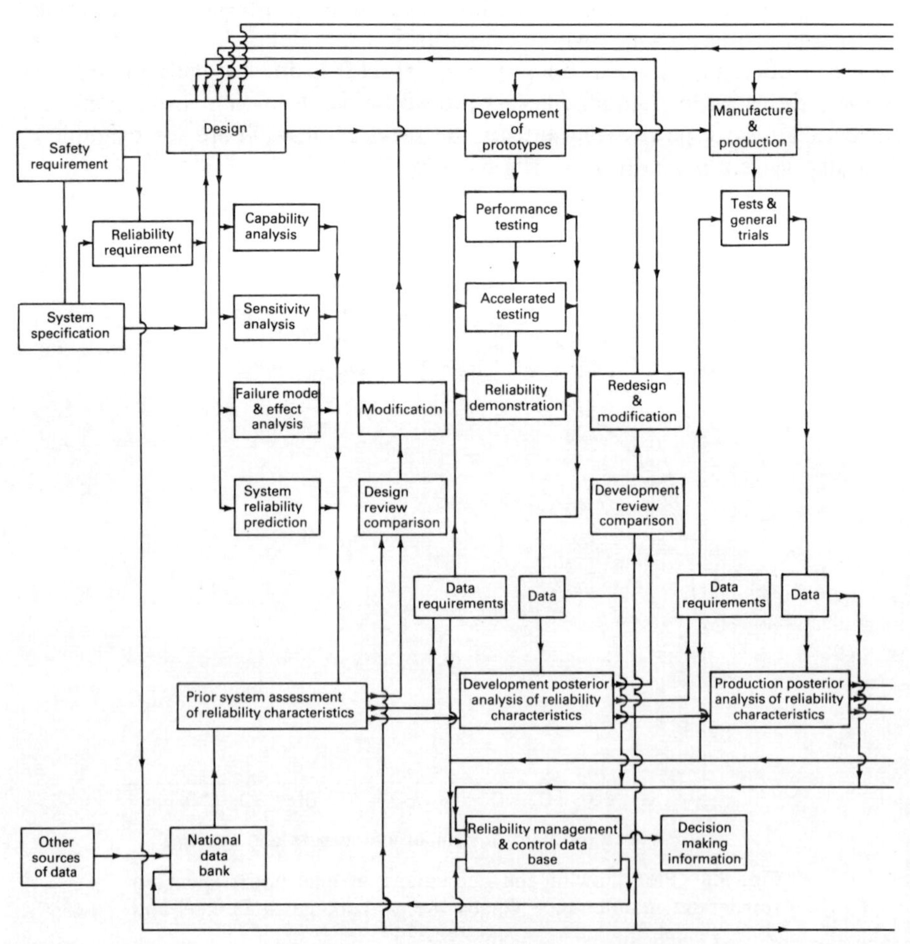

Fig. 5.7 Overall reliability programme

operation of the plant and in particular its protective system and equipment have been examined. The various techniques applicable to the different phases of the overall process have been investigated theoretically and experimentally using actual experience derived from the various activities.

The basic strategy which has emerged for protective systems dealing with high-risk situations is to evolve a safety and reliability procedure so that the overall growth programme may be 'tracked' for descriptors of the reliability characteristics of interest. Figure 5.7 shows in outline form the general process which has been conceived from the present study and investigation. The starting point being the safety requirement which sets up a set of criteria including a reliability requirement to be met by the protective system. An important part of the resulting strategy which is proposed is for a continuous assessment of those requirements laid down, not only qualitatively but quantitatively, to show at any time in the whole life cycle that they are being adequately met.

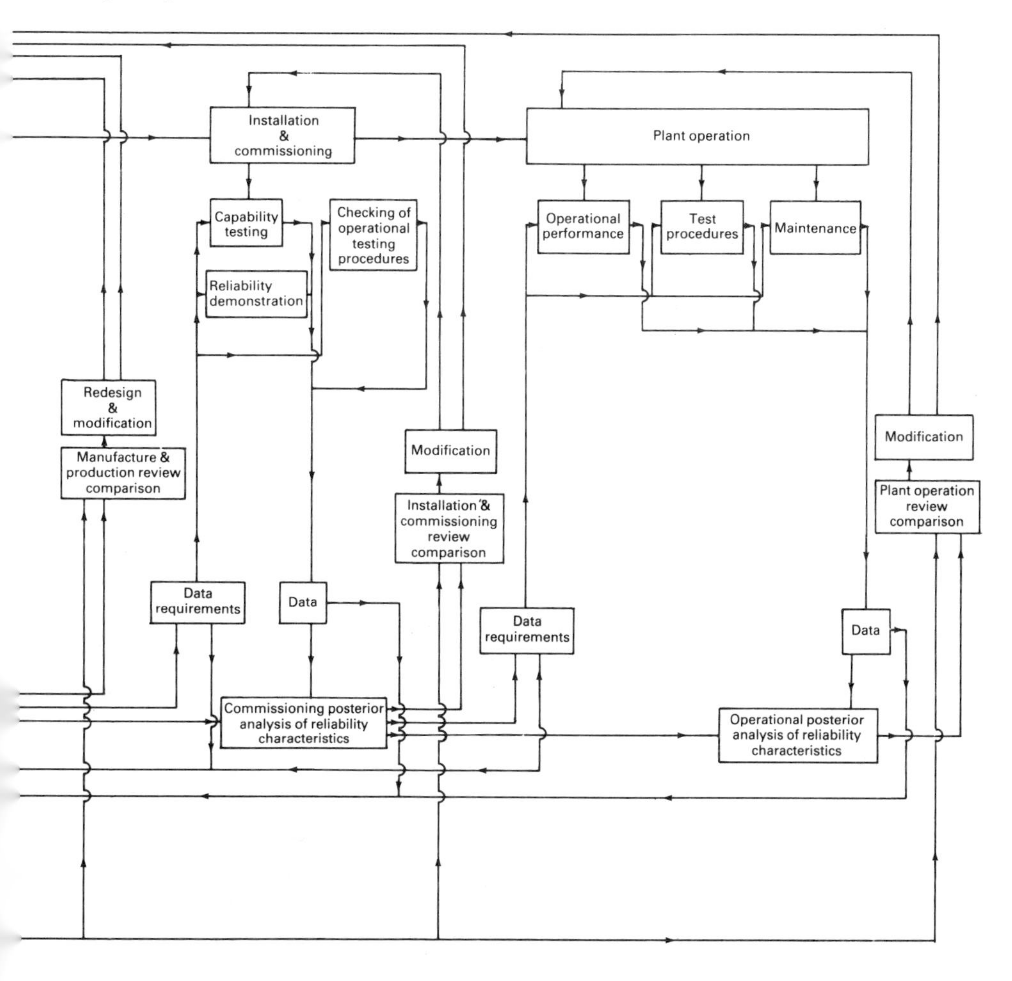

Figure 5.7 shows the general phases of specification, design and development, manufacture and production, installation and commissioning, and finally the ultimate objective of full operation. In practice and from the investigations carried out in the present work there is not always a clear demarcation between these various phases. It is considered important therefore that the interfaces should not produce distortion or leave gaps which can contribute to a lack of reliability even to the extent of the paper design being converted to a materialistic design of different reliability characteristics than those initially intended. This 'interface problem' has already been discussed in previous chapters. Hence the need for a reliability management and control data base as shown. Such problems require modern techniques for the handling of the information and data as mentioned in a previous section. Table 5.1 has already shown a sample print-out for a temperature protective channel (thermocouple and temperature trip amplifier) as occurring at the prior assessment phase. It shows the levels of assessment reached and the introduction of change by modification.

The reliability assessors, which as already discussed can benefit by being in an independent safety and reliability group, have a basic objective to create a prior assessment of the reliability characteristics and to update this prior with posterior information. This effectively leads to two tasks; first, the present measure of reliability which usually will be by inference in the type of protective system, secondly, to predict the reliability at some future time in the form of some type of projection. This leads to probabilistic statements of the type which have been discussed in much detail and the other feature which has clearly revealed itself is that of having to face the situation of reliability change. This change can be due to the perversity of nature itself or due to the general reliability growth programmes which have become part of a modern engineering concept. Hence one of the essential elements in the proposed reliability programme is to track suitable descriptors of reliability to follow its growth and that it is adequate.

Control is required to be imposed in order to manage the reliability programme through its various phases. Feedback is shown from each phase arising from a review comparison which has been named, design, development, manufacture and production, installation and commissioning, and operation. The overall decision-making information is from the Reliability Management and Control Data Base. Growth is used by the author to imply improving the reliability or improving the belief or confidence in the reliability achievement.

A basic tracking descriptor for reliability will be in terms of failure rate or failure probability and to monitor its change as a function of time. It will be noted in Fig. 5.7 that there is an output from a National Data Bank shown with a feedback from the Reliability Management Control Data Base. The operation of this 'loop' can assist particularly where generic data are concerned and comparisons can be made from time to time and trends investigated. However, with reference to the particular system being designed and processed through the various phases to operation it is clearly pertinent to have data specific to the system and its environment. Classical statistical approaches can be applied where appropriate but tend to lack the dynamic nature of other approaches and furthermore do not readily cater for small sample size data and the use of subjective or engineering judgement. The classical methods have been well established and have already been discussed in previous chapters and much more detailed discussion will be found in the general literature such as by Corcoran and Read[147], Lloyd and Liplow[148], and Stovall[149]. Problems arise due to the various formulae containing parameters which are not known and it becomes difficult from a statistical point of view to appropriately estimate the confidence statements for these paramters with reference to the observed data. However, it will be seen in earlier chapters where the methods have been quite adequate particularly where relatively large samples existed.

The Bayesian method has been mentioned previously and some of its limitations and assumptions quoted. In practice it is necessary to employ methods which can spot deterioration as well as growth in the reliability characteristics. Furthermore the system can be either in an unrepaired or repaired state. It has already been shown that the failure rates for the states can be predicted with some reasonable degree of precision. The techniques do exist and are viable for obtaining data as described in chapter 4 for the failure rates of these states which also leads if required to the probability of the system being repaired.

5.5 Common-Mode Failure

As discussed in previous chapters the prevention and minimization of a common-mode failure (CMF) throughout the whole cycle of design and operation is very significant in achieving highly reliable protective systems. The control of this type of failure involves an interplay between the overall reliability programme management and the various levels of the technical, engineering and operational aspects. In essence it is encompassed in the strategy described by Fig. 5.7 but requires a much more critical appraisal of the causes of the common-mode failure and the various defences which have been inbuilt into the overall reliability programme. Figure 5.8 shows the aspects which are involved in the common-mode failure causes and the related defences for protective systems and further research on some of these aspects is considered by Watson and Edwards[102] and CSNI Task Force[103]. Indicated in Fig. 5.8 is an overall CMF defensive strategy for protective systems, which incorporates the three main feedback loops by which CMF problems can be identified and alleviated.

The engineering needs for the appropriate management have been previously stressed which involve the appropriate level of design quality and review procedures coupled with quality control in the general process of construction. An independent reliability analysis, both qualitative and quantitative, must perform an important role in reducing CMF effects in both

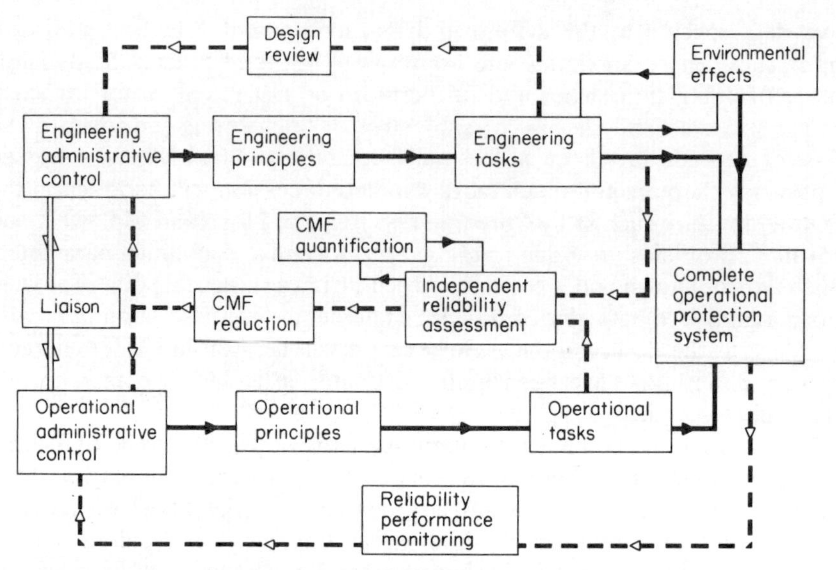

Fig. 5.8 Overall CMF defensive strategy

design and operational stages. Obviously, to complement these aspects, the appropriate level of operation management must be applied to maintenance, test and operational activities, and the monitoring of reliability performance should also be invoked to ensure that the reliability embodied in the design is achieved.

One of the important features of the Reliability Management and Control Data Base shown in Fig. 5.7 is to assist management decision-making on such factors. Since the events such as common-mode failure are rare it is essential that information on such rare events should be fed into the system which requires to be organized from various sources in a systematic fashion. Here the National Data Bank shown in Fig. 5.7 can play a very important part. Such data and general information can then be viewed against the overall safety requirements and reliability criteria which have been laid down and management decision-making can then be taken where action is required.

Some ideas along similar lines are also expressed in Fig. 5.9 which has been

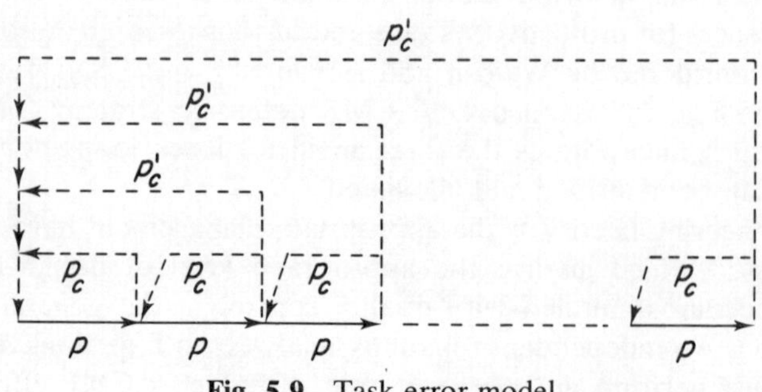

Fig. 5.9 Task error model

taken from the CSNI Task Force Report[103] and also given by Watson[191]. The original network approach for very complex tasks is still being researched but basically as design or more general work proceeds, checking takes place informally as well as formally. The solid line p arrows indicate stages of work and the p'_c dotted arrows represent the process of checking at various stages. It will be noted that the latter are indicated as a feedback function but it can be considered that a feedforward function may also be appropriate. If the assumption is made that basically these individual actions are largely independent and the p'_c symbols are taken as probabilities of error, then assuming the probabilities are small, the overall probability of failure is given by

$$p_a \doteqdot \{[(pp_c + pp_c)p'_c + pp_c]p'_c + \ldots + pp_c\} \cdot p'_c \tag{5.11}$$

$$\doteqdot pp_c p'_c + \text{higher order terms.}$$

From equation 5.11 it is seen that product pp_c represents the core behaviour activity of the whole process.

The early life characteristics of common-mode failures have been indicated by the CNSI Task Force report[103] and these are plotted in Fig. 5.10 for both design error and maintenance error caused common-mode failures. In Fig. 5.11 total safety-related occurrences in PWR plant in 1976 are plotted as a function of each and there are similarities with the Fig. 5.10 showing common-mode failures.

The CSNI Task Force report[103] gives a brief discussion of the possible similarities and differences between common-mode failures and other failures with Smith and Watson[129] pinpointing a definition of common cause failure.

An important question rests round 'Are common-mode failures rare events?'. The studies referred to indicate a mean frequency of 0.03 CMF per sub-system per year, i.e. several common-mode failures can be expected in a system life-time. Furthermore, it is considered that without any integrated design and operational common-mode failure defensive strategy, common-mode failure effects will generally determine the attainable system unreliability. This could be a few orders of magnitude worse than the unreliability due to independent failures alone. The general evidence which is being collected at the present time appears to substantiate this effect.

In general terms design and maintenance errors have been shown to be the most significant causes of common-mode failure and the research also indicates that there is a reduction for protective systems over approximately the first 6 years of operation to less than 20 per cent of the original rate. These are basically human error problems, involving procedures, supervision and detailed activities.

The proper application of inference and projection on technical aspects of the system design, construction or operation involves the various principles which have been researched in this present work. In the early design assessments of reliability great attention requires to be given to detail factors and the investigation of causes of failure at low hierarchical levels must be viewed

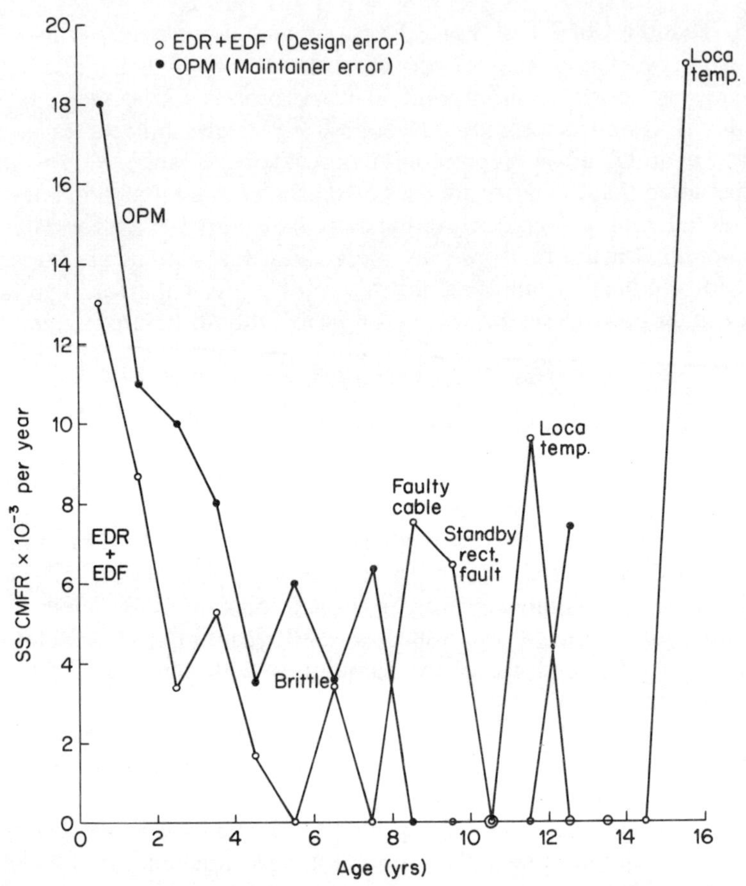

Fig. 5.10 APS and ECCS CMF rate/reactor age sub-system (SS) (per year) from criticality

collectively at the highest system level. Similarly, causes at the highest system level must be critically reviewed. The assumptions and uncertainties both in the basic concepts, analyses and procedures require to be resolved as far as possible by following a properly constructed reliability programme to observe the growth based on posterior information.

A technical illustration is that of employing diverse functional systems of protection and the use of diverse equipment design. In the highly reliable type of protective system which has been generally discussed it is a cardinal principle that such diversity is maintained for example in the physical sense of there being proper segregation maintained between 'independent' channels of protection and that activities involving human factors such as maintenance and testing should follow the basic strategies of the original design. Any modifications require very careful critical appraisal to ensure that the overall reliability programme will be maintained particularly with reference to rare events such as common-mode failure. Hence there is a real need for an integrated struc-

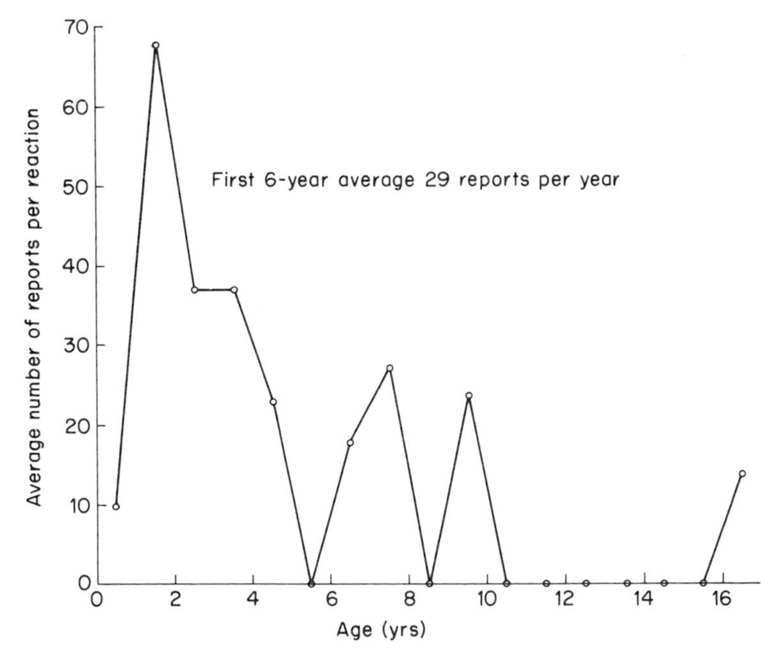

Fig. 5.11 Safety related occurrences in PWR power plant in 1976

ture which encompasses not only the original design concepts but the activities of maintenance, testing and operation. Such an integrated structure should ensure the elimination of interface problems for example which can distort or lose valuable information.

An example of the effects of CMF on the assessed reliability of a reactor protection system is given by the CNSI Task Force report[150]. A failure probability of 10^{-8} in a proof test interval of 1 month was calculated without consideration of CMF. By use of the β factor which is the ratio of CMF in a redundant channel to the total failures in the channel, the CMF effects were that the system failure probability was increased to 2×10^{-5}. The CNSI Task Force report[151] concluded from data collected that the overall LWR scram system unavailability is less than 10^{-4}. This report also considered CMF effects and these similar values give an indication of the limitation of the reliability of simple redundancy systems. The β factor method is further discussed in Appendices 8 and 9, pages 243 and 260, with applications to systems reliability analyses.

5.6 Some Aspects of an Integrated Structure

Clearly this integrated structure requires to be commenced and operated from the original early design stages and becomes more valuable with the passage of time. In fact it should basically be operated and data stored with analyses and references in the Reliability Management and Control Data Base so that plant management inherit not only a materialistic plant with various manuals and procedures but a tool which permits the greatest assistance in management decision-making.

The investigations carried out during this study have illustrated clearly that with the passage of the years, the original design groups have disbanded, manufacturers have ceased to exist and plant personnel have changed. New information, knowledge and techniques have emerged, hence from a safety point of view, the integrated approach to the reliability programme is effectively an updating process with the requirements for ready access to the present situation. Throughout the whole life cycle the creation of the protective system and its ultimate operation is 'speaking' in a 'language' which is describing its evolution. In an analogous way it is necessary to have a 'dictionary' for any spoken and written language which caters for change but enables an easy and ready reference to be available so it is necessary in a similar way to build up the reliability programme.

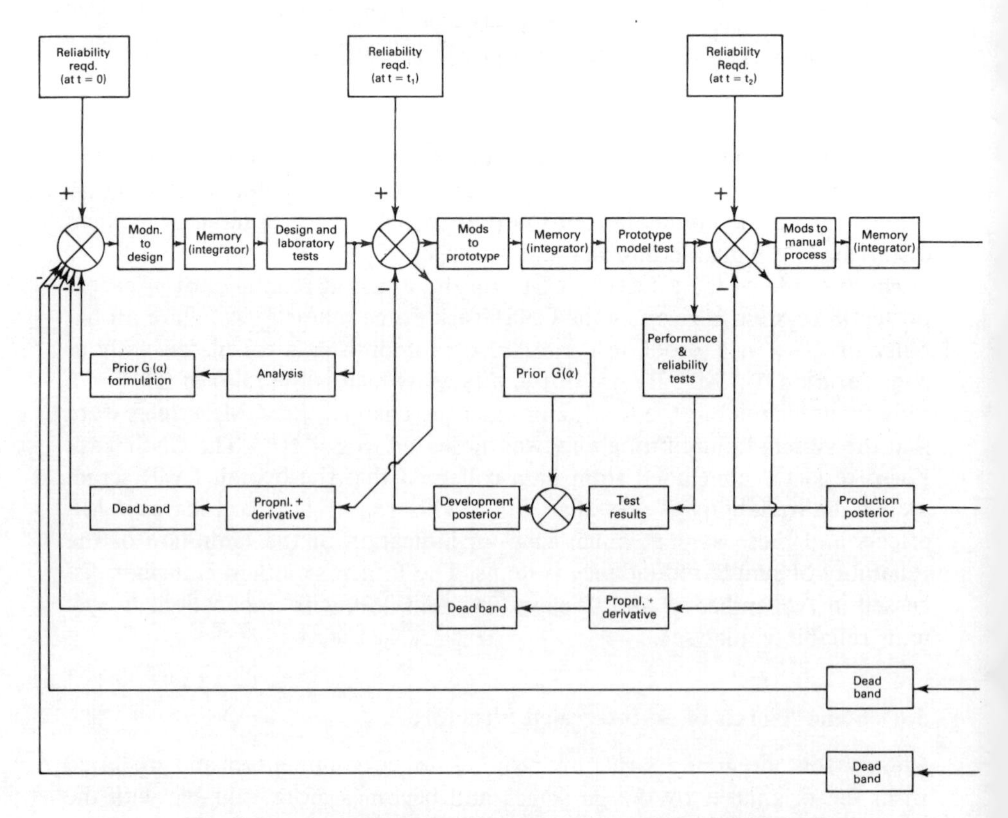

Fig. 5.12 Control system. Schematic for tracking failure rate

5.7 Reliability Control System Analogy of Integrated Structures

5.7.1 Control loops

The Fig. 5.7 may be viewed in an analogous way by considering the control system for tracking a moving target. The particular target being considered is that of failure-rate which is viewed as an index of reliability.

In the growth process the failure-rate being tracked is varying as a result of continuous elimination of systematic faults and general improvements. Developing the ideas postulated in Chapter 3 on the overall control of reliability leads to control loops basically as shown in Fig. 5.12. This shows the five separate phases of the safety life-cycle up to the ultimate operation of the plant. Each phase is identified as occurring at appropriate points in time t_0, t_1, t_2, t_3, and T where T is the start of the operational life of the plant and is the point at which the safety and reliability requirements have been specified.

Once the law governing the improvements to be expected to the equipment as the result of testing and modification is known then it is possible to calculate

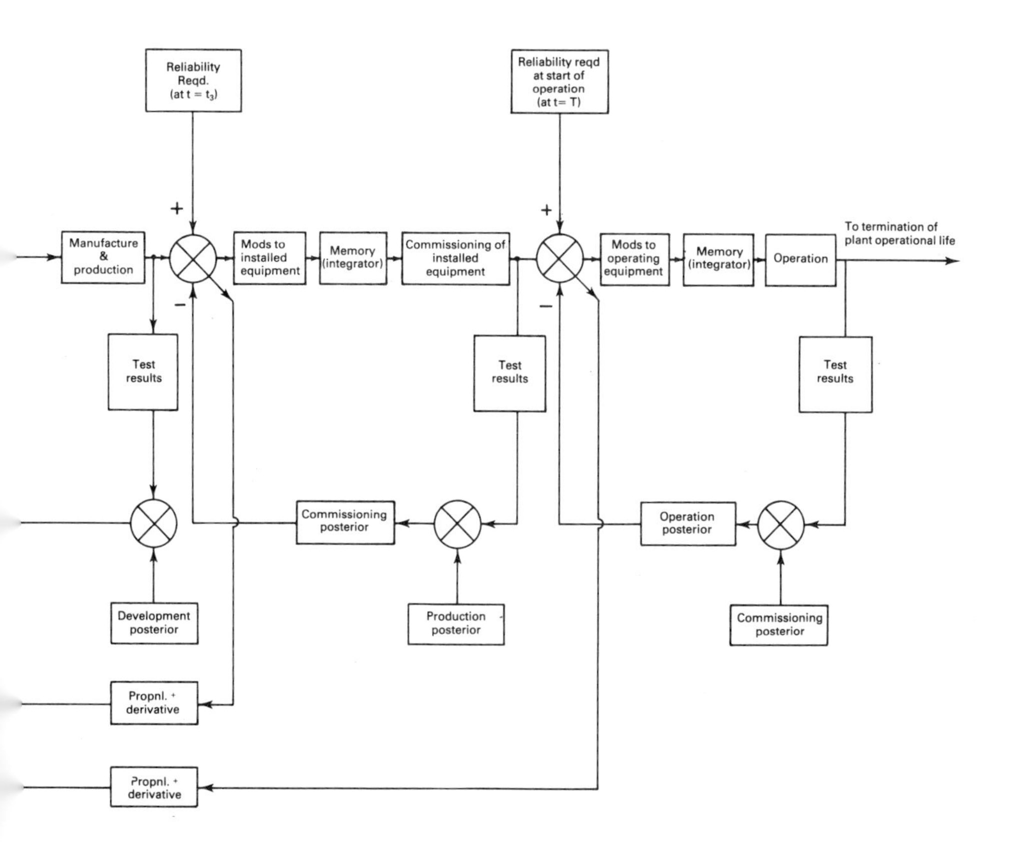

the reliability requirements expressed in terms of failure-rates at each of the foregoing times t_0, t_1, t_2 and t_3. Each control loop shown compares the failure-rate deduced from tests with that actually required at the appropriate points in time and results in two possible outcomes. These outcomes being

(a) modifications to the equipment under tests, and
(b) a major modification to the design of the equipment should a 'significant difference' exist between the actual and predicted failure-rates such that it would seem unlikely that the ultimate target failure-rate (or reliability) at $t = T$ will be met.

As an aid to making a decision for a major modification of the equipment design not only should the absolute values of failure-rates be compared but also their rates of change. Hence in the control loops indicated in Fig. 5.12 a proportional plus derivative term is shown. The 'significant difference' mentioned previously is represented by a dead band the value of which is influenced by the tolerance quoted in the original safety and reliability specification.

In order to retain the analogy with a control loop in which zero difference between the required and measured reliabilities results in no modifications as under the outcome (a) and yet leaves the equipment status unaltered; requires the introduction of a memory or integrator device as shown.

Because the test and operational results in each phase are of necessity based upon a limited sample the resulting confidence limits on the failure-rates could be extremely wide. However, applying the techniques involving a Bayesian approach as investigated in chapter 4, in which the prior has been updated by the previous accumulative test work and experience, then the resulting confidence limits will be much closer. In the comparison of the required and measured failure-rates the choice of the upper confidence limit ensures a conservative approach to the target reliability.

5.7.2 Dynamic behaviour

Although for practical application the Duane model has an attractive simplicity its application for control modelling has certain limitations, viz:

(a) At zero time it gives a zero mtbf which is not feasible.
(b) It implies continuous and indefinite growth which again is not feasible.

For these reasons the growth model developed by Lloyd *et al.*[152] is preferable. This subdivides the total number of failures F into random F_R and systematic F_S such that

$$F = F_R + F_S \tag{5.12}$$

where

$$F_R = \mathrm{A}t \tag{5.13}$$

and

$$F = \frac{\mathrm{B}}{\mathrm{C}}[1 - e^{-\mathrm{C}t}] \tag{5.14}$$

and A, B and C are characteristic constants.

It may be noted, from the above, that the number of systematic faults becomes asymptotic to B/C.

Differentiating F with respect to time gives the total failure-rate $\theta(t)$ as:

$$\theta(t) = \theta_R(t) + \theta_S(t) \tag{5.15}$$

where

$$\theta_R(t) = \mathrm{A} \tag{5.16}$$

is the failure-rate of the random faults and is assumed constant whilst the failure-rate of the systematic faults is given by:

$$\theta_S(t) = \mathrm{B}e^{-\mathrm{C}t} \tag{5.17}$$

The above exponential term represents the improvement in the systematic fault rate due to equipment modifications. It will be noted that $\theta_S(t)$ is asymptotic to zero at infinite time which is consistent with the finite number of systematic faults noted previously.

During a period when no modifications are made (as during a particular test or operational period) the systematic fault rate will remain constant at the value appertaining to the beginning of the interval.

At the beginning of the first interval, the total number of systematic faults as yet undetected must be B/C, the asymptotic value of equation 5.14, on the assumption that a systematic fault once detected and rectified does not re-occur. The fault rate at the beginning of the interval will be $\theta_S(0) = \mathrm{B}$ and hence for an interval of duration δ, there will be $\mathrm{B} \cdot \delta$ systematic faults arising. Once these have been rectified, the number of systematic faults remaining in the equipment reduces to $(\mathrm{B/C} - \mathrm{B}\delta)$. Consequently the failure-rate at the beginning of the next test interval will be reduced proportionally and will become

$$\theta_S(\delta) = \theta_S(0) \cdot \frac{\mathrm{B/C} - \mathrm{B}\delta}{\mathrm{B/C}} = \theta_S(0)[1 - \mathrm{C}\delta] \tag{5.18}$$

or a fractional fall of $\mathrm{C}\delta$ of its previous value.

By logical extension this argument can be applied to successive intervals giving the stepped response in measured failure-rates which approximates to the expected exponential as shown in Fig. 5.13.

The characteristic constants A, B and C can be derived from the test results although initially they may have to be estimated from experience with similar equipment.

For the initial period $t = 0$, $t = \delta$, the derived fault rate (random + systematic) is $\theta(0)$ and from equation 5.15

$$\theta(0) = \mathrm{A} + \mathrm{B} \tag{5.19}$$

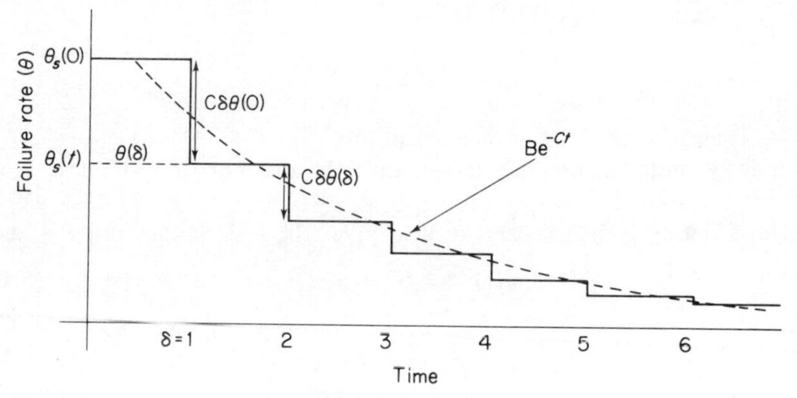

Fig. 5.13 Plot measured response and growth law for tracking parameter failure rate

At the end of this period, all the faults are rectified causing the systematic fault rate to fall by a fraction δC. For convenience, assume all the test intervals are of equal duration, then the observed fault rate during this second interval will be $\theta(\delta)$ given by:

$$\theta(\delta) = \mathrm{A} + \mathrm{B}(1 - \mathrm{C}\delta) \qquad 5.20$$

and for the third interval, a fault rate $\theta(2\delta)$ will be observed given by:

$$\theta(2\delta) = \mathrm{A} + \mathrm{B}(1 - \mathrm{C}\delta)^2 \qquad 5.21$$

From these, the values of A, B and C can be deduced in terms of the observed values of the failure-rates. The confidence limits on such a small number of test results could be extremely wide but can be much reduced by the application of Bayesian techniques as discussed in chapter 4.

With these values of A, B and C, the failure-rates to be achieved at any point in time in order that the target failure-rate will be achieved by the time the equipment becomes operational, can be calculated.

For example, if the design and safety specification calls for a reliability at time T equivalent to a failure-rate of $\theta_{\mathrm{SPEC}}(T)$, then recalling that the random failure-rate A is constant; the required systematic failure-rate $\theta_R(T)$ at time T and is given by:

$$\theta_R(T) = \theta_{\mathrm{SPEC}}(T) - \mathrm{A} \qquad (5.22)$$

But from equation 5.6

$$\theta_R(T) = \mathrm{B}_R\, e^{-CT} \qquad (5.23)$$

giving B_R the required value of B as

$$\mathrm{B}_R = \theta_R(T) \cdot e^{CT} \qquad (5.24)$$

So the required failure-rate at any time $t = t_1$, to give the specified failure-rate $\theta_{\mathrm{SPEC}}(T)$ at $t = T$ is given by:

$$\theta_R(t_1) = \mathrm{B}\, e^{-Ct_1} \qquad (5.25)$$

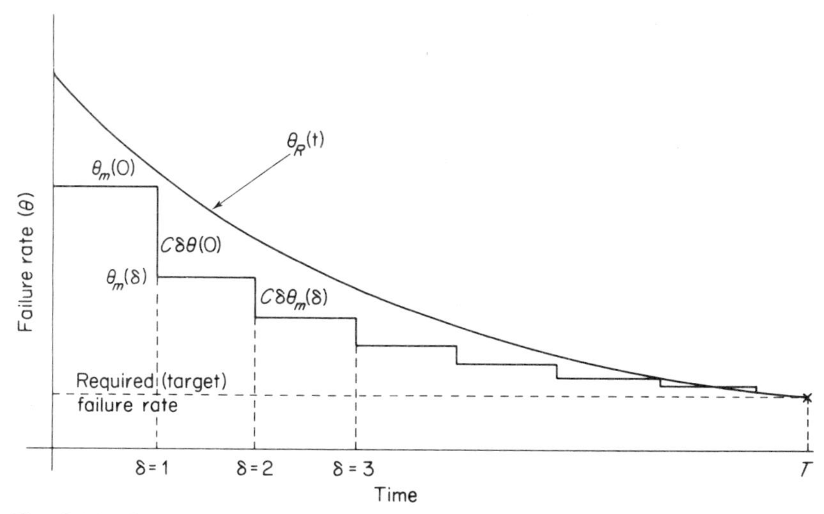

Fig. 5.14 Comparison of required failure rate $\theta_R(t)$ and measured failure rate $\theta_m(\delta)$

or

$$\theta_R(t_1) = [\theta_{SPEC}(T) - A]\, e^{C(T-t_1)} \tag{5.26}$$

which will be compared with the measured failure-rate $\theta_m(t_1)$ (as shown in Fig. 5.14) in order to assess whether the 'target' failure specified at $t = T$ is likely to be met. By making this comparison early in the life of the equipment, should it be decided that major modifications (shown in Fig. 5.12 as feedback to the design phase) are necessary, then sufficient time should be available.

5.7.3. Effect of time delays

The above assumes that any modifications to the equipment under test found necessary to remedy faults arising are carried out immediately, without interrupting the testing programme. In practice the equipment would be removed from test for the modification to be made before resuming testing.

If the time taken for modifications is δm *corresponding to a testing interval of* δ_T, then in the combined time interval $\Delta = \delta_T + \delta m$, only the fraction δ_T/Δ is usefully used for reliability growth.

Consequently in calculating $\theta_R(t)$ the failure-rate required at any time t to meet the target value at $t = T$, only a fraction δ_T/Δ of the actual time should be used.

Making the simplifying assumption (for the purposes of illustration) that all test intervals are equal throughout the equipment life and each requires the same modification period δm, then $\theta_R(t)$ of equation 5.26 becomes:

$$\theta_R(t_1) = [\theta_{SPEC}(T) - A]\, e^{-C(T-t_1)\frac{\delta_T}{\Delta}} \tag{5.27}$$

The effect of other major time delays can be examined in a similar manner by adjusting the time interval $(T - t_1)$ of $\theta_R(t_1)$ by the appropriate amount.

5.7.4 Loop stability

As can be seen by inspection of Fig. 5.12, the whole life cycle from initial design to final operation consists of five phases each with its own 'modification feedback' loop, but from the output of each phase there is also a 'design feedback' loop which encloses the modification feedback loops of previous phases. The interaction of such loops on the dynamic behaviour of the whole network remains to be examined in detail, but could conceivably give rise to an 'unstable' condition in which the control variable, failure-rate increases as a result of the interacting modifications with their different time responses.

Basically the Fig. 5.12 reduces to the form as shown in Fig. 5.15 and in Chestnut and Mayer[47], is shown the general solution for such interacting loops. The effective transfer functions which depend upon the delays due to modifications, design changes etc. could contribute towards the growth pattern deviating from the desired monotonic growth response pattern. Classically such analyses of a control system of this type containing time delays can have its stability analysis undertaken for example by Tustin's time series approach as described in Tustin[153].

5.7.5 Reliability control strategy

The whole cycle outlined and shown in Fig. 5.7 can be controlled and decision-making assisted by using the methods and strategies described. In the development of a piece of equipment, for example, time may be apportioned between the start of the design and the various stages up to commencing operation.

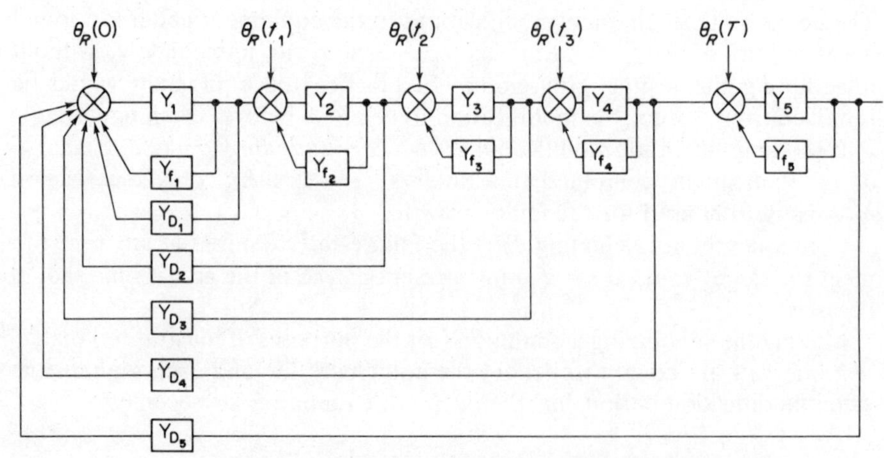

Fig. 5.15 Block diagram of equivalent tracking-control loops

At the start of the equipment's operational life the safety specification may call for a failure-to-danger rate which should not be exceeded. This applies particularly for protective equipment where the fail-safe/fail-danger ratio is important. The required $\theta_{SPEC}(T)$ is specified as in equation 5.26 and the combined random and systematic fault rate considered.

As discussed in BS5760[154] and in other sections of this book statistical approaches (e.g. Bayesian) can be used as a useful means for taking the designer's estimates from experience of similar equipment and combining them with the results of tests.

The required reliability index parameter, such as failure rate, can be calculated and compared at different times in the overall cycle with the observed values. Characteristic constants A, B and C can be further determined when the failure rates at different times become known. Such information can be derived from the equations 5.19, 5.20 and 5.21 as stated previously and used to calculate what the required failure rate should be at a given time in the programme cycle. Obviously if the difference between the required and observed values is too large it may lead to a major design change (a new Mark No.).

As more failure data are collected, the confidence limits on the failure-rates will converge and so too will those for the values of A, B and C permitting more accurate and reliable forecasts of the ultimate outcome of the growth in reliability.

This basic strategy can be illustrated in a simple way by assuming as an example a protective equipment which takes a 7 year period between the start of design and commencing operation. The first year is allocated for design and laboratory tests followed by 2 years allowed for prototype testing. This is followed by manufacture, installation and commissioning tests.

When the equipment starts its operational life the safety specification calls for an equipment failure-to-danger rate of < 0.1 faults per year or a MTBF of 10 years. However, these are fail-to-danger faults and for protective purposes the equipment is required to have a fail-safe/fail-danger ratio of at least 5 giving a revised failure-rate of 0.6 faults/year, which is the $\theta_{SPEC}(T)$ of equation 5.26. However, this is the combined random and systematic fault rate but by this time the faults should be almost entirely random in nature. Let it be assumed that the ratio of random/systematic faults has fallen to 5 : 1 by then giving

$$\theta_{SPEC}(T) - A = 0.1$$

Assume for this particular equipment that a typical value of $C = 0.5/$ year has been found to apply, which gives the initial value of θ_R the required failure-rate for systematic faults as

$$\theta_R(O) = Be^{-0.5\times 7} = 3.3 \text{ faults/year.}$$

The equipment is run for 6 months in the laboratory, during which time 5 faults develop. However, believing most of these are initial 'teething troubles' and combining this with experience of similar equipment, the designer expects the failure-rate to be 2.5 faults/year and to be between 1 and 10 with some degree of confidence which would have to be specified, e.g. 90% or 95%.

It would be necessary to consider the various statistical methods available but as previously discussed let us assume that this belief could be interpreted as a gamma prior of the form

$$f(\theta) = \frac{\theta^{\alpha-1}e^{-\theta/\beta}}{\beta^{\alpha}\Gamma(\alpha)} \tag{5.28}$$

This could be solved for the parameters α and β as described earlier. For the purposes of the example let it be assumed that some degree of confidence has been specified and that for convenience α and β have been determined and quoted to the nearest integer as $\alpha = 2$, $\beta = 1$, which gives a mean failure-rate of 2 faults/year.

Let the equipment be modified to correct the laboratory faults and then development models made during the next 6 months, which is at time t = 1.5 years. Development is started and after 6 months, 13 faults have developed in the 10 models involved, averaging 1.3 faults per model. Combining this with the design prior gives a new posterior of

$$\alpha_1 = 2 + 1.3 = 3.3$$

and

$$\beta_1 = \frac{1}{0.5 + 1} = 0.67$$

giving a new average failure-rate of 2.2/year. This can now be compared with the 'required' failure-rate at this time of $t = 2$ years which is

$$\begin{aligned} \theta_R(t) &= 0.1e^{0.5\times 5} \\ &= 1.2/\text{year} \end{aligned}$$

but adding the assumed 0.5/year random fault rate gives 1.7/year, which may be compared with that observed. In other words a 'tracking process' is in operation and decision making can now be undertaken as to the future strategy for suitable modifications and the continuation of tests.

This process of tracking throughout the safety life cycle can be repeated at each stage, permitting the values of the characteristics A, B & C to be further determined up to ultimate equipment operation.

Obviously it will be by a combination of engineering, statistical assessment and the appropriate disciplines which will allow reliability growth to be achieved in a controlled and safe way.

CHAPTER 6

Conclusions and Recommendations for Further Study

As a result of the work reported in this book two categories of recommendations arise, those which are applicable to reliability practitioners and those which most appropriately could be treated on a longer-term basis in an academic or similar institution.

6.1 Conclusions

The following conclusions can be drawn from the study.

6.1.1 Historical aspects

In this area the general outline of catastrophic events affecting philosophy was relatively easy to investigate but in any particular part of the spectrum of accidents it was difficult to understand and find details of the actual reliability policy involved. In fact the author was unable to find any extant treatise or comprehensive text relating to the area of the history of reliability as defined in this text.

The following general pattern regarding the evolution of legislation emerges in as much as that a given catastrophe (e.g. the sinking of the Titanic in 1912), see section 1.2.1.1, triggers technological developments to obviate these catastrophes reoccurring and this leads to these technological improvements being made mandatory by legislation.

The investigations into the plant-operating histories over periods up to 20 years or more have concentrated on the automatic protective systems and associated equipment. It clearly emerges that earlier methods of reliability assessment and various strategies tended to be of a deterministic nature but were extended to use probabilistic methodology for later designs. As a result of the analysis of the experience gained on certain nuclear-reactor protective systems (see chapter 4) it can be concluded that for the purposes of safety and reliability assessment it is insufficient to evaluate purely in qualitative terms. The use of quantitative terms for evaluation purposes necessitates the employment of greater discipline which in turn results in an improved understanding and greater precision. Furthermore, in order to communicate reliability criteria effectively it was necessary to evoke quantitative methods as discussed for example in section 1.2.3. It has been concluded historically that risk evaluations carried out at a high hierarchical level have led to the development of reliability technology using probabilistic techniques which are essential for the proper assessment of the high-integrity type of protective system. Furthermore, the formal and technical factors cannot be entirely divorced from the social and psychological factors as discussed in section 1.3.

An important strategy of preventing the continuation of a potentially dangerous sequence of events has emerged whereby the 'domino effect' is broken by the use of a highly reliable protective system.

6.1.2 Reliability methodology

In the methodology available and developed the study discusses various approaches which are available. The performance spectrum model which has been studied with its continuous variable aspects can readily cater for the two-state catastrophic or partial types of failures. Other types of models using fault-tree techniques for example, are better suited to the two-state type of failure but can be extended by repeated application to deal with partial failures. It is concluded that in dealing with the high reliability type of protective system that the appropriate methodology must of necessity use a process of synthesis due to the data not being available at the overall system level.

The deterministic approach, as discussed in chapter 2, makes a useful approximation to the assessment of the performance achievement of the protective system. This approach is considered a significant step in establishing the performance capability assuming no variability and forms a reference for the development of the reliability evaluation in the real world of variability.

There are many techniques which have emerged during the study for the theoretical prediction of the reliability of a particular design of protective system but the true model of reliability has revealed itself to be one of growth. This involves therefore not only the system at the conceptual design stage as particularly discussed in chapters 2 and 3 but also throughout the whole safety life cycle. Methods for dealing with this overall aspect have been developed from the Duane type of approach but it is concluded for a viable methodology both systematic and random faults must be described and analysed in order to give projection for the next phase of the life cycle as discussed in chapters 3 and 5.

A clear limitation in achieving high reliability and for its evaluation revolves round having a methodology which takes into account the common-mode failure of the complete sub-system or overall system. The investigations by the author and the developments given show that this aspect is very profound and searching and it arises throughout the book in various forms. Chapter 5 illustrates that in applying reliability methodology without taking into account common-mode failure can lead to probability of failure evaluations of protective systems on demand which are not consistent with the sparse data available. The treatment of 'rare events' extends the use of current techniques to their limits and raises some doubts regarding the validity of applying current probability concepts at the very low levels of occurrence inherent in some rare events.

6.1.3 Validation of applying reliability methodology

Particular attention has been given in the study with reference to the experimentation in applying various reliability techniques from an engineering point of view to automatic protective systems and associated equipment. Both classical and Bayesian techniques have been shown to be applicable depending upon the data available.

Where actual field or 'hard' data are available the important question of the correlation between theoretically predicted reliability characteristics, such as failure-rate, have been compared with actual results collected from field data. This gives rise to the conclusion that the majority of the predictions are within a factor of 2 : 1 as discussed in chapter 4 and this accuracy is adequate for current safety assessments. As the data become more sparse the Bayesian techniques enable the concept of belief to be introduced and it is concluded that this methodology can be most usefully applied by taking the theoretical predictions as priors and updating them with information obtained at a later date to give posterior estimates.

From the study carried out on actual operating plant protective systems and equipment it is found that the results obtained from using Bayesian techniques are in close agreement with those obtained by classical methods. In the growth of reliability from the conceptual design phase through the development phase to ultimate plant-system operation various methods of analysing reliability have been shown to be viable in chapters 4 and 5.

A difficult aspect of reliability estimation is the use of 'soft' or subjective data. The consensus method of approach using expert judgement for estimating the failure-rate of equipment has been experimentally investigated and gives results with encouraging good agreement when compared with actual field results, as described in chapter 4.

An important aspect in applying Bayesian techniques is to know the distribution of the prior and posterior. This aspect has yielded to satisfactory descriptions both for the hard and soft data cases studied. Implicit in the reliability methodology generally investigated is the conclusion that soft data can be used to extend the available hard data but the results and assumptions involved require careful consideration in any particular application. The methods illustrated have been developed and applied in order to remedy possible areas of deficiencies and further study would be advantageous.

Common-mode failure analysis has been considered and the limitations of the present methodology have been enumerated in chapters 3 and 5 in particular. However, it is concluded that the study shows that with careful attention to detail, techniques may be applied to take into account common-mode failure as discussed in chapter 5. Nevertheless, the author considers this is a weakness in many of the methods, specifically where independence has been assumed, as the practical results of the analysis of data for systems involved in this study indicate definite limitations of achievable reliabilities involving dependencies of the common-mode failure type.

Obtaining the appropriate reliability data from field experience requires the expenditure of considerable labour and time and a much more ready access to such data requires to be evolved. Paradoxically, as the data becomes more sparse greater effort is required to obtain a better understanding of the related factors. This extended the investigation to data available at a national and international level, as discussed in chapters 4 and 5 and it is concluded that the ready availability of appropriate reliability data requires further study and development of data-bank techniques. An additional conclusion is the need to review the ranges of available tables for the evaluation of some functions where high-reliability numbers are involved together with the associated techniques of calculation as for instance derived and discussed in the results given in chapter 4.

6.1.4 Reliability management and control

The study has indicated that from a total and overall point of view there are reliability parameters which can be monitored and measured throughout the complete life cycle of the protective system. Failure-rate has been derived in the study as a parameter or descriptor of reliability for the safety and reliability assessment process. On this basis a complete safety life-cycle approach has been developed in the study as described in chapter 5, and illustrated in Fig. 5.7.

From an engineering point of view and the experience gained in the study it can be concluded that right from the conceptual design stage it is not reasonable to assume that the reliability characteristics required will be right first time. Therefore for reliability management to be effective it must be forward looking as well as backward looking. This implies a need for means of projection for the descriptor such as failure-rate in order to know that the design is on the right trajectory throughout the life cycle.

Therefore particular attention has been paid to integrating the various reliability methods and strategies to 'track' the growth of reliability as given by the descriptor of failure-rate. In this context it has been concluded in the study that the analogy of a reliability control system involving the concepts of error and stability apply. These ideas have been developed and it is concluded that the general concepts of reliability stability can be applied. This involves a total modelling of both systematic and random failures as illustrated in Fig. 5.12 with reference to reliability requirements being set for each phase of the life cycle and derived from the overall reliability criterion given in the original design specification. A key aspect for proper decision making information for management and control is a highly organized Reliability Management and Control Data Base as illustrated in Fig. 5.7 and outlined in chapter 5.

6.1.5 Interfaces and human factors

Throughout the study have arisen interfaces of various kinds which give rise to distortion of information both as processed by the protective system and in relation to the general man–machine interface. It is concluded that sufficient attention in the past has not been given to communication techniques and the evaluation of human factors which pervade any reliability assessment of an automatic protective system.

The importance of the quantified reliability approach is to impose upon the assessor of such a system the need for a disciplined evaluation of the human factors. In the study various methods for evaluating the human influence factor have been investigated and developed as discussed in chapter 3 but it is the conclusion that currently this is one of the most pressing areas for further research. Furthermore, it represents one of the main contributors to be evaluated in common-mode failure particularly from a testing and maintenance point of view. However, the study does show that with careful application the techniques presented in this book can form the basis of a viable reliable assessment of an automatic protective system.

6.2 Recommendations for Practitioners

1 There is a need to avoid 're-inventing the wheel' to have a factual historical account prepared of the general subject of reliability. It requires to put together events and their relationship with human beings, technology, society and legislation. This could be accomplished as a long-term project by the co-ordinated effort of industry providing to an academic institution the appropriate information. An important contribution by industry would be to have diarists appointed on selected plants and in the associated design and manufacturing organizations to provide such information. The main contribution of the academic institution would not only be to prepare the published account but to undertake the appropriate analysis and present it in an easily accessible format.
2 A readily accessible up-dated reliability data source is required for the protective systems and equipment considered for different environments and industrial applications. Consideration should be given to creating in each industrial organization where high-reliability protective systems are used, an approved reliability data-collection centre. These centres should feed into a nationally-organized and sponsored focal point for the appropriate analysis and up-dated presentation of data for use.
3 Methods for dealing with the calculation and estimation from data giving rise to high-reliability numbers need to be reviewed and orientated for the

use of the engineer so that quick results can be obtained, for example by the use of modern hand calculators, where adequate tables are not available.

4 Greater attention needs to be focused upon the overall safety life cycle of the protective systems considered in this study in order to track descriptors of reliability such as failure-rate so that further developments for more effective reliability management and control can result.

6.3 Recommendations for Academic and Similar Institutions

1 To prepare an official history of the subject of quantitative reliability as related to the field of high-risk safety and technology (see also recommendation 6.2.1.

2 An investigation should be undertaken into the limitations of existing methods of reliability prediction for dealing with partial failure situations and a combined methodology created for example by combining aspects of the performance spectrum methods, fault-tree techniques and similar techniques.

3 Rare events and their analysis and prediction, such as common-mode failure, require to be studied with a view to developing an exhaustive treatment consistent with the data available and the needs of safety assessment evaluation.

4 Further research should be carried out in connection with the development and application of Bayesian techniques particularly with regard to the synthesis of these techniques with the 'consensus' and allied methods.

5 The development of the overall life-cycle reliability programme requires to be more fully researched to investigate the effectiveness of various descriptors of reliability for the purposes of control of the reliability of protective systems. Overall computer programmes should be prepared and integrated into a Reliability Management and Control Data Base to assist management in decision-making under actual industrial plant conditions.

6 Human factors should be researched at a high hierarchical level to formulate more clearly the relationship between social and psychological factors and technical factors affecting reliability and risk for the requirements applicable to high reliability protective systems. At a lower level such as that of the protective system itself and its associated equipment more defined and exhaustive methods require to be developed. Typically, these include the maintenance, testing and repair of protective systems. These methods require not only to be qualitative but also quantitative so that the human influence factors can be evaluated more precisely than at present as part of the general subject of reliability technology. Such an approach also requires to recognize related factors which involve aspects of human factors reliability data, communication techniques and interface problems.

APPENDIX 1

Extract from 'Vapour Clouds'

Table A1 Previous incidents involving vapour clouds

Place	Date	Material involved	Damage	Key
Hull, U.K.	1921	Hydrogen	Windows were broken in a 2-mile radius. The blast was felt in a 5-mile radius and tremors for 50 miles	CVE
Cleveland, USA	1944	LNG	136 killed, surrounding streets swept with burning gas. Windows broken, pavements raised up and manhole covers blown over buildings. Fire engine blown into the air	FB CVE
Ludwigshafen, Germany	1948	Dimethyl ether	245 killed, 2500 injured. Railcar ruptured alongside a dimethyl plant followed by an explosion and fire (Damage £8 million)	UVCE
Warren Petroleum Port, Newark, USA	1951	Propane	No record	UVCE
Wilsum, Germany	1952	Chlorine	7 people died when 15 tonnes were released from a storage tank	TOX
Whiting, Indiana, USA	1955		2 died, 30 injured following a detonation in a hydrofromer. Storage tanks punctured by shrapnel buried for 8 days (£8 million)	DET
New York, USA	1956	Ethylene	40 000 ft^3 of ethylene was released into the atmosphere causing aerial explosion	UVCE
Niagara Falls, New York State, USA	1958	Nitromethane	200 injured when a rail tanker car detonated forming a large crater (£0.5 million)	DET

Table A1 (continued)

Place	Date	Material involved	Damage	Key
Signal Hill, California, USA	1958		2 killed when vapour from a frothing tank which overflowed ignited and damaged 70% of the process area	FB
La Barre, Los Angeles, USA	1961	Chlorine	1 died in the cloud of 27.5 tonnes released from a rail tank car	TOX
Kentucky, USA	1962	Ethylene oxide	Explosion equal to 18 tonnes of TNT, 1 killed and 9 injured	UVCE
Berlin, New York, USA	1962	Propane	No record	UVCE
Louisiana, USA	1963	Ethylene	Fire of long duration	CVE
Texas, USA	1963	Propylene	£3 million damage from a fire and explosion in a low pressure polypropylene polymerization unit	CVE
Texas, USA	1964	Ethylene	Fire and explosion from a release of gaseous ethylene (£1.5 million)	FB CVE
Texas, USA	1964	Ethylene	2 died in a fire following a rupture of a high pressure ethylene pipe (£2 million)	UVCE
Massachusetts, USA	1964	VCM	Leaking sightglass ruptured on tightening under pressure. Escaping gas ignited and exploded. 7 killed, 40 injured (£2.5 million)	CVE
Louisiana, USA	1965	Ethylene	12 injured following a fire and explosion of ethylene from a ruptured pipe (£1.5 million)	CVE
Texas, USA	1965	Propylene	Pipeline failure in polypropylene polymerization plant caused £3 million damage in an explosion and fire	FB CVE
Feyzin, France	1966	Propane	Frozen valve during sampling from storage sphere allowed a vapour cloud to form and explode killing 16, injuring 63	UVCE
La Salle, Canada	1966	Styrene	11 died following an explosion after sight glass failure (£2 million)	CVE
West Germany	1966	Methane	3 killed, 83 injured	CVE
Santos, Brazil	1967	Coal gas	300 injured. 80 buildings of various sizes within a 2 km radius were either destroyed or damaged	CVE

Table A1 continued

Place	Date	Material involved	Damage	Key
Hawthorne, New Jersey, USA	1967		2 people killed, 16 injured. Explosion rocked buildings over a four-block area	
Buenos Aires, Argentina	1967	Propane	100 people injured. Fire destroyed 400 surrounding houses	
Antwerp, Netherlands	1967	VCM	4 people killed and 33 injured. Fire burned for 3 days	
Lake Charles, Louisiana, USA	1967	Isobutane	7 people died when a leaking 10 in gate valve liberated a cloud which exploded. Fires and secondary explosions continued for two weeks (£1 million)	UVCE
Bankstown, New South Wales	1967	Chlorine	Evacuation of large area of town. 5 people overcome by fumes	TOX
Perris, Netherlands	1968	Light hydrocarbons from stops tank	2 people killed, 75 injured. Blast shattered windows 1 mile away (£11 million)	UVCE
East Germany	1968	VCM	24 people killed	
Paris, France	1968	Petrochemical plant	400 people evacuated. Explosion rocked houses in area	CVE
Hull, UK	1968	Acetic acid	2 people killed, 13 injured	CVE
Rjukan, Norway	1968	Gas	Shop and car windows broken	CVE
Soldatna, Alaska, USA	1968	LPG	2 seriously injured	
Tamytown, USA	1968	Propane	3500 people evacuated	
Lievin, France	1968	Ammonia	(*AIChE* **112**, p17)	TOX
Teesside, UK	1969	Cyclohexane	2 killed, 23 injured	FB
Libya	1969	LNG	12 people injured	
Puerto la Cruz	1969	Light hydrocarbons	5 people killed. Extensive glass and ceiling damage in the town	
Long Beach, California, USA	1969	Petroleum	Lid of 600 gal tank was blown into suburban area. 1 person killed and 83 injured	CVE
Escombreras	1969	Petroleum	4 killed, 3 injured. Shock waves broke windows for miles around, 5000 people evacuated	UVCE
Repesa, Spain	1969	LPG	An LPG leak ignited causing a refinery fire which burned for 6 days	FB

Table A1 (continued)

Place	Date	Material involved	Damage	Key
Crete, Mebrasha, USA	1969	Ammonia	6 died after 64 tonnes of ammonia were released from a rail tank car	TOX
Basle, Switzerland	1969	Nitro-liquid	3 people killed, 28 injured. Blast shattered windows hundreds of yards away	DET
South Philadelphia, USA	1970	Oil refinery	5 people killed, 27 injured. Blast	
Mitcham, Surrey, UK	1970	Propane/ butane	Destruction to residential property nearby was wide-spread. Roofs were pierced, windows smashed, fences toppled and rooms burnt out. Two cars destroyed	CVE
St Thomas Island	1970	Natural Gas	Explosion rocked virtually the entire island. 25 injured	CVE
New Jersey, USA	1970	Oil refinery	40 people injured. Shock waves shattered windows over 60 sq mile area	UVCE
Port Hudson, Missouri, USA	1970	Propane	No fatalities, window breakage up to 18 km	UVCE
Crescent City, Illinios, USA	1970	Propane	Derailment of rail tank car, business section of town destroyed	BLEVE
Emmerich, Germany	1971	—	4 people killed, 4 injured. Numerous buildings in area damaged	CVE
Holland	1971	Butadiene	8 people killed, 21 injured	
Arkansas, USA	1971	Ammonia	Livestock and fish killed. Scorched leaves in 10 000 acres of forest	TOX
Holland	1972	Hydrogen	4 people killed, 4 injured	CVE
Brazil	1972	LPG	38 people killed, 75 injured. Shattered window panes in a 15 km area	UVCE
East St Louis, USA	1972	Propylene	230 people injured, window panes damaged up to 3 km from rail car shunting accident	UVCE
West Virginia, USA	1972	Gas	21 people killed, 20 injured. Entire island sealed off	CVE FB
Japan	1973	VCM	1 killed, 16 injured	UVCE
Lodi, N Jersey, USA	1973	Methanol	Relieved vapours from a reactor exploded. Evacuation of hundreds of people within a radius of several blocks. 7 died. (£1 million damage)	UVCE

Table A1 (continued)

Place	Date	Material involved	Damage	Key
Gladbeck, Ruhr, W Germany	1973	Cumol	1000 people evacuated	
Sheffield, UK	1973	Gas works	4 people died, 24 injured. Blast damaged buildings within a wide radius and blew in hundreds of windows. Cars were showered with debris and crushed by huge pieces of concrete	CVE
Saint-Acimand les Eaux, Nord, France	1973	Propane	LPG truck overturned, 4 people killed, 2 missing, 37 injured	UVCE
Tokuyama, Japan	1973	Ethylene	1 killed, 4 injured, (£12 million)	UVCE
California, USA	1973	VCM	55-gallon steel drums of chemicals were hurled into the air and came down on houses, fields and the bay. Thousands of windows were smashed and at least 8 small houses badly damaged. The shock was felt 50 miles away	
Cologne, Germany	1973	VCM	Flange rupture released 10 tonnes VCM	UVCE
New York, USA	1973	LPG	40 people killed	UVCE
Potchefstroom, South Africa	1973	Ammonia	38 tonnes released. Cloud (initially 20 m deep by 150 m diameter) drifted to adjacent town. 18 people killed, 6 outside boundary fence, 65 injured	TOX
Falkirk, UK	1973	Flammable liquid	Destruction of tar distillery	FB
Texas, USA	1974	Isoprene	12 people injured. Shattered windows in wide area	UVCE
Los Angeles, California, USA	1974	Organic peroxides	A leaking road tanker of organic peroxides exploded causing £25 million damage	DET
Beaumont, Texas, USA	1974	Isoprene	2 died and 10 people were injured after a vapour cloud explosion following a large spill of isoprene (£8 million)	UVCE
Czechoslovakia	1974	Ethylene	14 people killed, 79 injured	UVCE
Flixborough, UK	1974	Cyclohexane	28 people killed, 104 injured, 3000 people evacuated. River Trent closed to shipping. 100 homes damaged	UVCE

Table A1 (continued)

Place	Date	Material involved	Damage	Key
Rotterdam, Netherlands	1974	Petrochemicals	Large fire	FB
Rumania	1974	Ethylene	1 killed, 50 injured	UVCE
Nebraska, USA	1974	Chlorine	500 people evacuated. Clouds of poisonous fumes spread over area	TOX
Florida, USA	1974	Propane	Destroyed 2 warehouses. Crushed cars and broke windows in a 4-block area	UVCE
Wenatchee, Washington, USA	1974	Momomethyl-amine nitrate	2 died, 66 injured in rail tank car explosion	DET
Holland	1975	Ethylene	4 people killed and 35 injured	UVCE
Marseilles, France	1975	Ethylene	1 killed. 3 injured. Blast shattered windows in large area around complex	CVE
South Africa	1975	Methane	7 people killed, 7 injured. All gas supplies cut off to city for two days	
Antwerp, Belgium	1975	Ethylene	Ethylene leakage from compressors exploded. 6 killed, 13 injured. Widespread structural damage to the plant	UVCE
Philadelphia, USA	1975	Crude oil	Vapours from a storage tank exploded in a boiler house when filling a marine tanker. 8 died, 2 injured (£5 million)	CVE
Holland	1975	Propylene	14 people killed, 104 injured	UVCE
Seveso, Italy	1976	TCDD	Complete evacuation from area up until present time (1978)	TOX
Beek, Netherlands	1976	Naphtha	14 died, 30 injured when a leakage ignited and blast shattered windows of shops and houses (£10 million)	UVCE
Baton Rouge, Louisiana, USA	1976	Chlorine	Mississippi closed 50 miles north-ward. 10 000 people evacuated	TOX
Sandefjord, Norway	1976	'Flammable liquid'	Pipe rupture ignited, exploded, killing 6 people and caused £10 million damage	FB CVE
Brachead, UK	1977	Sodium chloride	Fire and explosion	DET
Mexico	1977	Ammonia	2 people died. 102 treated for poisoning. Gas entered sewer system	TOX

Table A1 (continued)

Place	Date	Material involved	Damage	Key
Umm Said, Qatar	1977	LPG	7 people died, many injured. Explosion scorched villages a mile away. Doha International Airport closed for two hours	FB
Mexico	1977	VCM	90 people injured	
Taiwan	1977	VCM	6 people killed, 10 injured	
Cassino, Italy	1977	Propane/ butane	1 killed, 9 injured	CVE FB
Jacksonville	1977	LPG	2000 people evacuated	
Gela, Italy	1977	Ethylene oxide/glycol	1 killed, 2 injured	CVE
India	1977	Hydrogen	20 people injured. Blast rocked nearby fertiliser plant, oil refinery and village	CVE
Italy	1977	Ethylene	3 people killed, 22 injured, shattered shop windows and doors. Car lifted several metres in air	UVCE
Colombia	1977	Ammonia	30 people killed and 22 injured. Nearby villagers suffered from effects of gas	TOX
Baltimore, USA	1978	Sulphur trioxide	Fumes drifted 10 miles away. More than 100 people treated for nausea	TOX
USA	1978	Grain dust		CVE
Waverley, Tennessee, USA	1978	Propane	12 dead and at least 50 injured when a derailed tank car exploded	BLEVE
Youngstown, Florida, USA	1978	Chlorine	8 people died and 50 were injured when gas escaped from a rail tanker involved in a crash	TOX

*Additional events

*Additional general event information given in *The Times Index*[186] and *Britannica Book of the Year*[187].

Key: TOX – Toxic; FB – Fireball; BLEVE – Boiling liquid expanding vapour explosion; CVE – Confined vapour explosion; UVCE – Unconfined vapour cloud explosion: DET – Detonation.

Reproduced from 'Vapour clouds', *Chemistry and Industry* No. 9, pp. 295–302, by permission of D. H. Slater, Technica

APPENDIX 2

Event and Fault Trees

Fault-tree analysis was developed in the early 1960s and was used in the aerospace industry principally for system safety analysis. It was applied to complex situations such as the Minute Man analysis and the analysis undertaken in the Space and Missile Organization (SAMSO). The basic concepts of this methodology are as described by Barlow and Lambert[161]. Later it was used in the Reactor Safety Study Wash-1400[21] which involved the use of event and fault trees as described by Vesely[162].

Definitions of event and fault trees may be given as follows.

(a) *Event tree*: Defines a logic method for identifying the various possible outcomes of a given event which is called the initiating event. The course of events is determined by the operation or failure of various systems.
(b) *Fault tree*: This is a graphical display to show how the basic component failures can lead to a pre-determined system failure state. It is used to determine the ways of failing and the likelihood of failure of the various systems identified in the event-tree accident paths.

The methodology uses basic logic and set-theory concepts of event trees and utilizes fault trees in event-tree models. An event tree begins with a defined initiating event as shown in Table A.2.1. where, for example, two safety systems are defined after the initiating event has occurred.

For the particular initiating event, the set of success and failure stated for each system is defined also, as shown in Table A.2.1.

The various system states are combined by use of the technique of decision-tree branching logic to obtain the various accident sequences

Table A.2.1. Event tree and system state definitions for Systems A and B

Initiating Event	System A	System B
	Success state	Success state
	Failure state	Failure state

Table A.2.2. Example of event tree branching

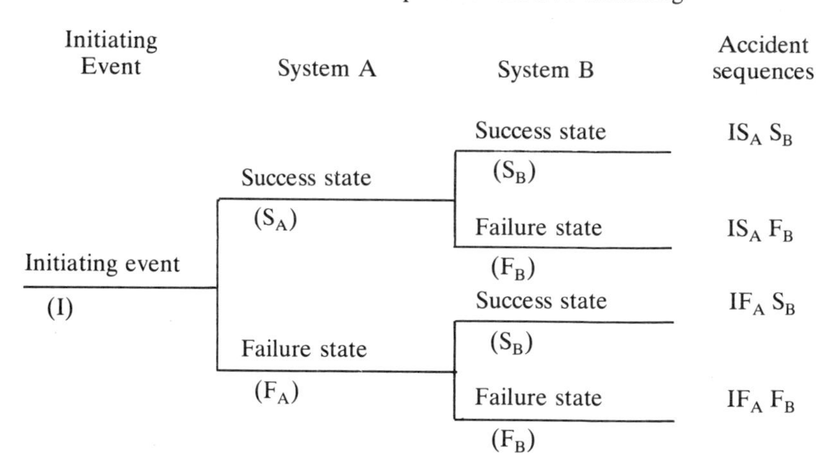

associated, such as $IS_A\ F_B$ with the particular initiating event as shown in Table A.2.2.

After each event sequence has been defined, the probabilities of system failure can be computed by the use of fault trees. Usually the overall failure probabilities for the system must be derived from data available for the failure rates which are available at the component level. This is because in this type of analysis the data are not available at the overall system level. Hence the system state definitions from the event tree are used to define 'top events' of the fault trees which are developed down to the component level.

Effectively the fault tree is a logic structure in which the system is analysed in a methodical manner to define the elements which contribute to the system failure probability. In this deductive process a failure is traced back to its basic causes. Table A.2.3 illustrates the associated fault-tree construction with the symbol ⌓ representing a fault AND gate and implies that the events above this gate will occur if all the lower input events occur. Other types of gates may be used such as an OR gate ⌓ as outlined by Cross[159].

In order to calculate the probability of failure of the system the minimum cut sets are listed and a cut set is defined as minimal if it cannot be reduced and still ensure the occurrence of the top event. Barlow and Lambert[161] describes this process and Fussell and Vesely[163] give a more detailed account.

Analysis by this method for complex systems and accident sequences may involve several thousand components and logic gates which lead to the requirement for computer calculations. Various computer codes exist for such calculations which involve the appropriate modelling techniques as described specifically by Shaw and White[164] for the PREP, KIT and SAMPLE computer codes. Brock and Cross[165] have more generally reviewed a number of computer programs which are available and have given an example of the application to an automatic protective system of the type discussed in this book.

Table A.2.3 Example of an accident sequence and associated fault-tree

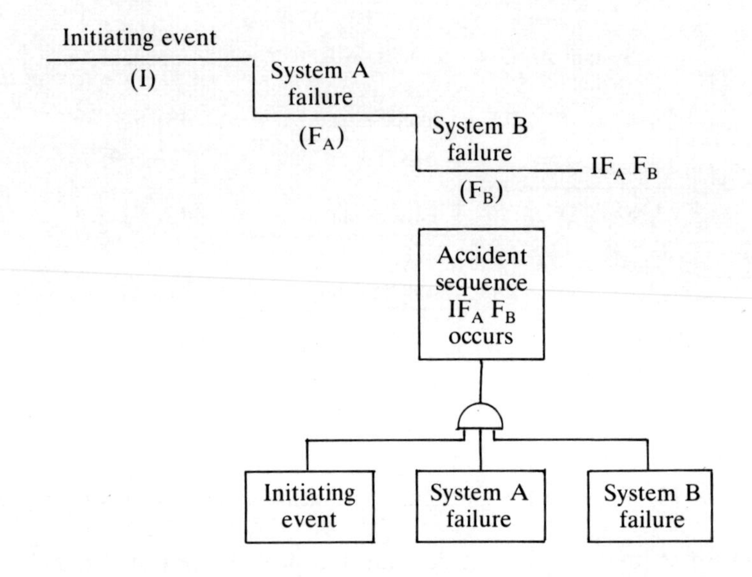

APPENDIX 3

Failure Modes and Effects Analysis

A.3.1 Differential Pressure Transmitter Example

Figure A.3.1 shows an example differential pressure transmitter which functions in the following basic manner. An increase in differential pressure applied across the measuring diaphragm causes the main beam to open the nozzle and flapper orifice against the tension of the balance spring. Opening of the nozzle and flapper reduces the air pressure on the feedback diaphragm due to the restrictor in the compressed air supply pipe. The feedback diaphragm then causes the auxiliary beam to release pressure on the main beam which has the effect of closing the nozzle and flapper again. This

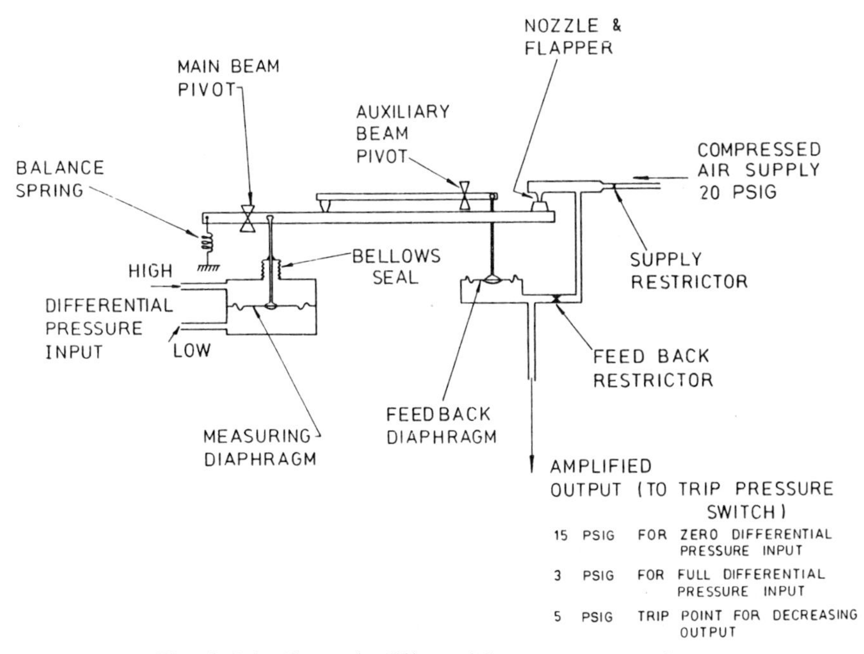

Fig. A.3.1 Example differential pressure transmitter

Table A.3.1 Summary Failure Rate Analysis of Differential Pressure Transmitter

Component	Fault to be assessed	Effect of fault	Fault category S or D	Failure rate %/1000 hr
Measuring diaphragm	Leaking or ruptured	Beam not actuated	D	0.8
Bellows seal	Leaking or ruptured	Signal in measuring cell reduced	D	0.5
Main beam pivot	Broken or loose	Nozzle opens	S	0.1
Balance spring	Broken	Nozzle opens	S	0.02
Nozzle and flapper	Blockage	Full output	D	0.6
	Breakage	Nozzle leaks	S	0.02
Supply restrictor	Blockage	Supply cut off	S	0.5
Feedback restrictor	Blockage	Output cut off	S	0.5
Feedback diaphragm	Leaking or ruptured	Output falls	S	0.8
Auxiliary beam pivot	Broken or loose	Feedback ineffective	D	0.1

Fail dangerous rate = 2.0%/1000 hr
Fail safe rate = 1.94%/1000 hr
Hence total failure rate = 3.94%/1000 hr

feedback mechanism results in the output pressure taking up a value which is proportional to the feedback differential pressure.

Table A.3.1 shows a summary analysis of the differential pressure transmitter to illustrate the method of approach. Faults which cause the output to remain high, even for low or zero differential pressure inputs, are fail-dangerous (D faults). All other faults are considered safe (S faults). The overall failure rate is seen to be 3.94 per cent/1000 hr with a failure-dangerous rate of 2.0 per cent/1000 hr and a fail-safe rate of 1.94 per cent/1000 hr.

A.3.2 Gamma Radiation Monitor Example

Figure A.3.2 shows an example electronic gamma radiation monitor of an early design and still in use, which functions in the following basic manner. Normally under conditions of low radiation the GM (Geiger-Muller) tube V1 is open circuit. Therefore, the trigger electrode of V2 (cold cathode trigger tube) is at zero voltage. When the radiation particle enters the GM tube the rarefied gas within the tube becomes ionized and the accelerating particles under the influence of the supply voltage of 600 V collide with other gas

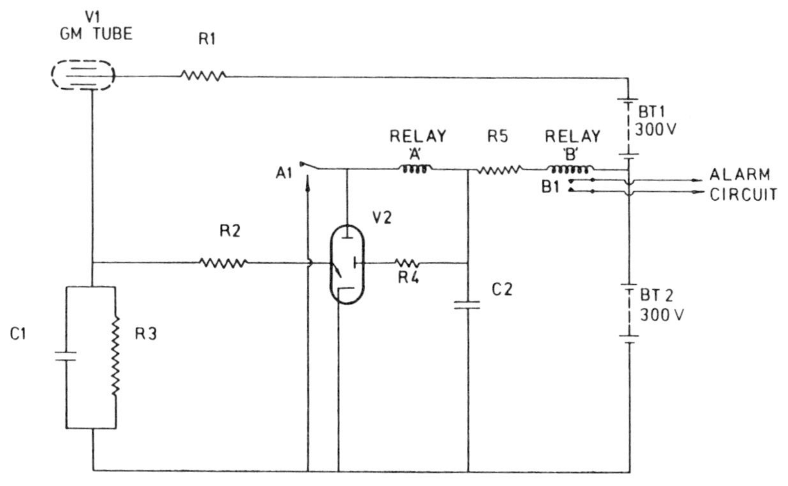

Fig. A.3.2 Example gamma-radiation monitor

particles producing further ionization. The effect is cumulative until virtually all of the gas within the tube becomes ionized. Hence for a single radiation particle entering the tube, the collisions result in 10^8 particles being generated thus this tube has an effective amplification of 10^8 times. Hence a pulse of current flows through the tube and charges up the capacitor C1 by a step of voltage. The GM tube then recovers and waits for the next radiation particle. Meanwhile the charge across capacitor C1 discharges slowly through the high resistor R3.

If the radiation is continuous then capacitor C1 will build up more rapidly than the resistor R3 can discharge it and in this way the voltage at the trigger electrode of V2 will build up. When this trigger voltage has been attained tube V2 fires. This results in an operating current through relays A and B, contact A1 closes and this maintains the current through the relays and causes V2 to turn off. The alarm contact B1 closes to raise the alarm.

One or two other aspects of the design can now be studied. The resistor R4 supplies the priming anode of V2 causing a very small local ionization to the cathode. Provided the tube V2 is primed in this way then the tube will fire within 2 msec. If the tube is not primed it may take up to 5 sec to trigger. For example, considering component V1 with fault mode open circuit the effect is 'no detection' which is a dangerous condition.

A general assessment of the failure rates of the example gamma radiation monitor is illustrated in Table A.3.2. Not only are there effects which are dangerous or safe, but neutral conditions can result. The overall failure rate is 3.4 per cent/1000 hr with a fail-dangerous rate of 0.84 per cent/1000 hr, fail-safe rate of 2.47 per cent/1000 hr and a neutral failure rate of 0.09 per cent/1000 hr. It may be observbed that another category, i.e. neutral, has entered into the analysis and for complete detailed analysis a more refined coding is found to be useful, as described in the next section.

Table A.3.2 Assesment of the failure rates of example gamma radiation monitor

Component	Fault Mode	Effect	Component %/1000 hr	Failure rate %/1000 hr Safe	Danger	Neutral
V1	o/c	Zero tube current	0.25		0.25	
	s/c	Trigger tube operates	0.25	0.25		
V2	anode o/c	Alarm inhibited	0.1		0.1	
	prime o/c	Alarm – 5 sec delay	0.1		0.1	
	trig. o/c	No alarm	0.1		0.1	
	cath. o/c	No alarm	0.1		0.1	
	a-p s/c	Tube damage – relays energized	0.1	0.1		
	a-t s/c	Relays energized	0.1	0.1		
	p-t s/c	Relays energized (tube fires)	0.1	0.1		
	p-c s/c	Tube will not fire (i.e. trigger)	0.1		0.1	
	t-c s/c	Tube will not fire	0.1		0.1	
	gas	Tube will not fire	0.5		0.5	
R1	o/c	Circuit inoperative	0.18		0.18	
	s/c	First pulse will trigger (background)	0.02	0.02		
R2	o/c	V2 will not trigger	0.18		0.18	
	s/c	V2 life shortened	0.02			0.02
R3	o/c	V2 will trigger on background	0.18	0.18		
	s/c	V2 will not trigger	0.02		0.02	
R4	o/c	V2 firing delayed by 5 sec	0.18		0.18	
	s/c	Tube damage Relay B energized	0.02	0.02		
R5	o/c	Relays will not energize	0.18		0.18	
	s/c	Excess current through relays	0.02			0.02
C1	o/c	No transfer energy to trigger V2	0.05		0.05	
	s/c	V2 will not trigger	0.05		0.05	
C2	o/c	Little effect	0.05			0.05
	s/c	Relay B energized	0.05	0.05		
A	coil o/c	Relays will not energize	0.024		0.024	
	coil s/c	Relays will not hold in	0.006		0.006	
	cont. o/c	Relays will not hold in	0.01		0.01	
	cont. s/c	Relays energized	0.01	0.01		
B	coil o/c	No alarm possible	0.024		0.024	
	coil s/c	Contacts B1 will not operate	0.006		0.006	
	cont. o/c	Alarm inhibited	0.01		0.01	

Table A.3.2 (continued)

Component	Fault Mode	Effect	Component %/1000 hr	Failure rate %/1000 hr Safe	Danger	Neutral
	cont. s/c	Alarm given	0.01	0.01		
BT1	no volts	Alarm inhibited	0.1		0.1	
BT2	no volts	Alarm inhibited	0.1		0.1	
TOTALS				0.84	2.47	0.09

A.3.3 Four-Character Coding System

A four-character code has been found of great assistance to represent each component fault. The first character describes the main effect of the fault and is designated, S, D. H or C where:

S is a fail-safe fault;
D is a fail-dangerous fault directly affecting the safety of the plant being controlled;
H is a fail-dangerous fault of an auxiliary function of the equipment which could indirectly affect the safety of the plant being controlled;
C is a significant change in calibration of any equipment function in the fail-dangerous direction.

The second character indicates which equipment function is affected by the fault. The functions are numbered 1, 2, 3 etc., the most important being first, usually the reactor trip function. The third character shows whether or not a warning is given when the fault has occurred where:

u designates an unrevealed fault;
r designates a revealed fault.

The fourth character describes which equipment function or facility reveals the fault (in the case of an 'r' designation in the third character). These are numbered 1, 2, 3 etc., and carry the same meanings as those defined for the second character with the addition of any other fault detecting facility.

By way of example, these numbers may be allocated as follows:

1 the main trip function;
2 self test function;
3 auxiliary warning device;
4 the meter or other visual indication of the measured parameter;
5 indicating lamps;
6 the routine testing or maintenance facilities.

Hence, on this classification, typical faults could be defined in the following way:

D1u- An unrevealed fail-dangerous fault of the main trip function.
S-r3 A fail-safe fault revealed by the auxiliary warning device.
H2r4 A potentially fail-dangerous fault in the self-test function revealed by the change in meter reading.

The D1u- type of fault is of prime concern in assessing any high integrity sensing channel.

A.3.4 Extract from a Standard and Code for Information Required on the Reactor Protection System on the Licensing Procedures in the Federal Republic of Germany*

1 CONCEPT INFORMATION

1.1 Analysis of incidents based on safety criteria†
1.2 List of physical variables and the admissible operational range of those variables, which produce the initiation criteria for the protection system
1.3 Concept for signalling operational readiness and functional capability of the reactor protection systems and safety devices
1.4 Correlation between incidents, initiation criteria and safety actions with regard to interlocking conditions
1.5 Considered failure causing occurrences and the counter-measures provided for ensuring proper function of the reactor protection system
1.6 Local installation and separation of the components of the reactor protection system
1.7 A summary description of the devices and systems provided for the implementation of the reactor protection conception
1.8 Wiring diagrams of the reactor protection system

2 BEGINNING OF BUILDING ERECTION

2.1 Drawings showing the routing lines of cables and electric wirings in so far as they may be relevant to the erection of buildings
2.2 Drawings of buildings showing location of the electronic components of the safety systems
2.3 Drawings showing the arrangement of penetrations through the containment and the reactor building

* Reproduced by permission of Gesellschaft für Reaktorsicherheit (GRS) mbH

† Nuclear Power Plant Safety Criteria, Federal Republic of Germany, The Federal Minister of the Interior, June 25, 1974
Criterion 6.1. sentence 4: 'Basically, at least two criteria for initiation of protective actions shall be available for any event to be controlled by the reactor protection system.'

3 BEGINNING OF MACHINERY INSTALLATION

3.1 Final concept of wiring diagrams of the reactor protection system
3.2 List of measuring points, wiring scheme and information on the arrangement of sensors and transducers together with the respective operational wiring diagrams of systems and their descriptions
3.3 Data sheets of sensors and transducers used on the measuring points of the reactor protection system
3.4 Data sheets showing the branching of all final control elements and drives activated by the reactor protection system. Data sheets are expected to show wiring diagrams and design data under operational and incident conditions
3.5 Concept of testing for sensors and transducers during specified normal plant operation
3.6 Block diagrams

4 BEGINNING OF INSTALLATION OF ELECTRICAL COMPONENTS

4.1 Detailed specification of the reactor protection system
4.2 Locations where the devices of the reactor protection system are to be installed including their distribution among the racks and information on the environmental conditions during specified normal operation and under incident conditions
4.3 Internal wiring diagrams and data sheets of the devices used in the reactor protection system
4.4 Documents concerning the power supply of the reactor protection system
4.5 List of alarm signals provided, showing method and location of registration as well as the location where the signals are reported
4.6 Basic specifications laying down the conditions for technical and acceptance tests of the reactor protection system
4.7 Binding signal and wiring diagrams of the reactor protection system (for measurement of variables, logic evaluation and open loop control)
4.8 Final operational logic diagrams of systems showing the attached devices of the measuring, open-loop and closed loop control systems, which are related to safety
4.9 Documents relating to the testability and to the scheduled test intervals of the systems and of the components of the reactor protection system
4.10 Information concerning the reliability (qualitative, if necessary) and the extent of satisfactory operational performance of the devices provided and information on the possibility of maintenance and repair

5 COMMISSIONING AND TEST OPERATION

5.1 Setting of limits for commissioning and test operation

5.2 Test schedule and test certificate forms for the functional tests of the reactor protection system
5.3 Schedule and test certificate forms for in-service tests of the reactor protection system outlining method and time intervals of the tests
5.4 Documentation of assembly, acceptance and functional tests

6 CONTINUOUS OPERATION

Limit value settings for continuous operation

APPENDIX 4

Comparative Reliability Case Study for Manual versus Automatic Control*

A.4.1 Introduction

The plant consisted of a basic steam-raising unit which comprised a boiler of typically 100 MW output generating 1500–2000 psi steam with a maximum throughput approaching 500 T/hr (the volumetric water equivalent of a block of ice the size of a TV set flashing over into high-pressure steam in each successive second). The steam is used for a variety of purposes, including heating, propulsion, rotating machinery, mixing, purging etc.; one large plant alone can consume up to £500 000 of steam per annum.

The boiler was fired with refinery waste oil and gas using dual-fuel burners mounted in groups of four, the groups being positioned at a series of levels over the height of the boiler. The plant system required to be flexible for two reasons:

(a) fluctuating throughout demands, dependent upon consumers,
(b) variations in quality and constituency of fuel.

The flexibility demanded a wide-ranging control capability. The designers presented three basic design philosophies, the first principally relating to manual operation, the second combining a balance of manual and automatic features and the third committing a heavy dependence on automation. Separate quantitative reliability assessments were carried out for the three designs taking into account the integrated human and hardware reliabilities. The analysis of the human component attempted to include the roles of the maintainer on the plant as well as those of the operator.

A.4.2 The Protective System

Notionally the plant could be divided into two distinct sections, (a) main plant hardware, and (b) control systems (human and hardware). The assessment reported in this case study dealt with only a particular subset of the

* Reproduced by permission of D. M. Hunns—extracted from Reference 93.

control-systems section, namely the protective system. However, the specified target success ratio depends upon the amalgamated contributions from all the plant sub-systems. Therefore an individual sub-system must achieve a success ratio significantly greater than the total system target requirement.

The protective system was designed to ensure a safe start-up sequence, the continuous monitoring of normal running conditions and safe automatic or manual shut-down of the boiler plant.

A.4.3 Assessment Process

The three designs, with the varying degree of automation, manual (M), automatic and manual (A–M), and automatic (M), were assessed with respect to,

(a) their capabilities to provide protection when all components were working up to specification; for example, taking the case of automatic vs. manual flame monitoring, account had to be taken of the faster response capability but the less dependable quality discrimination performance; another aspect was the need to objectively evaluate the influence of automation in sapping the capability of the operator to immediately exercise effective manual control should the automation fail; these capability aspects and others like them were identified in the analyses and estimates made of their quantitative contribution to performance.
(b) their reliabilities, probabilizing upon single (or combinations of) human and/or hardware failures which would act to partially or totally disable the protection function.

The total analysis was logically constructed using the matrix format similar to that shown in Fig. 3.20 (for the complete assessment over 200 pages of logic diagrams were produced).

The actual performance parameter predicted for each of the three systems, M, A–M and A, was mean accumulated days of 'outage' per operating year. The calculations used hardware failure rate data, originating from the Syrel Data Bank, together with human error probabilities which were extrapolated subjectively by two separate assessors making reference to a generalized data guide. The two assessors used a simple Delphi process to pool their judgments; in certain more critical cases a third assessor also contributed. Over 150 human error probabilities were quantified in this way a good level of agreement between judges was achieved throughout. The mean accumulated days of outage were calculated for each system postulating various testing strategies. The results presented in this case study are based on the finally decided 3-month test interval for all components in a system.

A.4.4 Assessment Results

For the purposes of assessment the sources of outage were dichotomized into

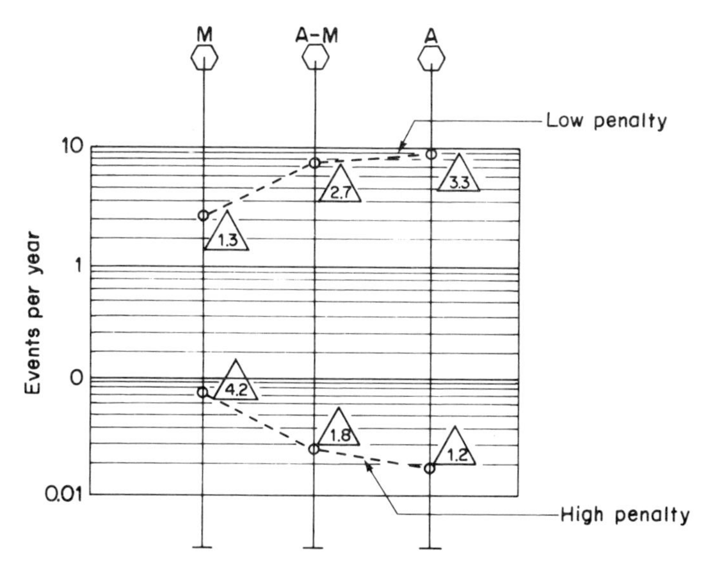

Fig. A.4.1 Outage event frequencies. △ = mean outage days/year

two categories, namely,

(a) low penalty—associated with failures and errors resulting in spurious trips of all or part of the plant—a pessimistic average of one-third of a day of outage per event was assessed.
(b) high penalty—where some form of detonation and/or destructive fire occurred necessitating extensive refurbishing operations—a mean of 60 days outage per event was assessed.

Figure A.4.1 summarizes the outage frequencies separately accruing from these two categories of outage for the three systems. As might be expected, increasing automation reduces the incidence of high penalty outage events (i.e. increases safety) but pays the price of increasing the frequency on the low-penalty side. The mean days of outage were calculated for each frequency (e.g. four low-penalty events per annum at one-third day outage/event produces a mean of 1.33 days accumulated outage per annum) and are given in the corresponding triangles in Fig. A.4.1. In Fig. A.4.2 these days of outage have been summated for each system. Thus the manual (M) system was predicted to produce a mean of 5.5 days of outage per annum whereas the semi-automatic (A–M) and the predominantly automatic (A) system were both predicted to produce mean outage of 4.5 days per annum. It will be noted that all three systems fall well within the 18 days/year (95 per cent success ratio) specified plant 'outage' limit—as illustrated in Hunns[93]. However, as can be seen from Fig. A.4.1 the minimum frequency (for the A–M and A systems) of the high hazard event was predicted to be approximately once per 50 years. This would seem to be unacceptably high from the point

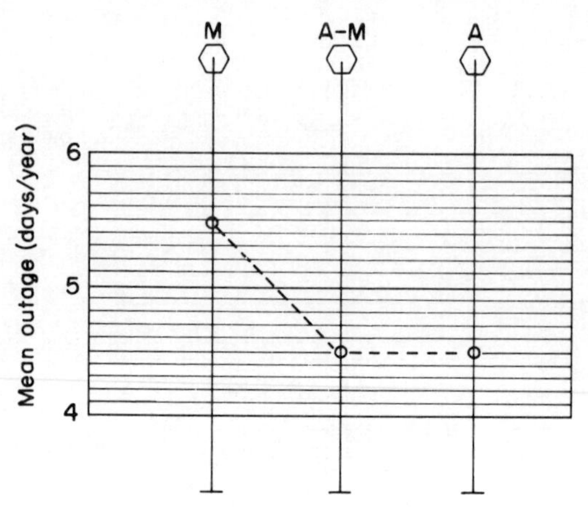

Fig. A.4.2 Mean outage

of view of operator risk. However, although damage to the internal structures of the plant could be extensive, it was estimated that only in a very low proportion of occasions would injury extend to operating personnel. Thus the A–M and A systems were agreed to provide an adequately low risk for the operating and maintenance staff.

The comparative reliability assessments together with the life-cycle costings analysis provide the very clear conclusion *viz.* that, in terms of reliability, in this case the semi-manual/semi automatic protective system is predicted to be the one to choose.

Although the case-study in this instance has dealt with only a sub-system, albeit an important one, of the plant, the same principles could be applied for the assessment of the complete plant performance in satisfaction of the specified success ratio.

APPENDIX 5

Mean failure rates for components and instruments

Electronic components Application — average industrial Temperature — 25°C	faults/10^6 hr
Resistors (fixed) all types	0.022
Resistors (variable) composition	0.068
Capacitors (fixed) electrolytic	0.38
Capacitors (fixed) paper, mica, glass	0.066
Capacitors (fixed) plastic	0.036
Capacitors (variable) all types	0.094
Transistors (PNP, NPN, low power)	0.05
Diodes, all types, low power	0.22
Transformers, AF	0.027
Ferrite cored devices	0.014
Electrical connectors all types	0.60
Panel meters (microammeters etc.)	4.6
Industrial instruments	
Temperature indicators (all)	20.0
Temperature trip amplifiers (electronic)	60.0
Pressure indicators (all)	8.0
Pressure transmitters (pneumatic)	22.0
Pressure transmitters (electronic)	8.5
Flow indicators (all)	16.0
Level sensors (all)	24.0
Signal convertors (electronic)	15.0
Recorders (electronic)	20.0

APPENDIX 6

Derivation of Bayes Theorem

Typically, Bayes Theorem may be written in the form:

$$P(\text{A/B}) = P(\text{A})\,\frac{P(\text{B/A})}{P(\text{B})}$$

where $P(\text{A})$ is the probability of A
and $P(\text{A/B})$ is the probability of A given B.

A derivation of this theorem may be given as follows:

If A and B are events then $P(\text{AB})$ is shown on the Venn Diagram (Fig. 6.1). The probability of (B/A) is clearly concerned with the 'heavy black' part of the diagram and is the ratio of the area (AB) to total area A, that is

$$P(\text{B/A}) = \frac{P(\text{AB})}{P(\text{A})} \qquad \text{(A6.1)}$$

By symmetry it may be shown that

$$P(\text{A/B}) = \frac{P(\text{AB})}{P(\text{B})} \qquad \text{(A6.2)}$$

Solving for $P(\text{AB})$ from equations 6.1 and 6.2 we have

$$P(\text{AB}) = P(\text{A})P(\text{B/A}) = P(\text{B})P(\text{A/B}) \qquad \text{(A6.3)}$$

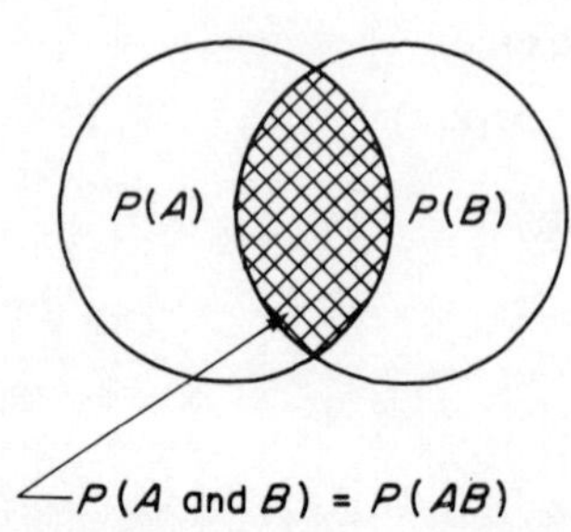

Fig. A.6.1 Venn diagram

giving Bayes Theorem

$$P(\mathrm{A/B}) = P(\mathrm{A})\frac{P(\mathrm{B/A})}{P(\mathrm{B})} \tag{A6.4}$$

where $P(\mathrm{A})$ = 'Prior' probability which is what is known before additional evidence becomes available.

$P(\mathrm{B/A})$ = 'Likelihood' which is how probable is the new evidence, assuming that the hypothesis A is true.

$P(\mathrm{B})$ = The probability of the evidence B alone.

$P(\mathrm{A/B})$ = The 'Posterior' probability of A now that evidence B has been added to the existing knowledge.

Various sources discuss this theorem in more detail, see for example, MacFarland[166], Mann *et al.*[167], and Crellin[168].

Deeley and Zimms[169] also by Easterling[170] which gives an extensive bibliography and similarly Martz and Waller[188] for an in-depth discussion of the subject.

APPENDIX 7

Proof of Theorem on the Asymptotic Behaviour of Gamma Percentiles*

The following proof of the theorem given in 4.20 is given for completeness and has been extracted from Waller *et al.*[122] Appendix C.

Theorem:

If, as $\alpha \to 0^+$, $\kappa(\alpha)$ is defined by

$$\frac{1}{\Gamma(\alpha)} \int_0^{\kappa(\alpha)} e^{-x} x^{\alpha-1} \, dx = p$$

and if $0 < p < 1$ then

$$\kappa(\alpha) \sim p^{1/\alpha}$$

where $f(\alpha) \sim g(\alpha)$ means $\lim\limits_{\alpha \to \alpha_0} \dfrac{f(\alpha)}{g(\alpha)} = 1$

Proof:

Consider

$$F(\alpha) = \frac{1}{\Gamma(\alpha)} \int_0^{\kappa} e^{-x} x^{\alpha-1} \, dx \tag{A7.1}$$

for small α and fixed $0 < \kappa < \infty$. Now

$$F(\alpha) = \frac{\displaystyle\int_0^{\kappa} e^{-x} x^{\alpha-1} \, dx}{\displaystyle\int_0^{\kappa} e^{-x} x^{\alpha-1} dx + \int_{\kappa}^{\alpha} e^{-x} x^{\alpha-1} \, dx} \tag{A7.2}$$

and

(i) $\lim\limits_{\alpha \to 0+} \displaystyle\int_0^{\kappa} e^{-x} x^{\alpha-1} \, dx = +\infty$

*Extracted from Reference 122 by permission of the Los Alamos National Laboratory and the authors.

(ii) there is an M such that, for $0 < \alpha < \alpha_1, \int_{\kappa}^{\infty} e^{-x}e^{\alpha-1}\,dx < M$.

Therefore, $F(\alpha) \to 1$ as $\alpha \to 0^+$. This emphasises, if $\kappa(\alpha)$ satisfies

$$\frac{1}{\Gamma(\alpha)} \int_0^{\kappa(\alpha)} e^{-x}x^{\alpha-1}\,dx = p \tag{A7.3}$$

then $\lim_{\alpha \to 0+} \kappa(\alpha) = 0$

Then, as $\alpha \to 0^+$

$$p = \frac{1}{\Gamma(\alpha)} \int_0^{\kappa(\alpha)} e^{-x}x^{\alpha-1}\,dx \sim \frac{1}{\Gamma(\alpha)} \int_0^{\kappa(\alpha)} x^{\alpha-1}\,dx \tag{A7.4}$$

$$= [\alpha\Gamma(\alpha)]^{-1}[\kappa(\alpha)]^{\alpha}$$

$$= [\Gamma(\alpha + 1)]^{\alpha}[\kappa(\alpha)]^{\alpha}$$

$$\sim [\kappa(\alpha)]^{\alpha}$$

Therefore, the conclusion, $\kappa(\alpha) \sim (1/p^{\alpha})$, holds.

APPENDIX 8

Failure Probability Evaluation of an Automatic Protective System on a PWR

by Mme A. CARNINO
Electricité de France
Direction Generale, 32 Rue de Monceau
75384 Paris Cedex 08, France.

A.8.1 Introduction

The example of assessment of the failure probability of a system is taken from a nuclear power plant—pressurized water reactor in France. The system under study is the automatic protective system whose main function is to shutdown automatically the reactor by stopping the nuclear fission reaction in the core, in case of any perturbating event.

A.8.2 System Description

Without describing the way a PWR functions, it is possible to characterize the function of the automatic protective system, also called the scram system, as producing automatically or even manually a fast decrease of power in the reactor when the integrity of the physical barriers containing the fission products and the fuel elements are threatened.

The scram system consists of the following elements:

(a) sensors which monitor permanently the evolution of the physical parameters representative of the functioning of the reactor, either in normal or abnormal state,
(b) equipments transforming the signals sent by sensors into measurable electrical currents (amplifiers etc.),
(c) comparators actuating a signal indicating that the physical parameters measured have exceeded the set value (threshold triggers),
(d) logic circuits grouping and processing the signal from the comparators and giving the scram signal,
(e) equipments called high-power components (by opposition with the

preceding ones called low or medium power) which from the scram signal directly actuate the mechanical equipments in order to scram the reactor by control rods insertion,

(f) control rods whose role is to stop the fission reaction in the core by fast insertion of a neutron poison.

As this automatic protective system is considered very important to the safety of the plant, it is possible to check its proper functioning while the reactor is on power—this allows the improvement of its availability. This is why; (a) test equipments, and (b) procedure defining the test policy, are considered as integral parts of the system.

A.8.3 System Actuation

The scram system is designed in order to respond to a long list of initiating events, characterized by some overridden parameters, and which are considered as potentially leading to abnormal consequences on the installation or even to accidental conditions. All these sequences are studied thoroughly in the Safety Report of the installation.

Here the purpose of the study will be limited to the assessment of the response (or non-response) of the system to a given initiating event. The study of the initiating event is not part of the scope of this assessment, this means that we assume a probability of 1 for the occurrence of the initiating event. The initiating event considered is a spurious withdrawal of the control rods, while the reactor is on power and with a reactivity insertion rate high. It belongs to the list of potential accident initiators taken into account for the safety design of the reactor. The direct effect of this initiating event would be to jeopardize the first barrier integrity (fuel elements cladding rupture). The study begins with the detection of the initiating event by the two measurements: neutron flux and temperature. This leads to the actuation signal of the scram system which induce the control rods to fall down into the reactor core.

The control-rod drive mechanisms reliability will not be part of the study as quantified numbers, but a qualitative reliability study will be performed.

The cumulation of the initiating event with other independent external initiators, such as earthquakes for instance, is not considered in the study.

A.8.4 Study of the Event Sequence

A.8.4.1 Description of the sequence

The spurious withdrawal of the control rods on power results in an increase of the thermal flux in the core. Until the opening of the discharge valves or the safety valves of the secondary circuit, the heat removal in the steam generators increases less quickly than the power generated in the primary circuit. The result is an increase of the primary circuit temperature.

If it is not stopped manually or automatically by an emergency shutdown, accident studies show that a boiling crisis could occur in the primary coolant. The reactor emergency shutdown system which must intervene is designed to inhibit this boiling by maintaining the *DNBR (Departure from Nucleate Boiling Ratio)*: ratio which gives the rate of thermal exchange between the fuel, cladding and the coolant above 1.3. Thus the risk of damaging the fuel rods and initiating a clad rupture is avoided.

A.8.4.2 Scram system response and structure

Within the framework of this study, the problem is to avoid exceeding a temperature and pressure in the primary coolant leading to a nucleate boiling in the core hottest parts. The phenomenon is not directly observable, therefore a set of measurement channels giving accessible values has been designed in order to prevent the phenomenon.

The structure adopted is a compromise between requirements of safety and those of functioning continuity. This compromise is worked out by a combination of redundancies (safety) and majority of vote systems (continuity of operation).

The protective system structure is schematically described in Fig. A.8.1. Upstream is the measurement system 'SIP' which processes the analog signals from the measurement sensors (pressure, flux, temperatures etc.). In the centre is the relay system 'RPR', consisting of vote elements providing the emergency shutdown order. The interface between SIP and RPR consists of threshold relays which convert the analog signals to logical signals 1 or 0.

The RPR consists of two identical logical trains in parallel.

Downstream the RPR is the system which actuates the emergency shutdown orders. It is composed of two circuit-breakers in series. If only one of the circuit-breakers receives an opening order, the power to the control-rod drive mechanism is switched. The control rods then fall by gravity.

The SIP structure is fairly logically deduced from its functions. The SIP consists of a number of channels, each of which has a particular protective role. There are twenty-seven channels. We do not include the manual emergency shutdown since it intervenes directly on the logical systems.

Nearly all these channels can be further divided into two, three or four identical sub-channels in parallel. Thence, an equipment redundancy has been built up. For an emergency shutdown to be actuated by a channel, it is then necessary that a majority of vote (1/2, 2/3, 2/4) of sub-channels should order it.

The channels structure can be represented by Fig. A.8.1.

The sensors, which deliver the analog signals processed by an analog processing block are in direct contact to the primary circuit. The output value of this processing block is compared to a set signal in a threshold relay whose output is binary.

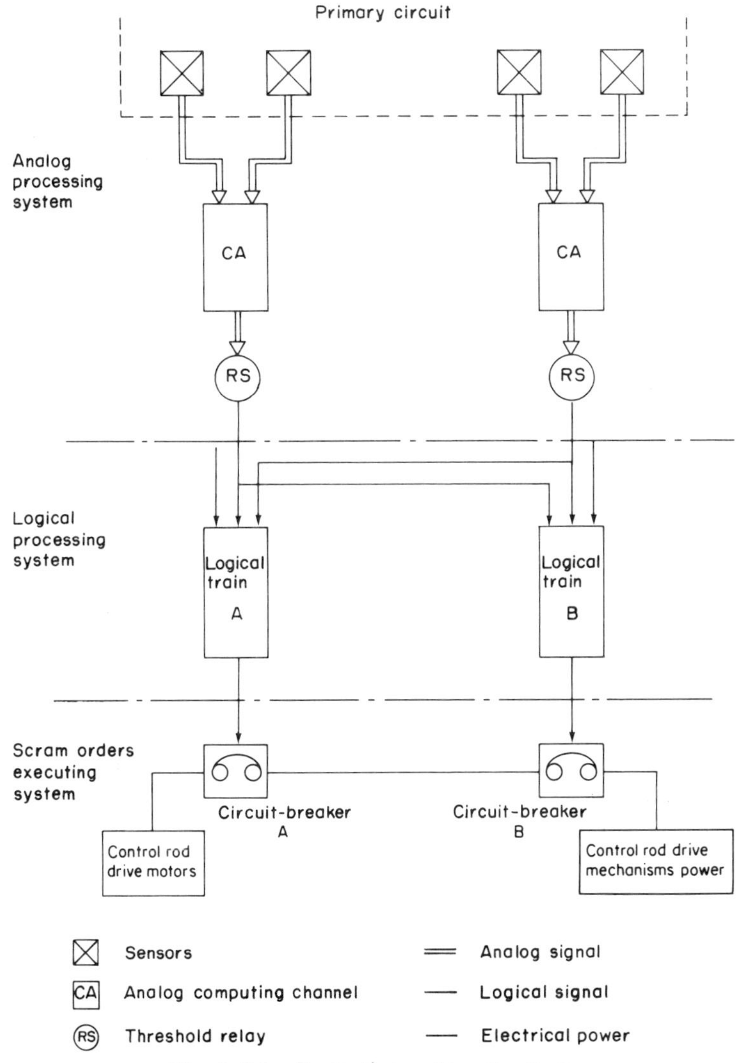

Fig. A.8.1 Protective system structure

Generally speaking, each sub-channel is part of a protective group and is therefore physically and electrically separated from the other sub-channels. In particular, the sub-channels of a same channel have independent power supplies; in the same way, the sensors are specific to each of them. On the contrary, there can be, inside the same protective group, sensors or items common to sub-channels of distinct channels. As a significant example of this, are the ΔT temperature ΔT power channels, where the temperature sensors are common for each sub-channel (Fig. A.8.4) of the same protective group. We will see later on that the existence of these common points has a great importance.

We can also note the presence of measurement output plugs and of test-signal input plugs. These devices allow the periodical testing of the channels.

Figure A.8.2 gives a more detailed diagram of the relay set (RPR) following the analog part (SIP).

Each logical signal from the SIP (three signals high ΔT temperature in the figure example) reaches two relays belonging respectively to trains A and B.

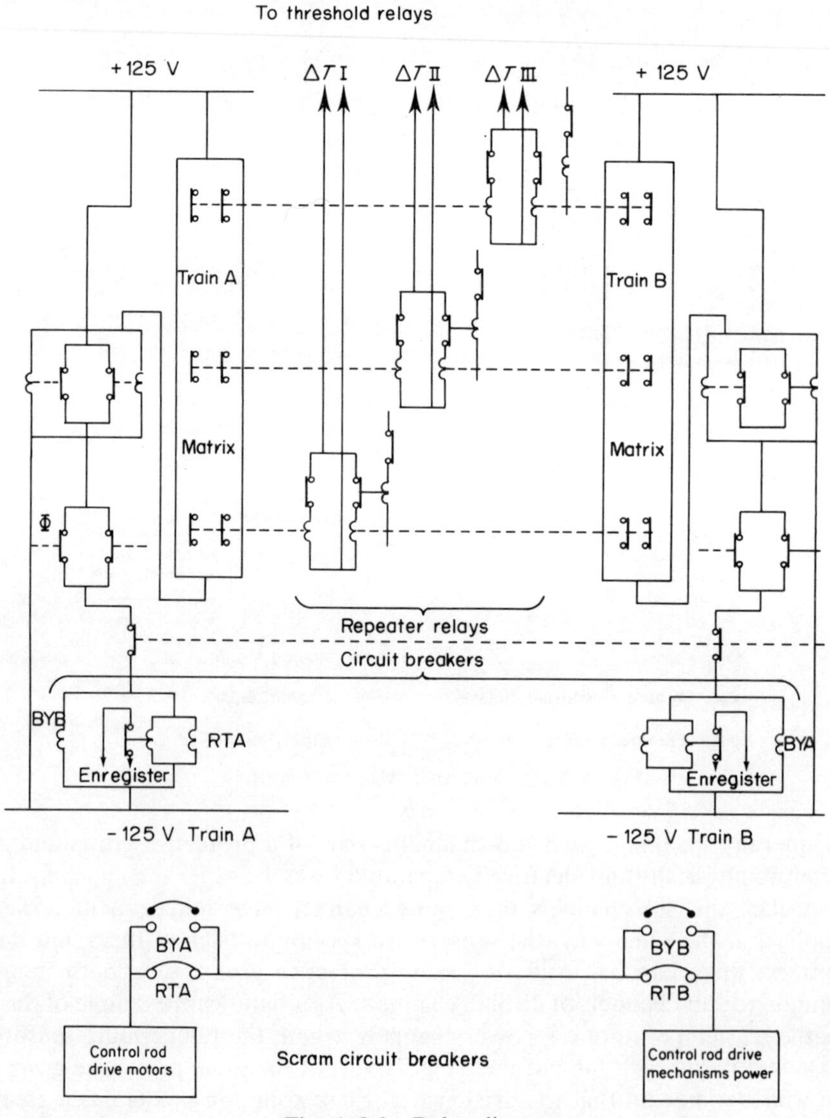

Fig. A.8.2 Relay diagram

It is important to point out that the emergency shutdown signal corresponds to a general lack of power of the whole RPR relay set.

The relay contacts cited above are organized into matrices which form the majority of vote elements (2/3 in the case of the example).

Each matrix is followed by 'matrix repeater' relays. The repeater relays of the different matrixes corresponding to the various emergency shutdown actuations have contacts connected in series to finally supply the circuit-breakers coils. These circuit-breakers trip the scram by switching off the electrical supply to the electromagnets which maintain the control rods in high position.

There is a test device for all these relays. This device is not shown in order to not overload the figure. It is not involved in the study since the possible failures lead to an undesired shutdown order, but never to an inhibition of the system. For the same reasons, the relays supplies are not taken into account.

A.8.4.3 Executor system structure

It consists of two circuit-breakers whose coils are fed respectively by trains A and B and the contacts cabled in series to supply the control-rod electromagnets. The by-pass circuit-breakers used to test the main circuit-breakers will only be mentioned since, by instruction, they are constantly maintained in the 'disconnected' position and, moreover, when the testing, for example, of breaker A is performed the maintaining of the corresponding by-pass breaker is ensured through that contacts series of train B.

A.8.5 Reliability Assessment of the System

A.8.5.1 Scope

Considering what has been said so far, the problem is limited to the following terms.

The reactor being at its nominal power, an event causes the spurious withdrawal of some control rods. The subsequent reactivity insertion velocity and fuel burn up are such that the sequence start is normally detected early enough by the measurement of at least two different physical parameters:

(a) measurement of the high level neutron flux;
(b) measurement and calculation of a high ΔT temperature (temperature variation of the coolant between the core inlet and outlet).

The neutron flux sensors (long ionization chambers) are four in number and distributed around the vessel at 90° from one another. The scram decision is taken when at least two of the four measurements have exceeded the safety threshold.

The ΔT temperature is calculated for each loop and therefore we get three redundant pieces of information combined in a 2/3 system.

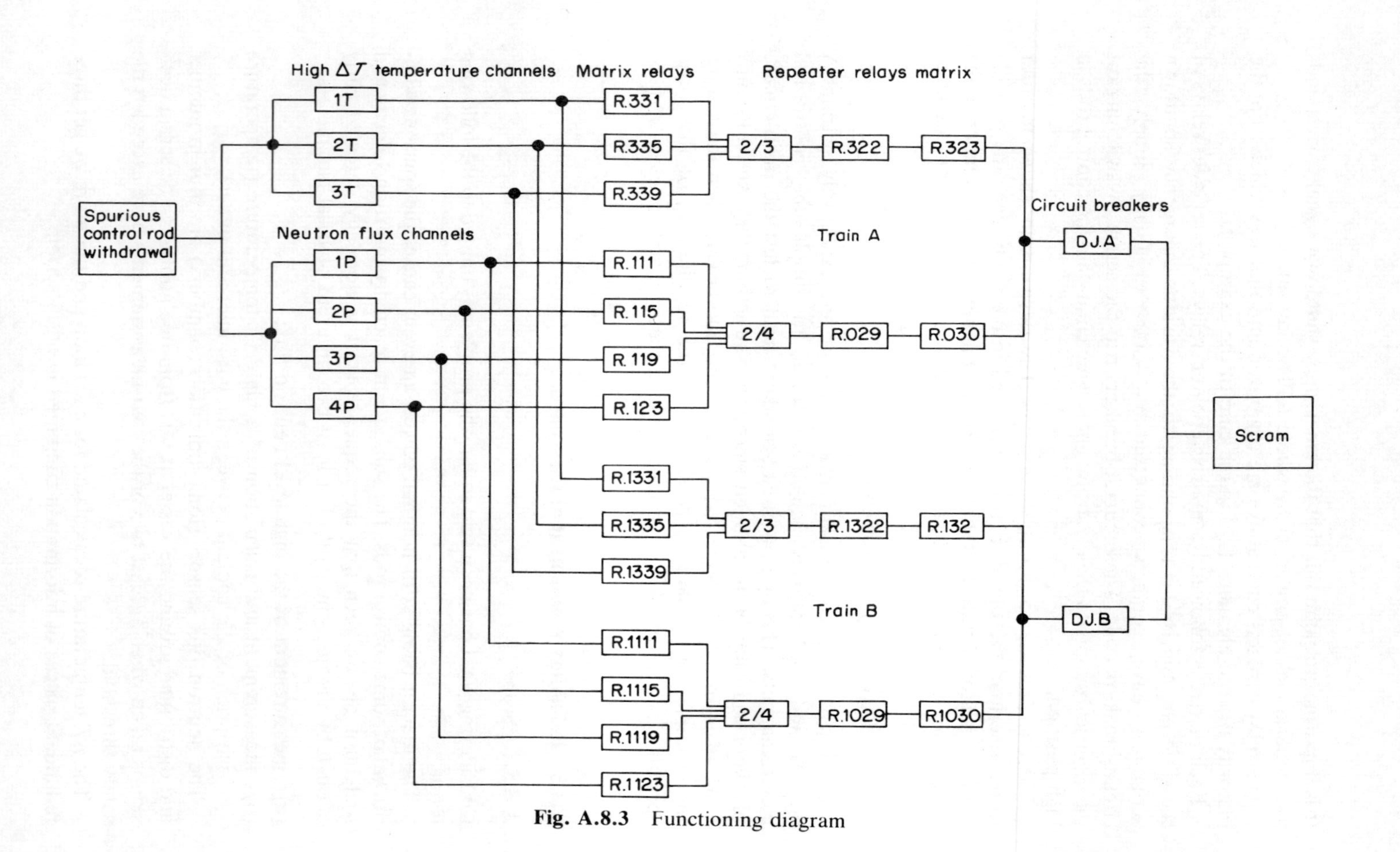

Fig. A.8.3 Functioning diagram

This being the case, one attempts to evaluate the probability that the control rods do not fall in response to the event 'spurious withdrawal'.

A.8.5.2 Functioning diagram

Figure A.8.3 schematizes, from left to right, the sequence resulting from the occurrence of the initiating event followed by an effective scram.

We think that the diagram is clear enough and does not require comments which would be useless and redundant with all that has already been said.

A.8.5.3 Fault tree

The fault tree has been established from the preceding diagram of operation (Fig. A.8.4).

The logical part of the system, particularly complex, has been studied with the CHAMBOR computing code. This code (after transposing the conventional wiring diagrams into equivalent diagrams directly usable by a computer) gives two advantages:

(a) checking that the diagram contains no error, either in design or in recopy, thanks to the program ability to simulate a dynamic operation of the system;
(b) determining the critical paths.

This diagram takes into account the dependencies. We notice (for example) that the 1*T* train (leaves 52 and 55) may affect simultaneously trains A and B.

The numbers below the leaves correspond to a numbering used in the PATREC code which solves fault trees.

A.8.5.4 Data

The problem of the numerical data is always a difficult problem to solve. The data here come from: (a) the WASH 1400 report[21], (b) statistics from 'Electricité de France', and (c) the files established by the DSN on the 'Phenix' reactor.

Three types of difficulties must be pointed out.

(a) The data are presented either in the form of a failure rate, or in the form of a probability of failure on demand; this led us to present the results in a form which, though quite significant, is nevertheless peculiar. This could be done thanks to the wide capabilities of the PATREC/RCM code.
(b) The data are always global, that is they include safe and unsafe failures. Within the framework of this study, only the safe failures are useful. A ratio of 1/10 between unsafe and global failures is generally admitted. It was agreed to apply this coefficient to the relays and the circuit-

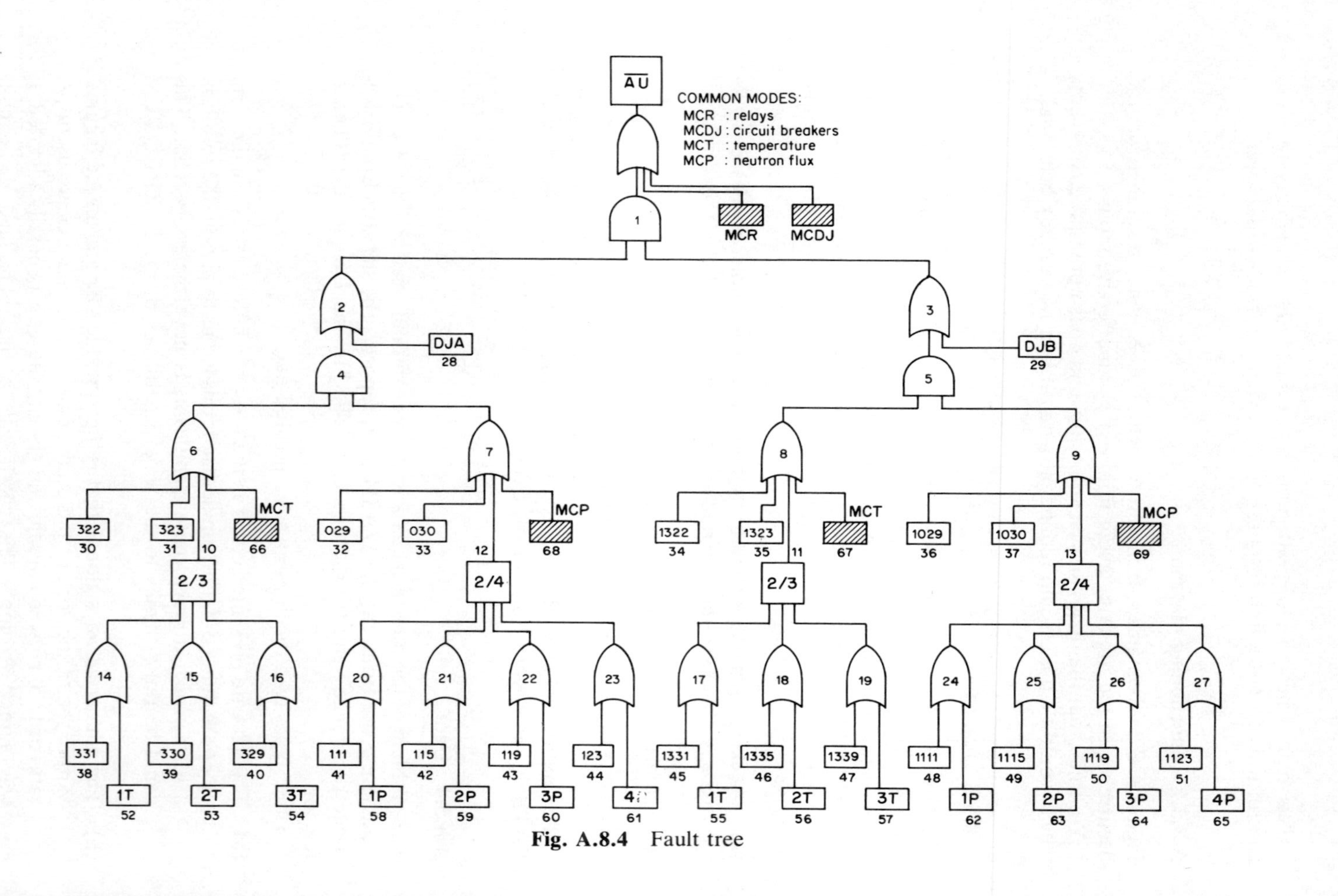

Fig. A.8.4 Fault tree

Table A.8.1

Device designation	Failure rate λ First estimation $\frac{\text{UF}}{\text{global failure}} = 1/10$	Failure rate λ Second estimation $\frac{\text{UF}}{\text{global failure}} = 1/2$	Probability of failure on demand
ΔT temperature channel	8×10^{-6}/hr	4×10^{-5}/hr	
Neutron power channel	9.6×10^{-6}/hr	4.8×10^{-5}/hr	
Relay	1.4×10^{-7}/hr	1.4×10^{-7}/hr	
Circuit breaker			3×10^{-5}/days

UF : unsafe failure

$$\beta = 0; 10^{-2}; 10^{-1}; 1 = \frac{\lambda \text{ (or probability) common mode}}{\lambda \text{ (or probability) self mode}}$$

breakers. On the contrary, there was divergent opinions concerning the analog part of the system; the calculations have therefore been made for two values of the ratio: 1/10 and 1/2.

(c) We do not have a very clear idea of the value to be assigned to the common mode failures. Some authors recommend a ratio $\beta = 1/10$ between common mode and self mode failure rates. As this value is questionable, we have made the calculations for four values of the β ratio: 0 (no common mode failure), 10^{-2}, 10^{-1}, 1.

Considering what has been said above, the values shown by Table A.8.1 have been adopted in the calculations.

A.8.5.5 Calculations and results

The calculations have been made from the fault tree (Fig. A.8.4) with the data defined above and with the help of the PATREC/RCM code, which is a very efficient code taking into account in particular the dependencies.

The results are given in Tables A.8.2 and A.8.3 respectively corresponding to the two values of the ratio λ unsafe failures/λ global failures (1/10 and 1/2) adopted for the analog trains (Figs. A.8.5 and A.8.6).

The values in the 'Resulting probability' column of Table A.8.3 must be interpreted as follows:

At the time when a test of the system is performed, we establish that it is in a good operating condition. After a T *period equal to 1 month (720 hr), a second test is performed. The number written in column 5 gives the probability that this second test reveals a failure in the system operation.*

Therefore, a simplifying assumption has been made that the system is

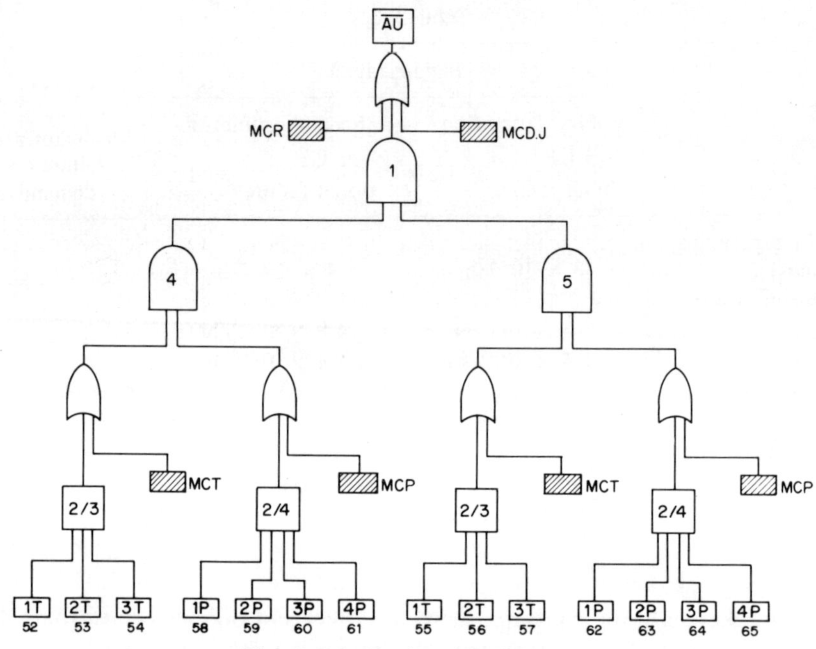

Fig. A.8.5 Reduced fault tree

Table A.8.2 Case 1

	TCM + PCM	RCM	BCM	Resulting probability
0	0	0	0	1×10^{-9}
10^{-2}	1.1×10^{-8}	1×10^{-6}	3×10^{-7}	1.3×10^{-6}
10^{-1}	4.6×10^{-7}	1×10^{-5}	3×10^{-6}	1.3×10^{-5}
1	4×10^{-5}	1×10^{-4}	3×10^{-5}	1.7×10^{-4}

TCM = temperature channels common mode
PCM = neutron flux channels common mode
RCM = relay common mode
BCM = circuit breaker mode

Table A.8.3 Case 2

(1)	TCM + PCM (2)	RCM (3)	BCM (4)	Resulting probability (5)
0	0	0	0	3.6×10^{-7}
10^{-2}	4.4×10^{-7}	1×10^{-6}	3×10^{-7}	2.1×10^{-6}
10^{-1}	1.6×10^{-5}	1×10^{-5}	3×10^{-6}	3×10^{-5}
1	1×10^{-3}	1×10^{-4}	3×10^{-5}	1.1×10^{-3}

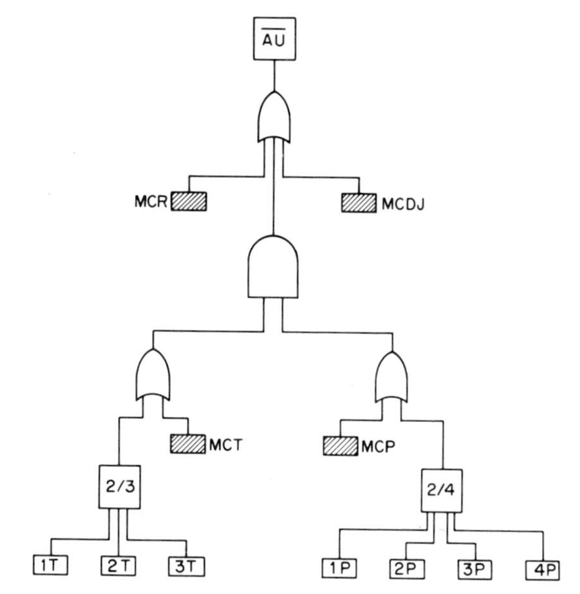

Fig. A.8.6 Reduced fault tree

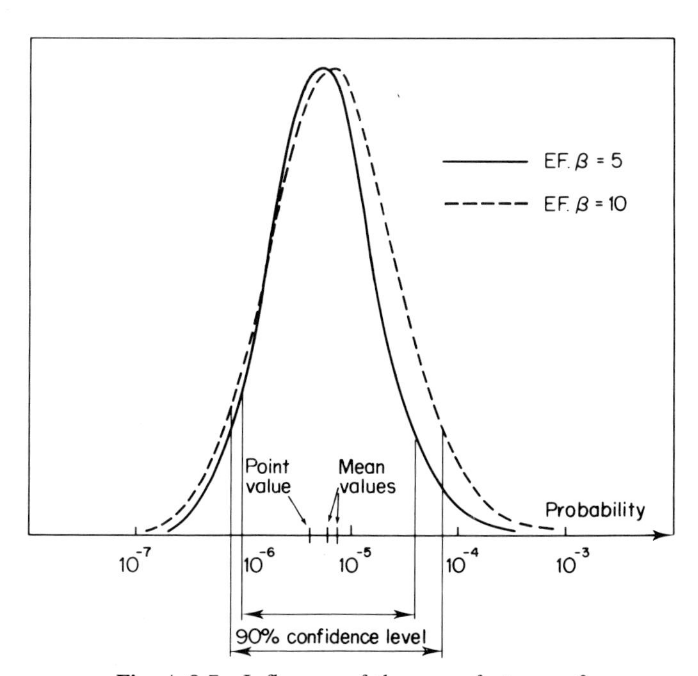

Fig. A.8.7 Influence of the error factor on β

tested in its whole and instantaneously every month. In reality, things are different and, though it is true to say that the system is tested every month in its whole, one might take into consideration that, in fact, the different parts of the system are sequentially tested. A separate study will be carried through on this subject.

The results bring to light that in the two cases the common mode failures are predominant, as could be expected. The relay common mode failure is predominant, except in Case 2 for $\beta = 1$ (unlikely case). We can also notice that in case 1, for $\beta = 0$, the circuit-breakers are predominant (9×10^{-10} for the circuit-breakers compared with a total probability of 1×10^{-9}).

A.8.5.6 Credibility of the results

Given the uncertainty on the validity of the numerical values which were used, it was interesting to evaluate the uncertainty factor affecting the global results, by taking into account the uncertainty factors specific to the component failure rates.

Calculations have been made with the following values (taken from WASH-1400[21]):

analog trains *uncertainty factor* 10
relays and circuit-breakers *uncertainty factor* 3
Data: those from Case 2 (§ A.8.5.5) for $\beta = 10^{-1}$

For these calculations, the PATREC/MC code has been used (version of the PATREC code using a Monte-Carlo method).

A first run was used to simplify the tree of Fig. A.8.4 by identifying the predominant cut sets. So, if we neglect the cut sets lower by a factor 10^4 to the cut sets kept, we come to the simplified tree of Fig. A.8.5.

The tree of Fig. A.8.5 shows a perfect symmetry in relation to the circuit 'ET' numbered 1. This permits (using the properties of the Boolean algebra) a second reduction of the tree which is shown in Fig. A.8.6.

Fig. A.8.7 shows the dispersion of the global result value related to the dispersion of the input parameters, after 10 000 runs for the two β values.

As can be seen, the curve looks like a lognormal law. With this assumption, for a 90 per cent confidence level, the code provides the following values:

lower limit 5%	1.045×10^{-5}
upper limit 95%	5.663×10^{-4}
adjusted mean value	7.694×10^{-5}

That is an uncertainty factor of about 7.

If we remember that the deterministic value precedently found was of 3×10^{-5}, we find out that the results are quite consistent:

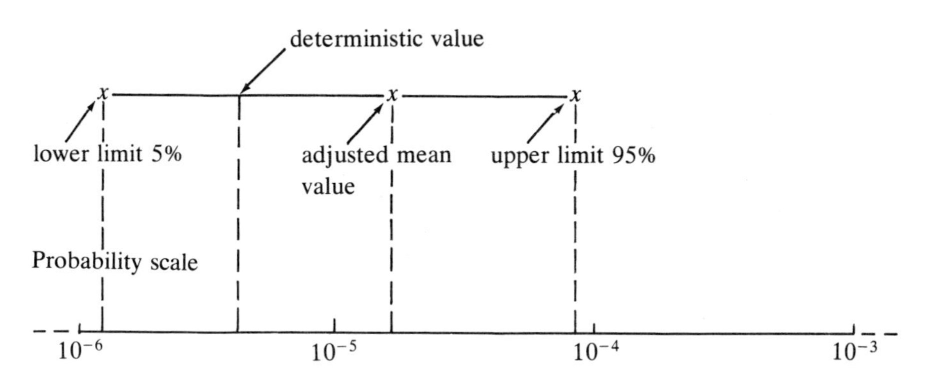

A.8.6 Analysis of Control-Rod Drive Mechanisms

It was only possible to analyse the probability of the control rods to fall down into the core from a qualitative point of view. A fault tree was drawn concerning the elementary mechanical failures which could lead to a blockage of the control rods. But it was not possible to allocate failure rates to these mechanical failures, even after the analysis of events that had led to mechanical blockage of one control rod from incidents reporting all over the world (Fig. A.8.8).

Therefore the analysis of mechanical control-rod blockage was not included in the quantitative analysis.

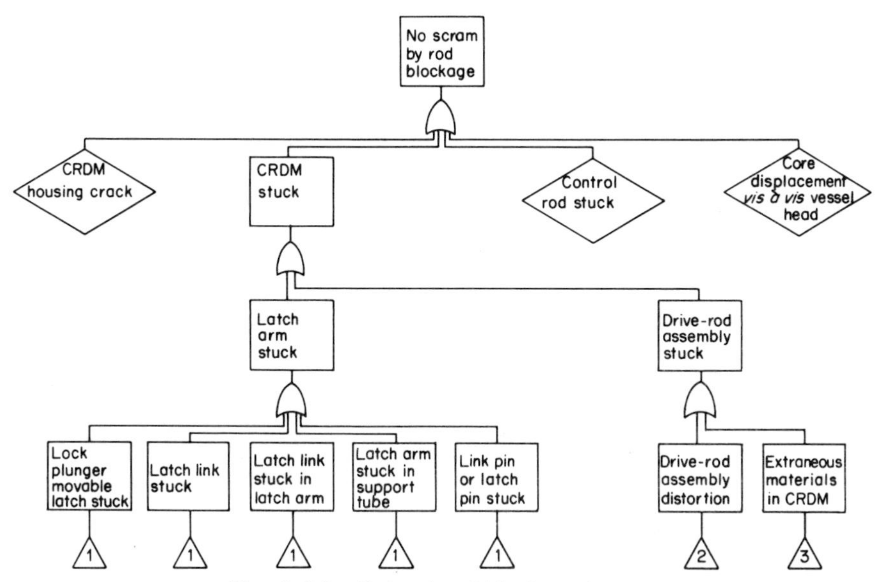

Fig. A.8.8 Control-rod blockage fault tree

A.8.7 Test of the System and Human Factors

Although we did not include human factors consideration we performed a trial and limited quantification for the test of the system where human factors could have induced unavailability of the system prior to the demand from the initiating event.

The goal to obtain the probability of a failure to scram the reactor following a specific initiating transient—uncontrolled withdrawal of a control-rod cluster. During the study of the reliability of the equipment in the system considered, it became apparent that the human factors which could influence the behaviour of the system might originate from breakdowns that left the system unavailable during and after periodic testing, that is to say, before the demand studied.

The following approach was used. According to the results of an evaluation of reliability of the emergency shutdown system of the reactor following an uncontrolled control-rod withdrawal, the failure rate for one signal channel was estimated to be 5×10^{-5}/hr, or 3.6×10^{-2} per test interval. For the operator who tests this signal channel to render it unavailable by mistake, he must make an unnoticed fail-dangerous error with a probability of less than 3.5×10^{-2} per test. Because, in fact, there are several channels to be tested, these errors must also be common-mode errors affecting all the channels, which brings the probability back to 3×10^{-5} per test.

One potential source of human error was voluntarily set aside because it would have necessitated a study far too long in comparison with the scale of the exercise performed. This was the question of external sources of interference which could have led to manipulation errors on other systems whose control-system cabinets were beside the one studied. Similarly ignored was any interference with the system studied due to errors committed during periodic testing of these neighbouring cabinets.

This approach thus led us to:

(a) break down the task 'execute the test procedure' into subtasks or elementary actions,
(b) observe how the operators really perform the task,
(c) do an error/consequence analysis taking into account the recovery factors.

The task analysis yielded a sequence of 186 actions, and also identified ambiguities, uncertainties, and potential errors and their effects. By direct observation of execution of the task the consequences of certain errors were identified, along with the factors affecting recovery of these errors when this was possible. The final analysis thus allowed us to separate out from the 186 actions the possible errors that would render the channel unavailable:

(a) organization and setup of the test team (organizational changes by the plant management in a period of stress, strikes, illness, beginners),
(b) choice of the correct cabinet for the test,

(c) equipment needed for the test (its calibration, its maintenance),
(d) the calibration itself (earlier undetected errors, incorrect reading etc.),
(e) reviewer's assessment of the contents of the records from the periodic tests,
(f) restoration of the channel to the normal operating mode after testing and verification in the control room,
(g) final verification of the status of the channels in the control room.

In fact, human error data was not available for each of these actions identified as being significant. After analysing the sources of data and assessing operating events—which only gave global statistics without allowing identification of the denominator—another solution was chosen to meet the objective that has been fixed.

If one adopts an error rate of 10^{-2} per action for each elementary action, and if one assumes a ratio of 0.5 for fail-dangerous to total rate of error, the decomposition into 186 actions gives (by employing an approximation frequently used to obtain an estimate of the collective reliability of several pieces of equipment):

$$186 \times 10^{-2} \times 0.5 = 0.93$$

This result is obviously unrealistic and far too high. In fact, the estimate represents the errors committed if the procedure is not respected, and not the rate of errors leading to unavailability of a channel. However, it shows well the danger that there is in decomposing a task into subtasks and elementary actions and quantifying directly, without taking into account dependencies and situation and performance shaping factors which modify the evaluation completely.

If one takes instead only the actions that can lead to a fail-dangerous fault in the channel (eight actions), the failure rate—which is the goal sought—is then 0.86×10^{-2}/test, using the high failure rate of 10^{-2} per action.

After this analysis of the errors which could render *one* channel unavailable, a similar analysis was done to identify common-mode errors which would lead to the unavailability of the whole set of measurement channels considered. Three of them were identified:

(a) personal assignment: a change in the make-up of the test team, planning of the test, etc., planning of parallel activities that could interfere with the test studied.
(b) equipment needed for the test: calibration, maintenance, possible modifications, etc.
(c) functional 'improvements' in the procedure to lessen inconvenience in parts of the task, improvisations, etc.

Concerning common-mode actions which affect the set of measurement channels considered, the quantification could not be continued in the exercise because it required supplementary information having to do with the personnel, the teams and their qualification, and the equipment needed for the test.

However, there are several lessons to be drawn from this trial of quantification about the approach that one should adopt when quantifying human error in the execution of a task:

(a) clearly define the objective sought,
(b) analyse (as for reliability) the nature of the errors their effects on the system and their impact on the objective,
(c) take into account in the analysis recovery (or detection) factors for the errors committed,
(d) quantify what comes out of this analysis either by estimation using pessimistic data, or by observation and a more limited exercise on only those errors that were identified as significant,
(e) do not quantify directly by decomposition into elementary tasks without taking account of interdependencies, and situation and performance shaping factors.

Due to the difficulties discussed here, the quantification of potential human error was not included in the global failure probability of the system under study. It shows that further research is needed for quantification of human errors in reliability calculations in order to give credit to such error probability determination.

A.8.8 Conclusion

This preliminary study is a decision experiment and constitutes a good illustration of the present knowledge in the field of probabilistic estimation of the quality of a complex system. It brings to light the field in which research is further necessary.

The study has been limited to the scram system actuation after a spurious withdrawal of some control rods (reactor being on power). The external events have not been taken into account.

However, a wider study would not have brought more to the purpose pursued, which has to concretize the possibilities and the drawbacks in lacks of the method by studying a real case.

The fault-tree analysis has once more shown its value. For example, we have noted that it was possible without expensive modifications to greatly improve the efficiency of the measurements processing by a continuous search for eventual discordance.

Doubts remain on the numerical data. The failure rates are generally given in a global form (safe failures + unsafe failures) while we are only interested in the unsafe failures. The factor β related to the common mode failures is not well known:

$$\left(\beta = \frac{\text{probability of common-mode failure}}{\text{probability of self-mode failure}}\right)$$

We do not yet know exactly how to introduce the human factor in the calculations. Simplifying assumptions had to be adopted for the test procedures.

The parameter calculations show the importance of these factors on the global result.

A sensitivity study performed with the help of PATREC/MC code in a case which (in the present state of the art) seemed to us realistic, gives the following results for the non-operating probability of the emergency shutdown system after a 1-month period:

adjusted mean value	7.7×10^{-5}
lower limit 5%	1.0×10^{-5}
upper limit 95%	5.6×10^{-4}

This means an uncertainty factor of about 7 for a confidence level of 90 per cent.

We can therefore see the difficulties related to the establishment of a good-quality probabilistic criterion and the need to develop research in the adequate fields.

Further reading

The problems of rare events in the reliability analysis of Nuclear Power Plants, OECD–CSNI Report SIN DOC (78) 33.

PATREC, A computer code for fault tree calculations, Rapport DSN no. 235 (e) by A. Carnino *et al.*, septembre 1978.

Donnés de faibilité extraites de l'exploitation des centrales by A. Carnino, J. F. Greppo, *NUCLEX-BALE*, octobre 1975.

Défaillance du système de protection d'un réacteur nucléaire considérée en tant qu'évènement rare by A. Carnino, B. Gachot, Rapport DSN no. 103, juin 1976.

Composantes humaines dans les phénomènes initiateurs d'accidents nucléaires by A. Carnino, A. Raggenbass, Rapport DSN no. 140, AIEA Vienne, mars 1977.

Etude sous l'angle fiabiliste des mécanismes de commande des grappes des réacteurs PWR by R. Carnino, Internal report, juin 1978.

Cours Fiabilité, Etudes probabilistes de sûreté by A. Carnino, Rapport DSN no. 89, février 1976.

La fiabilité dans une méthode d'analyse de sûreté by A. Carnino, R. Quénée, Rapport DSN no. 14, *AIEA Julich*, février 1973.

The problems of rare events in the reliability analysis of Nuclear Power Plants by A. Carnino, J. Royen, M. Stephens, Congres of Newport Beach, mai 1978.

APPENDIX 9
HTGR Core Auxiliary Cooling System (CACS)—Reliability Prediction

by V. JOKSIMOVICH
NUS Corporation, 16885 West Bernardo Drive,
Suite 250, San Diego,
California 92127, U.S.A.

A.9.1 Introduction

The power conversion loops (or main loops) are able to remove decay heat under most accident conditions. However, the HTGR plants sold commercially in early 1970s also had a dedicated, fully redundant decay-heat removal system that was separate from and independent of the power-conversion system. This separation was intended to increase the resistance against common-cause failures. Inherent features of the HTGR with its associated predictabilities, enabled the design of this core auxiliary cooling system (CACS) to be kept simple. Its sole function was to remove reactor decay heat under the full range of operating conditions and postulated design basis accident environments. There was no requirement to inject coolant since adequate heat removal can be obtained with the vessel depressurized after a postulated leak in the primary coolant system boundary. There was hardly any ambiguity with regard to which cooling system can be operated at which temperature and pressure etc. As a result of the slow thermal response of the HTGR core, interruptions in core-cooling system operation of several hours can be tolerated without damage to the fuel or significant release of radioactivity. Reliability of this system was evaluated in the HTGR risk assessment study named AIPA (Accident Initiation and Progression Analysis). The material presented here is based on the prediction in the study.

A.9.2 Brief Description of CACS

The CACS system consists of three independent cooling loops. Each loop consists of an electric motor-driven circulator which takes cold helium coolant

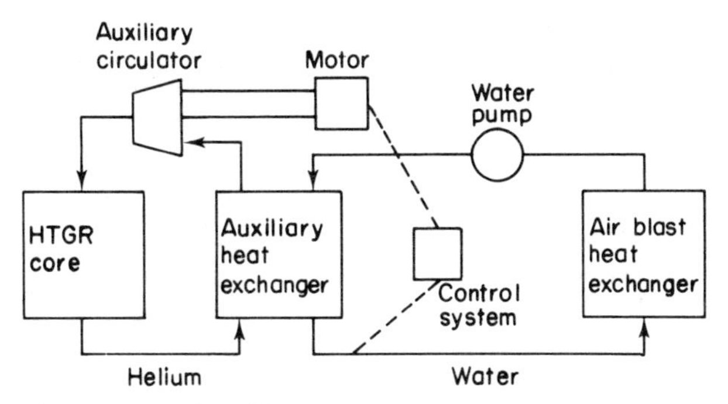

Fig. A.9.1 Simplified diagram of CACS heat-transport loop

from the exit of the auxiliary heat exchanger (CAHE) and circulates it back to the upper plenum. The coolant then flows down through the reactor core where it is heated up and into the lower plenum. From the lower plenum the hot helium flows up through the CAHE to complete the loop. The auxiliary heat exchanger is cooled by the auxiliary cooling-water supply as displayed in Figs. A.9.1 and A.9.2. The loop cooler (air-blast heat exchanger (ABHE)) removes heat from the CACS cooling water and rejects that heat to its ultimate heat sink, the surrounding atmosphere. To accommodate this function hot water from CAHE enters ABHE and flows through metal tubes, rejecting heat to the metal. The tubes pass the heat to air that is blown by large fans over the tubes.

A.9.3 Start-up Reliability Prediction

Since the CACS is a standby system, consideration has to be given to whether or not the system can be brought into operation before operating failures need to be considered. In order to bring a CACS loop into operation from

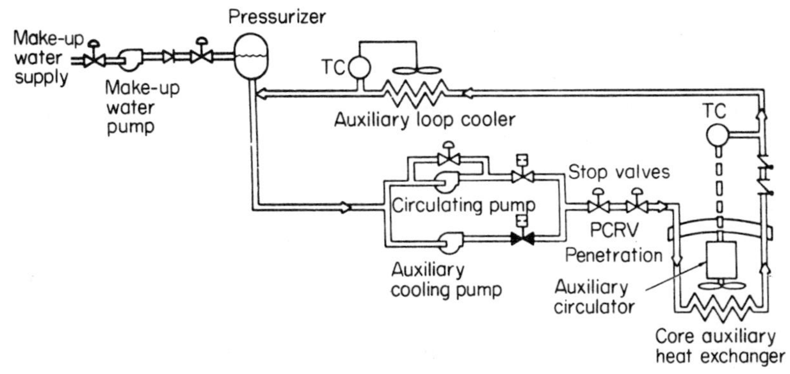

Fig. A.9.2 Major components in core auxiliary cooling system loop

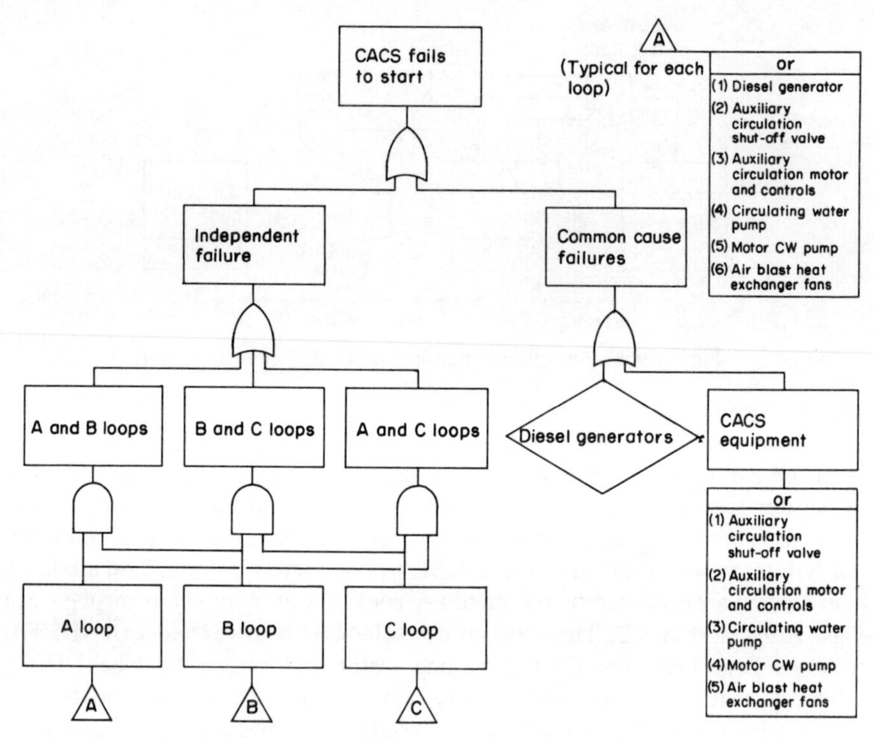

Fig. A.9.3 Fault tree for CACS start-up given loss of off-site power

stand-by following loss of main-loop cooling, the following components are susceptible to a failure to start: diesel generator, auxiliary helium shut-off valve, auxiliary circulator motor and controls, circulating water pump, motor cooling-water pump, and the air-blast heat exchanger fans. Successful start of each of these items is required for a successful start-up of each loop. System success is defined when at least two of the three loops start following a reactor trip from full power. The fault tree describing this is displayed in Fig. A.9.3.

A.9.3.1 Methodology

Using the fault tree methodology discussed in Phase 1 of the AIPA Study[171] the probability of CACS start-up failure is given by:

$$P_{\mathrm{START}} = 3q^2 - 2q^3 + \sum_{j=1}^{6} \beta_j Q_j \tag{A.9.1}$$

where $q = \sum_{j=1} (1 - \beta_j)Q_j$ (A.9.2)

Q_j = failure-to-start probability for CACS component j,
j = common cause failure fraction for component j,

$$j = \begin{cases} 1 & \text{diesel generator,} \\ 2 & \text{auxiliary circulator shutoff valve,} \\ 3 & \text{circulator motor and controls,} \\ 4 & \text{CAHE cooling water pump,} \\ 5 & \text{motor-component cooling-water pump,} \\ 6 & \text{air-blast heat exchanger fan.} \end{cases}$$

The first two terms in equation A.9.1 represent the contributions to system failure from independent combinations of component failures (a two-out-of-three system) and the last term represents the contribution to system common-cause failures.

A.9.3.2 Quantification

Equation A.9.1 is quantified employing relevant reliability data assessments for components similar to those employed in the CACS. These assessments are summarized in Table A.9.1.

In arriving at these assessments it should be observed that definition of system success, as reflected in the plant design criteria was for the system to start up and assume full heat load within 20 min. Normally, however, when all components are functioning satisfactorily, start-up would occur in less than 5 min. Thus, if a component such as a diesel generator did not operate successfully on the first automatic start attempt there would be sufficient time for several restart attempts. To reflect this capability it was assumed in Phase I of the AIPA Study that the component failure probabilities (i.e. the Q_j in equation A.9.1), at 20 min are an order of magnitude less than the probability of failure on the first automatic start attempt with the exception of the auxiliary circulator motor and shut-off valve, which are located inside the pressure vessel and would not be repairable. This resulted in $Q_1 = 2 \times 10^{-3}$ which was in good agreement with results of a special pre-start-up testing program at Fort St. Vrain during which a total of 355 manual diesel generator starts were attempted.

Subsequent to completion of Phase 1, the operating experience of diesel generators in nuclear power plants was examined in greater detail to re-evaluate validity of the assumption related to credit given for time delay. Crooks and Vissing[173] summarize the failures that occurred at twenty-nine nuclear plants during the period October 1959 until October 1973. Although most of the failures occurred during surveillance testing, some occurred during actual demand situations.

The experience data given by Crooks and Vissing[173] are summarized as follows:

(a) During the above-mentioned period, there were a minimum of 2940 diesel generator start attempts.
(b) There were forty-three occurrences of failure of individual (i.e. independent) diesel generator units to start and accept the load (forty-one during surveillance tests, two in actual demand situations).

Table A.9.1 Reliability data for quantification of CACS start-up

Component symbol j	Description	Component failure-to-start probability Q_j*	Common cause factor β_j*	References
1	Diesel generator	6×10^{-3} (2)	1.7×10^{-1} (3)	Joksimovich *et al.*[172]
2	Auxiliary circulator shut-off valve	3×10^{-4} (10)	2.2×10^{-1} (3)	Scaletta[175]
3	Auxiliary circulator, motor and controls	3×10^{-4} (3)	1.4×10^{-1} (3)	Assessment for pumps in RSS Appendix III (WASH-1400[174]) reduced by factor of 3
4	CAHE cooling-water pump	3×10^{-4} (3)	1.4×10^{-1} (3)	Assessment for pumps in WASH-1400[174] reduced by factor of 3
5	Motor cooling-water pump	3×10^{-4} (3)	1.4×10^{-1} (3)	Assessment for pumps in WASH-1400[174] reduced by factor of 3
6	Air-blast heat-exchanger fan	3×10^{-4} (3)	1.4×10^{-1} (3)	Assessment for pumps in WASH-1400[174] reduced by factor of 3

*Values in parentheses are lognormally distributed uncertainty factors

(c) There were three occurrences of common-cause failures that affected three, two and two diesel generator units respectively, giving a total of seven diesel generator units failed due to a common cause failure.

Counting the total number of diesel generator failures from (b) and (c) above, the total number of diesel generator startups from (a), the point estimate of the failure-to-start probability per diesel unit, become: $\lambda_D = 1.7 \times 10^{-2}$ and $\beta_D = 0.14$.

The probability of at least two of the three diesel generator units failing to start on demand is assessed using the formula derived in Volume IV of the AIPA Study[171]

$$P(> 2DG) \simeq 3[(1 - \beta)\lambda]^2 - 2[(1 - \beta)\lambda]^3 + \beta\lambda \qquad \text{(A.9.3)}$$

Using above stated values in equation A.9.3 the case of no credit for time dependence gives

$$P\left\{\geqslant 2DG, \begin{array}{c}\text{fast}\\ \text{autostart}\end{array}\right\} = 3.2 \times 10^{-3} \qquad \text{(A.9.4)}$$

which is a factor of 10 greater than that assessed in Phase I of the AIPA Study.

The failure summaries in the report referred to earlier were examined to determine whether the fault was actually corrected and the diesel successfully restarted within 20 min. In many cases, such as the failures that occurred during surveillance testing, no attempt was made to quickly correct the fault since there was no incentive, i.e. not an emergency situation. In these cases a judgment was made as to whether or not the fault could be reasonably expected to be corrected within 20 min during an actual demand situation. Marginal cases were assigned a fifty-fifty chance of successful restart within 20 min. Based on this examination, the following results were obtained:

(a) Of the forty-three occurrences of independent diesel generator failures, twenty-four can be expected to be capable of repair within 20 min with a high probability, and eight were found to be marginal. Therefore, $28 = 24 + 8(0.5)$ was assumed to be the expected number of cases of independent failures that would be corrected within 20 min in an actual emergency situation.

(b) Of the three occurrences of common cause failures to start, two were actually corrected well within 20 min. Since one event that could not be repaired quickly affected three units, a total of four of seven diesel generator failures due to common cause were repaired within 20 min.

Eliminating the failures corrected within 20 min, the revised point estimates of λ_d and β_d were obtained as:

$$\lambda_d = \frac{50 - (28 + 4)}{2940} = 6.2 \times 10^{-3}$$

and

$$\beta = \frac{7 - 4}{50 - (28 + 4)} = 0.17$$

Using these values in equation A.9.3, the updated assessment for failure to start two or more diesels within 20 min becomes:

$$P[\geqslant 2DG \text{ in } 20 \text{ min}] = 1.1 \times 10^{-3} \qquad (A.9.5)$$

which is a factor of about 7 greater than the original AIPA Study assessment and a factor of about 2.5 less than the value for fast autostart.

Therefore, it was found that the experience data on actual diesel generator demand situations in nuclear power plants support the view that credit should be given for time dependence in the assessment of diesel generator systems. A review of experience data reveals that there have been six incidents (four British and two US) of loss of off-site power followed by common-cause failure of the on-site emergency power system (i.e. six occurrences of loss of all AC power). Summaries of these events have been included as Table 7.1 in Joksimovich *et al.*[172] All of these events were repaired or recovered within 50 min and four out of six were repaired within 5–10 min. This also corroborates a factor of 3 credit for 20-min time delay type of situations.

A.9.3.3 Results

The uncertainty distribution have been evaluated via Monte Carlo simulation (Cairns and Fleming[176]) employing equation A.9.1 and the input distributions

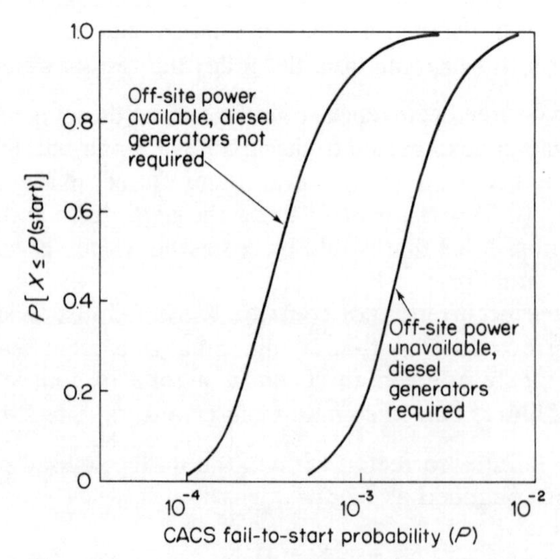

Fig. A.9.4 Uncertainty distribution for CACS fail-to-start probability

quoted in Table A.9.1 for two cases. One case applies to the loss of off-site power (LOSP) sequences since the CACS depends on successful operation of diesel generators for these sequences. The second case applies to the loss of main-loop cooling (LMLC) for which off-site power is assumed available, and hence, diesel generators are not required for successful CACS startup. The results, plotted in Fig. A.9.4, for the medians, upper and lower bounds are:

Case 1. *Off-site power available*

$$P_{\mathrm{START}} = \begin{cases} 1 \times 10^{-3}, \text{ upper bound}, \\ 3 \times 10^{-4}, \text{ median}, \\ 1 \times 10^{-4}, \text{ lower bound}. \end{cases} \tag{A.9.6}$$

Case 2. *Off-site power unavailable*

$$P_{\mathrm{START}} = \begin{cases} 5 \times 10^{-3}, \text{ upper bound}, \\ 2 \times 10^{-3}, \text{ median}, \\ 7 \times 10^{-4}, \text{ lower bound}, \end{cases} \tag{A.9.7}$$

A.9.4 Operation Until Main Loops are Restored—Reliability Prediction

The probability assessments that CACS operates until MLCS is restored assumed two separate plant states which in the study have been modelled by two distinct event trees named LOSP and LMLP respectively. Each of the two event trees used different methods to model the repair distribution.

In the LOSP event tree the repair distribution was treated explicitly, whereas the repair distribution for LMLC was treated at the fault-tree level. Since these different approaches result in different probability models, these different models are presented separately here, beginning with the model for LMLC. Complexities added to the model have been warranted since the results could be affected by an order of magnitude or even more.

A.9.4.1 Methodology

Given that the CACS starts, it need only operate until such time as the MLCS is restored since failures that occur following MLCS restoration would not lead to loss of forced circulation, the condition that is of interest in safety evaluations. The actual time of MLCS restoration, on the other hand, is not known deterministically and for this reason is treated as a random variable uniquely characterized by an assumed probability distribution. The probability of CACS failure before MLCS restoration is given by:

$$P_{\mathrm{RUN}} = \int_0^L \mathrm{pdf}_c\{x\} F_m\{x\}\, \mathrm{d}x \tag{A.9.8}$$

where $\mathrm{pdf}_c\{x\} \equiv$ probability density for time of CACS system failure,
$F_m\{x\} \equiv$ probability MLCS is not restored by time x,

$L \equiv$ maximum time the CACS must operate to prevent eventual radioactivity release from the reactor core (~1000 hr)

It is convenient to evaluate P during different time intervals following the initiating event:

$$P = \int_0^{t_1} \text{pdf}_c x F_m x \, dx + \int_{t_1}^{t_2} \text{pdf}_c\{x\} F_m\{x\} \, dx + \ldots + \int_{t_{n-1}}^{L} \text{pdf}_c\{x\} F_m\{x\} \, dx \quad (A.9.9)$$

If it is assumed that repair of the MLCS occurs at a constant repair rate, $F_m x$ can be expressed in the simple exponential form

$$F_m x = \exp(-\mu_m x) \quad (A.9.10)$$

where $M_\mu \equiv$ repair rate of the MLCS, inverse of the mean repair time.

It is convenient to derive $\text{pdf}_c\{x\}$ in terms of the expressions for system reliability $R_c\{x\}$, which were developed in Joksimovich *et al.*[171] using the identity:

$$\text{pdf}_c\{x\} - \frac{d}{dx} [R_c x] \quad (A.9.11)$$

During the first several hours of operation ($t_1 \approx 10$ hr) following reactor trip from full power, two of the three loops must operate for system success and thereafter only one is required. Using the expressions developed by Joksimovich *et al.*[171] for the reliability of non-repairable redundant systems subject to independent and common cause failures, we have

$$R_c\{x\} = \begin{cases} 3e^{-(2-\beta)\lambda x} & 0 \leqslant x \;\; 10 \text{ hr}, \\ 3e^{-\lambda x} - 3e^{-(2-\beta)-(3-2\beta)\lambda x} & 10 \leqslant x \;\; 1000 \text{ hr}, \; x \geqslant 1000 \text{ hr} \\ 1 & \end{cases} \quad (A.9.12)$$

where λ and β are the failure rate and common cause factor for each CACS cooling loop, respectively. The density function $\text{pdf}_c\{x\}$ can now be readily obtained with use of equation A.9.11 and constituents in equation A.9.9 derived as demonstrated in Joksimovich *et al.*[172]

A.9.4.2 Loop failure rate

Quantification can now proceed by quantifying the independent variables λ, β, and M_μ. However, since all three of these variables are reliability parameters for rather complex systems as opposed to relatively simple components, it is necessary to develop additional probability models to facilitate quantification. The term λ is the failure rate for an entire CACS cooling loop, major components of which are shown in Figs. A.9.2 and A.9.3, and β is the corresponding common-cause failure fraction. A probability

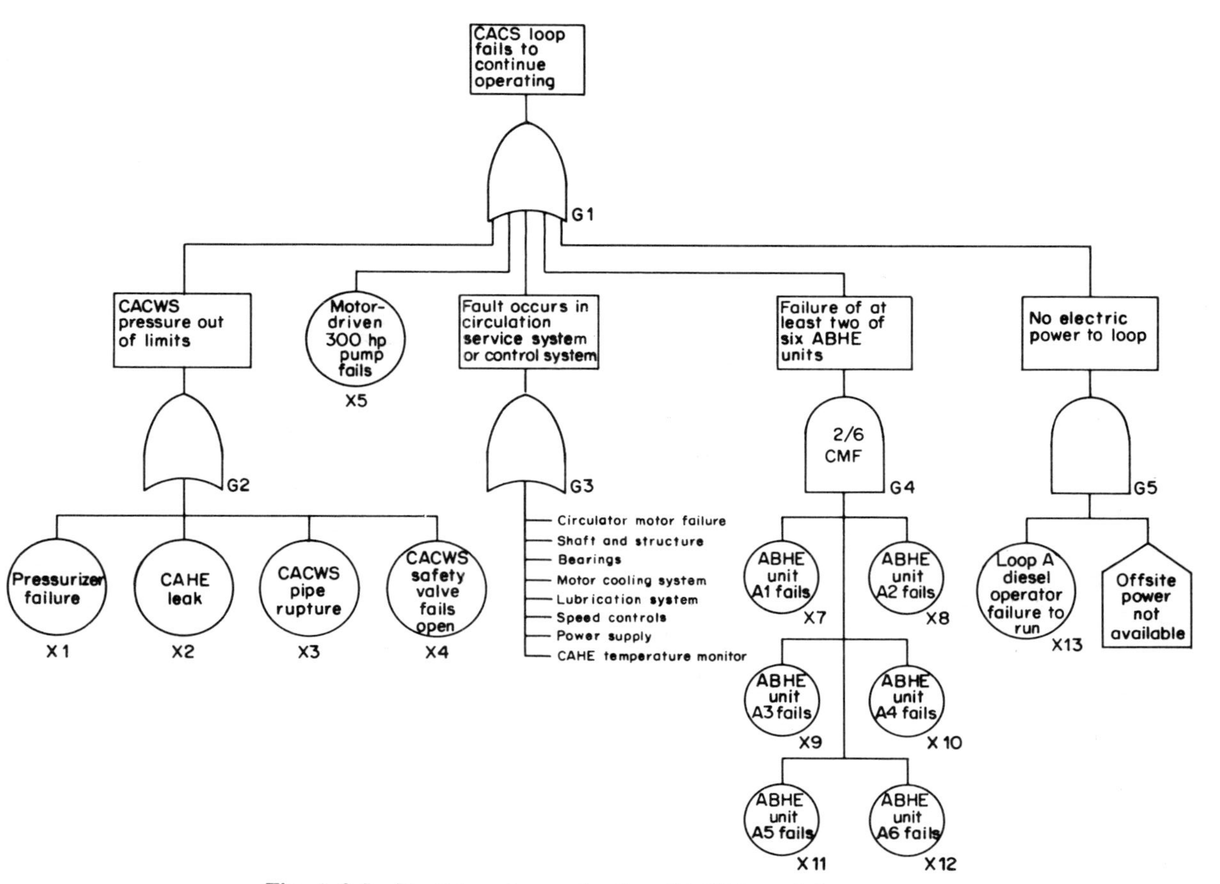

Fig. A.9.5 Fault tree for evaluating CACS loop failure rate

model for λ can be developed using the CACS cooling-loop fault tree presented in Fig. A.9.5. The CACS loop failure-rate is given by

$$\lambda = \lambda_{X1} + \lambda_{X2} + \lambda_{X3} + \lambda_{X4} + \lambda_{X5} + \lambda_{G3} + \lambda_{G4} + \lambda_{X13} \qquad \text{(A.9.13)}$$

Event G4 requires multiple failures (at least two of six) of repairable air-blast heat-exchanger units to result in CACS loop failure.

This failure-rate can be approximated as

$$\lambda_{G4} \simeq \binom{6}{2}[(1 - \beta_{X7})\lambda_{X7}]^2\tau_{X7} + \beta_{X7}\tau_{X7} \qquad \text{(A.9.14)}$$

Combining equations A.9.13 and A.9.14:

$$\lambda = \sum_{j=1}^{5} \lambda_{xj} + \lambda_{G3} + \binom{6}{2}[(1 - \beta_{X7})\lambda_{X7}]^2\tau_{X7} + \beta_{X7}\lambda_{X7} + \lambda_{X7} + \lambda_{X13} \qquad \text{(A.9.15)}$$

The input data used to quantify λ are presented in Table A.9.2. Note that X13 only applies to sequences in the LOSP event tree and not to the LMLC tree. Operating experience with electrically driven gas circulators in Magnox reactors has been used to quantify λ_{G3} (Cave and Gow[179]). The results in the form of uncertainty distributions are plotted in Fig. A.9.6 for the cases of off-site power availability and off-site power unavailability and are summarized as follows:

Case 1. *Off-site power available (LMLC event tree)*

$$\lambda = \begin{cases} 1.7 \times 10^{-3}/\text{hr, upper bound,} \\ 2.8 \times 10^{-4}/\text{hr, median,} \\ 8.7 \times 10^{-5}/\text{hr, lower bound.} \end{cases} \qquad \text{(A.9.16)}$$

Case 2. *Off-site power unavailable (LOSP event tree)*

$$\lambda = \begin{cases} 4.8 \times 10^{-3}/\text{hr, upper bound,} \\ 7.2 \times 10^{-4}/\text{hr, median,} \\ 2.0 \times 10^{-4}/\text{hr, lower bound.} \end{cases} \qquad \text{(A.9.17)}$$

A.9.4.3 Loop common cause factor

A model to evaluate β, the equivalent common cause factor for a CACS cooling loop, can be generated by considering that the common-cause failure-rate of the CACS system consists of contributions from the common-cause failure-rates of each CACS component. Hence,

$$\beta = \frac{1}{\lambda}\left[\sum_{j=1}^{5} \beta_{Xj}\lambda_{Xj} + \beta_{G3}\lambda_{G3} + \beta_{X7}\left\{\binom{6}{2}[(1 - \beta_{X7})\lambda_{X7}]^2\tau + \beta_{X7}\lambda_{X7} + \beta_{X13}\lambda_{X13}\right]\right. \qquad \text{(A.9.18)}$$

In other words, β is estimated as the average of the individual β_{Xj} weighted

Table A.9.2 Reliability data for quantification of CACS operation

Fault tree symbol	Description	Failure rate λ*	Common cause factor β*	Mean repair time τ (hr)	References
X1	CACWS pressurizer rupture	3×10^{-8}/hr (10)	2×10^{-1} (3)	168	Melvin and Maxwell[177]
X2	CAHE leak	3×10^{-5}/hr (10)	2×10^{-3} (4)	100	Hannaman[178]
X3	CACWS pipe rupture	3×10^{-7}/hr (100)	2×10^{-2} (10)	30	Hannaman[178]
X4	CACWS safety valve opens prematurely	1×10^{-5}/hr (3)	2.3×10^{-1} (3)	24	WASH-1400[174]
X5	CACWS motor-driven pump fails	3×10^{-5}/hr (10)	2×10^{-2} (10)	24	WASH-1400[174]
G3	Fault in circulator, motor, service, or power supply	3.7×10^{-5}/hr (5)	8×10^{-3} (5)	193	λ based on 85 failures in 264 circulator-years, β based on 0 CMF in 85 failures; these values and that for τ taken from WASH-1400[174]
G4	Air-blast heat-exchanger fans	1.6×10^{-4}/hr (10)	2×10^{-1} (3) within loop 2×10^{-2} (10) among loops	24	Hannaman[178]
X13	Diesel generator	3×10^{-4}/hr (10)	1.3×10^{-1} (3)	21	Cave and Gow[179]

*Values in parentheses are lognormally distributed uncertainty factors.

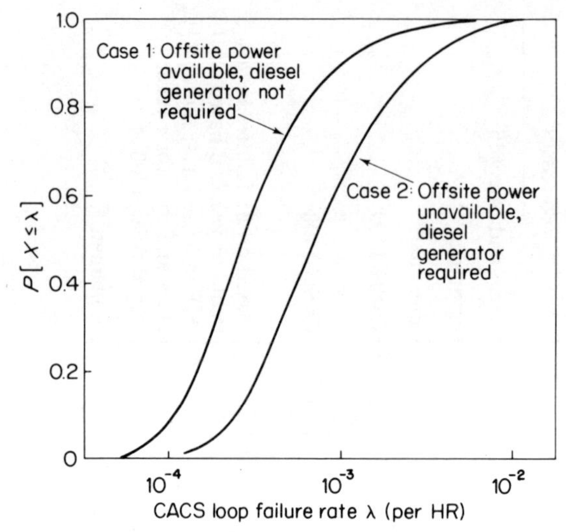

Fig. A.9.6 Uncertainty distribution results for CACS loop failure rate, λ

by the fraction of the total loop failure-rate attributable to component j (i.e. weighted by $\lambda j/\lambda$). Note that the two common-cause factors are specified for X7, one of common cause failures within a cooling loop, β_{X7}, and one for those among loops, β'_{X7}.

The results in the form of uncertainty distributions are displayed in Fig.

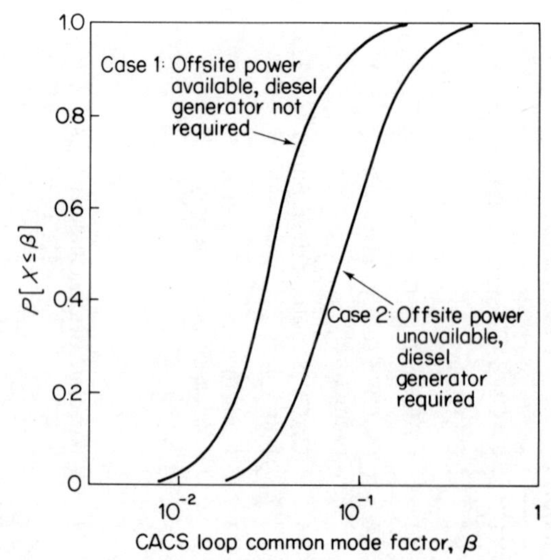

Fig. A.9.7 Uncertainty distribution results for CACS common-mode factor

A.9.7 and summarized as follows:

Case 1. *Off-site power available (LMLC event tree)*

$$\beta = \begin{cases} 9.5 \times 10^{-2}, \text{ upper bound,} \\ 3.3 \times 10^{-2}, \text{ median} \\ 1.4 \times 10^{-2}, \text{ lower bound.} \end{cases} \tag{A.9.19}$$

Case 2. *Off-site power unavailable (LOSP event tree)*

$$\beta = \begin{cases} 2.5 \times 10^{-1}, \text{ upper bound,} \\ 8.0 \times 10^{-2}, \text{ median,} \\ 2.8 \times 10^{-2}, \text{ lower bound.} \end{cases} \tag{A.9.20}$$

A.9.4.3 MLCS repair rate

The remaining variable to be quantified is μ_m, the repair rate of the MLCS. The probability model for μ_m is obtained by noting that the total unavailability of the MLCS can be approximated as the sum of the unavailabilities of the nominal cut sets of the MLCS fault tree* (Joksimovich *et al.*[172], Fig. 4.4).

$$\lambda_m \tau_m = \sum_{j=1}^{N} \lambda_j \tau_j \tag{A.9.21}$$

where λ_m = MLCS failure rate
$\tau_m = 1/\mu_m$
λ_j = failure rate of minimal cut set j of MLCS fault tree
τ_j = mean time to repair MLCS for jth failure mode.

* In view of complexity this fault tree is not reproduced.

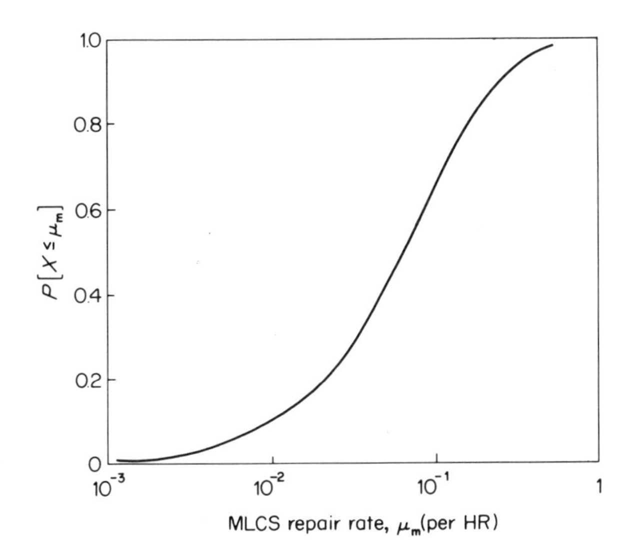

Fig. A.9.8 Uncertainty distribution results for MLCS repair rate

Table A.9.3 LMLC event tree: probability of CACS failure before restoration of MLCS

Time interval of CACS failure since initiating event (hr)	Upper bound (95th percentile)	Median (50th percentile)	Lower bound (5th percentile)
100 – 1000	2×10^{-3}	4×10^{-7}	$< 10^{-9}$
10–100	2×10^{-3}	7×10^{-5}	2×10^{-7}
0–10	8×10^{-4}	8×10^{-5}	7×10^{-6}

The uncertainty distribution evaluated is displayed in Fig. A.9.8 and summarized as follows:

$$\mu_m = \begin{cases} 3 \times 10^{-1}/\text{hr, upper bound,} \\ 6 \times 10^{-2}/\text{hr, median,} \\ 5 \times 10^{-3}/\text{hr, lower bound.} \end{cases} \tag{A.9.22}$$

A.9.4.4 Results—LMLC event tree

Now that uncertainty distributions have been evaluated for λ, β and μ_m, sufficient information is available to complete the quantification of P for the LMLC event tree. The final results are summarized in Table A.9.3.

A.9.4.5 Methodology and results—LOSP event tree

In the LOSP event tree, the initiating event-repair distribution is presented explicitly by the assignment of probabilities for restoration in different time intervals following the initiating event. Thus, for each sequence in the event tree a finite time interval for off-site power restoration is specified, although the exact time of restoration within each interval is uncertain and, in fact, has a uniform (rectangular) time to repair distribution. That the unconditional probability of failure of the CACS is an arbitrary time interval (t_1, t_2) is given by

$$P_{\text{RUN}} = R_c(t_1) - R_c(t_2) \tag{A.9.23}$$

where $R_c(t_1)$ is the probability that no CACS failure occurs by t_1.

If the time of MLCS restoration is known to be uniformly distributed in the time interval (t_1, t_2) and, hence, the mission time of the CACS is also uniformly distributed in (t_1, t_2), it can be shown that the probability of CACS failure in the time interval (t_1, t_2) before the MLCS is restored is given by:

$$P_{\text{RUN}} = R_c(t_1) - R_c(t_1 + x) \tag{A.9.24}$$

where x is uniformly distributed in the interval $(0, t_2 - t_1)$. The results are summarized in Table A.9.4.

Table A.9.4 LOSP event tree: Probability of CACS failure before restoration of MLCS

Time interval of LOSP restoration (hr)	Time interval of CACS failure (hr)	Point estimate	Upper bound (95th percentile)	Median (50th percentile)	Lower bound (5th percentile)
0–3	0–3	8.9×10^{-5}	8.2×10^{-4}	7.7×10^{-5}	4.5×10^{-6}
3–30	0–3	1.8×10^{-4}	1.4×10^{-3}	1.8×10^{-4}	2.5×10^{-5}
3–30	3–30	7.8×10^{-4}	6.6×10^{-3}	6.6×10^{-4}	4.0×10^{-5}
30–150	3–30	1.8×10^{-4}	1.4×10^{-3}	1.8×10^{-4}	2.5×10^{-5}
30–150	3–30	1.6×10^{-3}	1.1×10^{-2}	1.5×10^{-3}	2.1×10^{-4}
30–150	30–150	3.6×10^{-3}	4.1×10^{-2}	4.3×10^{-3}	1.6×10^{-4}

A.9.5 Comparison of Reliability Data with some Other Sources

The failure-rate data employed in the AIPA Study pertinent to reliability predictions of the CACS were compared against some other sources, i.e. IREP (Interim Reliability Evaluation Program), LER (Licensee Event Reports) and NPRDS (Nuclear Plant Reliability Data System) (Sullivan and

Table A.9.5 Comparison of reliability data

Components	Units	AIPA	IREP	LER	NPRDS
Unique to HTGR					
Circulator shutoff valve	Demand	3×10^{-4}	NA	NA	NA
Pipe					
Core auxiliary cooling water system	Reactor/Year	3×10^{-3}	8.5×10^{-7}	—	—
Heat exchangers					
CAHE	Hour	3×10^{-5}	—	—	8.5×10^{-6}
Valves					
Core auxiliary cooling-water system	Hour	1×10^{-5}	—	3×10^{-6}	—
Diesel generators					
Fail to start	Demand	6×10^{-3}	—	2.4×10^{-2}	1.5×10^{-3}
Fail to run	Hour	3×10^{-4}	—	2.6×10^{-2}	5×10^{-3}
Pumps					
Core auxiliary cooling-water system motor driven	Hour	3×10^{-5}	3×10^{-5}	1.8×10^{-6}	4.4×10^{-6}
Core auxiliary heat-exchanger cooling water	Demand	3×10^{-4}	3×10^{-3}	6×10^{-4}	4×10^{-4}
Motor cooling water	Demand	3×10^{-4}	3×10^{-3}	6×10^{-4}	4×10^{-4}
Electric motor					
Auxiliary circulator motors and controls	Demand	3×10^{-4}	—	1.6×10^{-3}	2×10^{-4}
Air-blast heat-exchanger fan	Hour	1.6×10^{-4}	—	—	4×10^{-6}

Poloski[180], Poloski and Sullivan[181], Hubble and Miller[182], Miller and Hubble[183], Murphy[184], and Monnie[185]. This comparison is summarized in Table A.9.5. The general conclusion can be derived that the AIPA study failure rates were equal or conservative in all cases except for the diesel generator. A detailed discussion related to rationale behind the diesel generator data is offered in section A.9.3.2.

A.9.6 Acknowledgement

Invaluable contributions of K. N. Fleming and G. W. Hannaman in the AIPA Study made this work possible.

References for Main Text

1. Heinrich, H. W. (1959). *Industrial Accident Prevention* (4th Edition). McGraw-Hill, New York.
2. The Health and Safety at Work Act, 1974, HMSO, London, UK.
3. The Occupational Safety and Health at Work Act of 1970, USA, *National Safety News*, **103**(2). 36–37, February 1971.
4. A Programme of Nuclear Power, Cmnd 9389, HMSO, London, UK, February 1955.
5. Haddon, W., Jr, Suchman, A. E., and Klein, D. (1964). *Accident Research: Methods and Approaches*. Harper and Row, New York, p. 28.
6. An Investigation of Potential Hazards from Operations in the Canvey Island/Thurrock Area. (1978) HMSO, London, UK.
7. Slater, D. H. (1978). Vapour Clouds. *Chemistry and Industry*, no. 9, 295–302.
8. Thom, R. (1975). *Structural reliability and morphogenesis—An outline of a general theory of models*. Institute des Haute Etudes Scientifiques, Bures-sur-Yvette. (English translation by D. H. Fowler and C. H. Waddington, published by Benjamin, Advanced Book Program, Massachusetts, USA, 1975.)
9. Thompson, J. M. T. (1975). Experiments in catastrophe, *Nature* **254**.
10. Thompson, J. M. T. and Hunt, G. W. (1974). *Dangers of Structural Optimization*. Vol. 1. *Engineering Optimization*, Gordon and Breach Science Publishers, UK, pp. 99–110.
11. Chilver, H. (1975). Wider implications of catastrophic theory, *Nature* **254**.
12. McCutcheon, J. E. (1963). *The Hartley Colliery Disaster, 1862*. G. McCutcheon, 2 Ambleside Avenue, Seaham, Co. Durham.
13. Clanzy, T. K. and Luxmore, C. (1975). Mechanical and electrical engineering safety requirements for the transport of men and materials in shafts and under ground. *Symposium on the Transport of Men and Materials Underground, organised by The Association of Mining, Electrical and Mechanical Engineers, Harrogate*, UK 28–30 October 1975.
14. Moore, R. V. (ed.) (1971). *Nuclear Power*. Cambridge University Press, pp. 152, 153, and 157.

15. Moore, R. V. (1961). Introductory paper, *Proceedings of the Symposium on the Dounreay Fast Reactor*, Institute of Mechanical Engineers.
16. Marsham, T. N. (1968). *The Place of the Fast Reactor in Commercial Development—British Nuclear Power Plant*. British Nuclear Export Executive, British Engineering Exhibition, Copenhagen.
17. *The Second Nuclear Power Programme*. (1964). Cmnd 2335, HMSO, London, UK.
18. Silverleaf, D. J. and Weeks, R. J. (1968). *The Performance of the Central Electricity Generating Board Nuclear Power Stations—British Nuclear Power Plant.* Nuclear Export Executive, British Engineering Exhibition, Copenhagen.
19. Farmer, F. R. (1967). Siting criteria—A new approach. SM-89/34, *IAEA Symposium on the Containment and Siting of Nuclear Power Reactors*, Vienna.
20. Siddall, E. (1963). Reliability of reactor control systems, *Nuclear Safety*, **4**(4).
21. *Reactor Safety Study, An Assessment of Accidental Risks in US Commercial Nuclear Power Plant*. WASH-1400 (NUREG 75/014) October 1975.
22. Minutes of the Risk Assessment Review Group Meeting, Washington DC, 5–6 October 1977, US Nuclear Regulatory Commission.
23. Daffer, R. (1975). Flixborough shows need for a rethink, *Financial Times*, 27 June.
24. *Royal Commission on Environmental Pollution*, (Chairman Sir Brian Flowers) *Sixth Report on Nuclear Power and the Environment*. HMSO, Cmnd 6618, London, UK, September 1976.
25. Stewart, R. M. and Hensley, G. (1971). *High Integrity Systems on Hazardous Chemical Plants*. CREST Meeting on the Applicability of Quantitative Reliability Analysis of Complex Systems and Nuclear Plants in its relation to Safety. Munich, 26–28 May.
26. Lowrance, W. W. (1976). *Of Acceptable Risk—Science and the Determination of Safety*. W. Kaufman, Los Altos, California.
27. Slovic, P., Fischoff, B., and Lichenstein, S. *Cognitive Processes and Societal Risk Taking*, Oregon Research Institute, Eugene, Oregon, USA.
28. Otway, H. J., Panner, P. D., and Linnerooth, J. (1975). *Social Values in Risk Acceptance*. International Institute for Applied Systems Analysis, IIASA, RM-75-5, November.
29. Otway, H. J. and Cohen, J. J. (1975). *Revealed Preferences. Comments on the STARR Benefit—Risk Relationships*. International Institute for Applied Systems Analysis. IIASA, RM-75-5 March.
30. Singh-Wadura, K. (1978). *Risk and Rare Events in Perspective—A Literature Review*. GRS-A-135.
31. Treversky, A. and Kohnman, D. (1974). Judgement under uncertainty. Heuristics and biases, *Science*, **185**, 1124–31.

32. Green, C. H. and Brown, R. *Life Safety. What is it worth and how much is it worth?* CP 52/78, Building Research Establishment, Garston, Herts., UK.
33. Bowen, J. H. (1976). Individual risk versus public risk criteria. UKAEA SRD, Warrington, England. *Chemical Engineering Process*.
34. BØE, C., Dr-Ing. *Some Views on Risk and Risk Acceptance*. The Royal Norwegian Council for Scientific and Industrial Research, Oslo, Norway.
35. Firth, A. (1967). Project management, *Symposium of Engineering Aspects of the Winfrith SGHWR*, Institute of Mechanical Engineers.
36. Hill, A. (1973), Reliability and maintainability activities utility application, *Canadian Nuclear Association Symposium on Reliability and Maintainability*, Sheridan Park, Ontario, 19 September 1973.
37. Lakner, A. A. and Anderson, R. T. (1978). Analytical approach to determining optimum reliability and maintainability requirements, *The Radio and Electronic Engineer*, **48**, 7/8.
38. Green, A. E. and Bourne, A. J. (1976) *Safety Assessment with Reference to Automatic Protective Systems for Nuclear Reactors*. UKAEA Report AHSB(S) R117, Parts I, II, and III (reprinted).
39. Joyce, G. H. (1950). *Principles of Logic*. Longman, London, UK.
40. Neumann, J. von and Morgenstern, O. (1947). *Theory and Games and Economic Behaviour*. 2nd Edn, Princetown University Press, Princetown, NJ.
41. Laplace, P. S. (1814). *Essai Philosophique sur les Probabilities*, Paris. (Translation New York: Dover 1951.)
42. Savage, L. J. (1954). *The Foundation of Statistics*. John Wiley, New York.
43. Kaufman, G. and Thomas, H. *Modern Decision Analysis. Edited Selected Readings*. Penquin Books, Harmondsworth.
44. Beattie, J. R. and Bell, G. D. (1975). A possible standard risk for large accidental releases. *Symposium on Principles and Standards of Reactor Safety*, IAEA, Jülich, Germany, 5–9 February.
45. Penland, J. R. (1975). A formation for risk assessment and allocation. *Proceedings of Annual Reliability and Maintainability Symposium*, Washington DC, USA, January.
46. Goode, H. H. and Machol, R. E. (1957). *System Engineering*, McGraw-Hill, New York, p. 306.
47. Chestnut, H. and Mayer, R. W. (1951–55). *Servomechanisms and Regulating System Design*. Vol I, John Wiley, New York.
48. Holbrook, J. G. (1959). *Laplace Transforms for Electronic Engineers*. Pergamon Press, London, UK.
49. Green, A. E. and Bourne, A. J. (1972). *Reliability Technology*, John Wiley, London UK.
50. Eames, A. R. (1971). *Principles of Reliability for Nuclear Reactor Control and Instrumentation Systems*. UKAEA Report, SRD R1.

51. Deverell, C. S. *Techniques of Communication in Business*. Gee and Company, London, UK.
52. Ablitt, J. F. (1960). *Culcheth*, Unpublished work.
53. Green, A. E. (1972). Quantitative assessments of system reliability. *4th Symposium on Chemical Process Hazards with special reference to Plant Design*. Institute of Chemical Engineers Symposium Series No. 33, London, UK.
54. Topping, J. (1962). *Errors of Observation and Their Treatment*. (3rd edn.), Chapman and Hall for the Institute of Physics and The Physical Society.
55. Tables of Normal Probability Functions. National Bureau of Standards, Applied Mathematics Series No. 23, US Government Printing Office, Washington, DC, 1953.
56. Keller, A. Z. Private Communication, 15 March 1979.
57. Kendal, M. G. and Stuart, A. (1958 and 1961). *The Advanced Theory of Statistics*. Vols 1 and 2. Charles Griffin.
58. Bourne, A. J. (1967). A criterion for the reliability assessment of protective systems. *Control*, **11**, 112, pp. 495–8, October.
59. Gangloff, W. C. (1977). 'What If's' on Nuclear Plants. *IEEE Spectrum*, June.
60. Barlow, R. E. and Fussell, J. B. (eds) (1974). *Proceedings of Conference on Reliability and Fault Tree Analysis* held at the University of California, September 3–7. Society for Industrial and Applied Mathematics: 33 South 17 St. (1975). Philadelphia, Penn, 19103, USA.
61. Kontoleon, J. M., Lynn, J. W., and Green, A. E. (1975). Diakoptical Reliability Analysis of Transistorised Systems. *Symposium on Reliability of Nuclear Power Plants*. Innsbruck, Austria, April 1975.
62. Kontoleon, J. M., Lynn, J. W., and Green, A. E. (1975). *Application of Diakoptical Analysis in System Reliability Assessment*. NCSR Report R6. UKAEA.
63. Decomposition as a tool for Solving Large-Scale Problems. NATO Advanced Study Institute, July 17–26 1972, Pembroke College, Cambridge, England. Proc. Published by North Holland Publishing Company.
64. Balasubrannanian, N. V., Lynn, J. W., and Sen Gupta, D. P. (1970). *Differential Forms of Electromagnetic Networks*. Butterworth, London, UK.
65. Kron, G. (1963). *Diakoptics—The Piecewise Solution of Large-Scale Systems*. MacDonald, London, UK.
66. Lynn, J. W. (1963). *Tensors in Electrical Engineering*. Edward Arnold, London, UK.
67. Green, A. E. (1968). Assessment of sensing channels for high integrity protective systems, *Instrument Practice*, **22**, 2.
68. Eames, A. R. (1966). Reliability assessment of protective systems, *Nuclear Engineering*, II, 118.

69. Green, A. E. (1969–70). Reliability prediction, *Proceedings of Institution of Mechanical Engineering*, **184**, Part 3B.
70. Safety Assessment Principles for Nuclear Power Reactors. Report para 124, p. 14, 1982. HM Nuclear Installations Inspectorate, Health and Safety Executive, UK.
71. Kletz, T. A. (1971). *Hazard Analysis—A Quantitative Approach to Safety*. Institute of Chemical Engineers Symposium Series, pp. 75–81, Institute Chemical Engineers, London, UK.
72. Lawley, H. G. (1974). Operability studies in hazard analysis, *Chemical Engineering Process*, **70**, 4.
73. Adkins, L. A. *et al*. Development of all Weather Landing System Reliability Analysis and Criteria for Category III Airborne Systems, Phase III. Report FAA-RD-67-20 and Report AD-655-240 Lockheed Georgia Co., Marietta, Georgia, USA.
74. Mackworth, Sir David. *Is Man Necessary*. Unreferenced document.
75. Embrey, D. W. (1978). *Ergonomics and Human Reliability Considerations in Reactor Safety and Reliability*. UKAEA Reactor Safety Course, Harwell, Lecture No. 24, September 1978.
76. Singleton, W. G., Easterby, R. S., and Whitfield, D. (1967). *Proceedings of Conference on The Human Operator in Complex Systems*. Taylor and Francis, London, UK.
77. McCormick, E. J. (1970). *Human Factors Engineering*. (3rd edn), McGraw-Hill, New York, USA.
78. De Greene, K. B. (ed.) (1970). *Systems Psychology*. McGraw-Hill, New York, USA.
79. Meister, D. (1971). *Human Factors: Theory and Practice*. John Wiley, New York, USA.
80. Sheridan, T. B. (1962). *Human Operator in Control Instrumentation*. Vol. 1. *Progress in Control Engineering*, Edited by MacMillan, R. H., Higgins, T. J. and Naslin, P. Academic Press, New York, USA.
81. Morgan, C. T., Cook, J. S., Chapinis, A. and Lund, M. W. (Editors) (1963). Human Engineering Guide to Equipment Design, pp. 223–237, Human Dynamics. McGraw-Hill, New York, USA.
82. Green, A. E., Marshall, J., and Murphy, T. (1968). Preliminary Investigations into the Time Response of Operators, UKAEA Internal Document.
83. Maynard, H. B., Stegemarten, G. T., and Schwab, J. L. (1948). *Method-time Measurement*. McGraw-Hill, New York, USA.
84. Ablitt, J. F. (1969). *A Quantitative Approach to the Evaluation of the Safety Function of Operators on Nuclear Reactors*, UKAEA Report AHSB(S) R160.
85. Green, A. E. (1970). Safety assessment of automatic and manual protective systems for reactors, *Instrument Practice*, February.
86. Rasmussen, J. and Timmermann, P. Measurement of Reponse Times of Human Operators in Control Rooms, Report published by Research Establishment, Danish Atomic Energy Commission, Risø, Denmark.

87. Embrey, D. E. *Human Reliability in Complex Systems—An Overview*. Report NCSR R10, The National Centre of Systems Reliability, UKAEA.
88. *An Index of Electronic Equipment Operability*; in five related documents, published by American Institute for Research, AD697161, Pittsburgh, Pennsylvania, USA, 1962.
89. Swain, A. D. *A Method for Performing, a Human Factors Reliability Analysis*. Monograph SCR-685, Sandia Corporation, Albuquerque, USA.
90. Swain, A. D. *Human Reliability Assessment in Nuclear Plants*. Monograph; SCR-69-1346, Sandia Corporation, Albuquerque, USA.
91. Regulinski, T. L. (Guest Editor) (1973). *Special issue on human performance reliability. IEEE Transactions on reliability,* **R-22**, 3, August.
92. Report by the CSNI Group of Experts on Human Error Analysis on Quantification, Task Force on Problems of Rare Events in the Reliability Analysis on Nuclear Power Plants, Report No. SINDOC(78)74. OECD, 1978.
93. Hunns, D. M. (1980). *How Much Automation*. UKAEA/NCSR Report 25.
94. Fichiner, N., Becker, K., and Bashir, M. (1977). *Cataloque and Classification of Technical Safety Standards Rules and Regulations for Nuclear Power Reactors and Nuclear Fuel Cycle Facilities.* Report EVR 5849 EN, Commission of the European Communities, Directorate—General, Scientific and Technical Information Management Batiment Jean Monnet, Luxembourg.
95. *Safety Codes and Guides, Edition 4*/75, Compilation of information on the reactor protection required for examination purposes in the licensing procedures of nuclear power plant. Issued by the Institut für Reaktorsicherheit der Technischen, Uberwachungsvereinee V. D-5 Koln, 1, Glockengasse, 2.
96. *Safety Assessment Criteria for Nuclear Power Reactors and Associated Plant*, Issue No 2, April 1977. HM Nuclear Installations Inspectorate, Health and Safety Executive, London, UK.
97. Aitken, K. (1977). Assessment of high integrity protective system for loss of electric power, *Nuclear Safety*, **18**, 2, March.
98. Green, A. E. (1977). *Reliability Assessment of Nuclear Systems with Reference to Safety and Availability*. Presented at the Conference on Nuclear Systems Reliability Engineering and Risk Assessment, June 20–24 1977, Gatlinburg, Tennessee, sponsored by The University of Tennessee, EPRI and ERDA, USA.
99. Dhillon, B. S. and Singh, C. (1981). *Engineering Reliability: New Techniques and Applications*. John Wiley, New York, USA.
100. Green, A. E. (1974). *The Reliability Assessment of Industrial Plant Systems*. Symposium on Major Loss Prevention, the Netherlands, May.
101. Epler, E. P. (1969). Common-mode failure considerations in the design of systems for protection and control, *Nuclear Safety*, **10**(1), 38–45.

102. Watson, I. A. and Edwards, G. T. (1979). *A Study of Common Mode Failures*. UKAEA Report SRD 146.
103. CSNI Task Force on Rare Events. Research Sub-Group on Common-Mode Failures. Report SINDOC(78)41, OECD, 1978.
104. CSNI Task Force on Rare Events. Group of Experts on Human Error Analysis and Quantification. Report SINDOC(78)84. OECD, 1978.
105. CSNI Task Force on Rare Events. Communication Techniques Working Group. Report SINDOC(78)86. OECD, 1978.
106. MacFarlane, A. G. J. (1970). *Dynamical System Models*. George G. Harrap, London, UK.
107. Keller, A. Z. (ed.) (1975). *Uncertainty in Risk and Reliability. Appraisal in Management*. Adam Hilger, London.
108. Lüsser, R. (1956). *The Notorious Unreliability of the Complex Equipment*. Redstone Arsenal, Ala., Att R&D Division, Rel. Branch.
109. Green, A. E. (1973). *Safety Assessment of Systems.* IAEA Symposium on Principles and Standards of Reactor Safety, Julich, Germany, 5–9 February.
110. Shooman, M. L. and Sinkar, S. (1977). *Generation of Reliability and Safety Data by Analysis of Expert Opinion*. Proceedings Annual Reliability and Maintainability Symposium, Philadelphia, USA.
111. Booth, L. E. *et al*. (1976). The Delphi Procedure to IEEE Project 500 (Reliability Data Manual for Nuclear Power Plants). *Proceedings 3rd Annual Reliability Engineering Conference for the Electric Power Industry*. IEEE and ASQC, Montreal, September.
112. *IEEE Guide to the Collection and Presentation of Electrical, Electronic and Sensing Component Reliability Data for Nuclear-Power Generating Stations*. Distributed in co-operation with Wiley-Interscience, a division of John Wiley and Sons, Inc, IEEE Std 500–1977.
113. Goarin, R., Monnier, B., and Quenee, R. (1978). An electronic reliability data bank. *The Journal of the Institution of Electronic and Radio Engineers*, **48**, 7/8, July/August.
114. Moore, J. C. (1966). Research reactor fault analysis, Parts 1 and 2, *Nuclear Engineering*, **11**, 118 and 119 March and April.
115. Eames, A. R. *Data Store Requirements Arising out of Reliability Analysis*. UKAEA Report AHSB(S)138.
116. Ablitt, J. F. *An Introduction to the 'SYREL' Reliability Data Bank*. Report No SRS/GR/14, Systems Reliability Service, UKAEA, Culcheth.
117. Fothergill, C. D. H. (1973). *The Analysis and Presentation of Derived Reliability Data from a Computerised Data Store*. Proceedings of Seminar on Reliability Data Bank, Jarvakrog, Stockholm, Sweden, 15–17 October.
118. Eames, A. R. and Fothergill, C. D. H. (1972). *Some Reliability Characteristics for Operating Plant*. Lecture given at Birmingham University–Techniques in Tetrotechnology at the Lucas Institute for Engineering Production, November 1972.

119. Moss, T. R. (1978). *Developments in the SRS Reliability Data Bank*. Lecture given to SRS Course on an Introduction to Reliability Assessment—Theory and Practice, Liverpool University, September 1978.
120. Snaith, E. R. (1979). *Can Reliability Predictions be Validated?* Second National Reliability Conference, Birmingham, UK. Sponsored by the National Centre of Systems Reliability and the Institute of Quality Assurance, March 1979.
121. Crellin, C. L. Private Communication, 13 November 1978.
122. Waller, R. A. Johnson, M. M., Waterman, M. S., and Martz, H. F. Jr. (1977). *Gamma Prior Distribution Selection for Bayesian Analysis of Failure Rate and Reliability*. Los Alamos Scientific Laboratory, Report LA-6879-MS.
123. Calder Hall Reactors, *The Engineer*, 30 November 1973.
124. *Calder Works Nuclear Power Plant Symposium*. The British Energy Conference, London, 22 and 23 November 1956.
125. Stretch, K. L. (1956). 'Commissioning of Calder Hall No 1 reactor, *The Engineer*, 30 November.
126. Green, A. E. (1956). The reactor control system at Calder Hall, *The Metropolitan-Vickers Gazette*, December 1956 and also Metropolitan-Vickers Special Publication 7657/1.
127. Ghalib, S. A. and Bowen, J. H. (1956). Paper No 15, *Equipment for Control of the Reactor Symposium on Calder Works Nuclear Power Plant*, British Nuclear Energy Conference, 1956.
128. Green, A. E. (1956). Unpublished work (under contract for the UKAEA).
129. Smith, A. M. and Watson, I. A. (1980). Common Cause Failures—A Dilemma in Perspective. *Proceedings Annual Reliability and Maintainability Symposium*, San Francisco, USA.
130. Murdoch, J. and Barnes, J. A. (1970). *Statistical Tables (for Science, Management and Business Studies)* (2nd edn), p. 17, Macmillan, London, UK.
131. Moore, R. V. and Holmes, J. E. R. (1968). The SGHWR System, *Proceedings of the Conference at the Institution of Civil Engineers*, 14–16 May 1968, on *Steam Generating and Other Heavy Water Reactors*. British Nuclear Energy Society, 1–7 Great George Street, London, UK.
132. Bradley, N., Dawson, D. J., and Johnson, F. G. (1968). Engineering design of SGHWRs. Proceedings at the Institution of Civil Engineers, 14–16 May 1968, on *Steam Generating and Other Heavy Water Reactors*. British Nuclear Energy Society, 1–7 Great George Street, London, UK.
133. Wray, D., Butterfield, M. H., and McMillan, R. N. H. (1968). Control of SGHWRs. Proceedings of the Conference at the Institution of Civil Engineers, 14–16 May 1968, on *Steam Generating and Other Heavy Water Reactors*. British Nuclear Energy Society, 1–7 Great George Street, London, UK.

134. Keller, A. Z. Private Communication, 3 April 1979.
135. *Report on the Work and Activities of the Plant Availability Study Group*, published by the National Centre of Systems Reliability of the UKAEA. NCSR R8.
136. Palmer, R. G. and Platt, A. (1961). *Fast Reactors*. Temple Press, London, UK, pp. 5–7.
137. Bainbridge, G. R. (1969). *The Fast Breeder Reactor Programme in the United Kingdom*. UKAEA Report 1925 (R) 1970, presented at the IAEA International Survey Course, Vienna, September 1969.
138. Hensley, G. (1979). *From System Assessment to Long Term Safety Assurance*. Second National Reliability Conference, Birmingham. Sponsored by the National Centre of Systems Reliability and the Institute of Quality Assurance. March 1979.
139. Selby, J. D. and Miller, S. G. (1970). *Reliability Planning and Management*, ASQC/SRE Seminar, Niagara Falls.
140. BS 4200: 1971 Guide on the Reliability of Electronic Equipment and Parts used therein. Part 6—Feedback of reliability information on equipment. British Standards Institution, 1971.
141. Crow, L. H. (1975). On tracking reliability growth. *Proceedings of the 1975 Annual Reliability Symposium*, pp. 438–443.
142. Mead, P. (1978). The role of testing and growth techniques in enhancing reliability. *The Radio and Electronic Engineer*, **48**, 7/8.
143. Brandt, H. W., Crellin, G. L., Graham, J., Mullunzi, A. C. Simpson, D. E., and Smith, A. M. (1978). Reliability testing of reactor shutdown systems in support of low-risk design. ENS/ANS Topical Meeting on Nuclear Power Reactor Safety, Brussels, Belguim October 1978.
144. Marsh, W. and Ferguson, H. R. M. S. (1970). The Development of a System for Collecting and Analysing Service Failures, *Nucl. Eng. and Design*, **13**, 2.
145. Harris, C. M. (1968). The Pareto Distribution as a Queue Service Discipline, *Operational Research*, **16**, pp. 307–313.
146. Moore, J. C. *Experience in the Use of Automatic Protective Systems*. AHSB Reactor Safety Course, Lecture No 21, Figure 2, UKAEA Harwell (UKAEA Internal Document RRL64/985).
147. Corcoran, W. J. and Read, R. (1966). *Comparison of Some Reliability Growth Estimation Schemes*. United Technology Center, Sunnyvale, California, UTC 2140-ITR, 15 November.
148. Lloyd, D. K. and Lipow, M. (1962). *Reliability: Management, Methods and Mathematics*. Englewood Cliffs, NJ, Prentice-Hall.
149. Stovall, Frank A. (1973). Reliability management, *IEEE Transactions on Reliability*, **R-22**, 4.
150. CSNI Task Force on Rare Events in the Reliability of Analysis of Nuclear Power Plants. Report SINDOC(78)82, OECD, 1978.
151. CSNI Task Force on Problems of Rare Events in the Reliability

Analysis of Nuclear Power Plants, Committee on Safety of Nuclear Installations, Report SINDOC(78)83, OECD, 1978.

152. Lloyd, D. A., Staley, J. E., and Sutcliffe, P. S. (1977). *A Theoretical Study of Methods for Analysing Reliability Growth*. Ministry of Defence (PE), Report No GR/77/076-01, prepared by Smith's Industries Ltd.
153. Tustin, A. (1947). A method of analysing the behaviour of linear systems in terms of time series. *IEE Journal*, **94** (IIA), 1.
154. BS5760: Guide to the Reliability of Systems, Equipment and Components. Part 2 Reliability Technology. British Standards Institution.
155. Green, A. E. (1981). *The Reliability Assessment of Automatic Protective Systems for the Safety of Nuclear Reactors*. Proceedings Annual Reliability and Maintainability Symposium, Philadelphia, USA.
156. Green, A. E. (1980). Systems Reliability/Structural Reliability—Interaction and Differences in Approach, *Nuclear Engineering and Design*, **60**, 1. North Holland Publication Company, Amsterdam.
157. Swain, A. D. and Guttmann, H. E. (1980). *Handbook of Human Reliability Analysis with Emphasis on Nuclear Power Plant Applications*. NUREG/CR-1278, Sandia Laboratories, USA (Draft Copy). October 1980.
158. Rassmussen, J. and Rouse, W. B. (eds) (1981). Human Detection and *Diagnosis of System Failures*. Plenum Press, New York, USA.
159. Green, A. E. (ed.) (1982). *High Risk Safety Technology* John Wiley, New York, USA.
160. Green, A. E. (1982). Maintainability versus disposibility. *Nuclear Engineering and Design* North Holland Publication Company, Amsterdam, August.

161 to 188. See References for Appendices page 280.

189. McRuer, D. T., Graham, D, Krendel, E. S. and Reisener, W. (1965). *Human Pilot Dynamics in Compensatory Systems*. AFFDL Technical Report No. 65–15, July, US Air Force Flight Dynamics Lab;, WPAFB, Ohio.
190. Birmingham, H. P. and Taylor, F. V. (1954). *A Human Engineering Approach to the Design of Man-Operated Continuous Control Systems*. Report No. NRL-4333, April, Naval Research Lab, Washington, DC.
191. Watson, I. A. (1982). The rare event dilemma and common cause failures. *Proceedings Annual Reliability and Maintainability Symposium*, Los Angeles, USA.

References for Appendices

161. Barlow, R. E. and Lambert, H. E. (1975). *Introduction to Fault Tree Analysis, Reliability and Fault Tree Analysis, Theoretical and Safety Assessment*. Siam, Philadelphia, USA, pp. 7–35.
162. Veseley, W. E. (1975). *Reliability Techniques used in the Rasmussen Study. Reliability and Fault Tree Analysis, Theoretical and Safety Assessment*, Siam, Philadelphia, USA, pp. 775–803.
163. Fussell, J. B. and Veseley, W. E. (1972). A new methodology for obtaining cut sets. *American Nuclear Society Transactions*, **15**, 262–263.
164. Shaw, P. and White, P. F. An Appraisal of the PREP, KIT and SAMPLE Computer Codes for the Evaluation of the Reliability Characteristics of Engineered Systems. UKAEA Report SRD R57.
165. Brock, P. and Cross, A. Computer Programs for Reliability Analysis. Lecture No. 27, Safety Course Notes, Education Centre, UKAEA, Harwell.
166. Macfarland, W. J. (1968). *Use of Bayes Theorem in its Discrete Formulation for Estimation Purposes*. 7th Reliability & Maintainability Conference, San Francisco, 14–17 July.
167. Mann, N. R., Schafer, R. E., and Singpurwalla, N. D. (1974). *Method of Statistical Analysis of Reliability and Life Data*. John Wiley, New York, USA.
168. Crellin, G. L. (1972). The Philosophy and Mathematics of Bayes' Equation. *IEEE Transactions on Reliability*, **R-21**, 3, August.
169. Deeley, J. J. and Zimmer, W. J. (1968). *Bayesian and Classical Confidence Intervals in the Experimental Base.* 7th Reliability and Maintainability Conference, 14–17 July 1968, San Francisco, USA.
170. Easterling, R. G. (1972). A Personal View of the Bayesian Controversy in Reliability and Statistics. *IEEE Transactions on Reliability*, **R-21**, 3 August.
171. V. Joksimovich, *et al* (1975). *HTGR Accident Initiation and Progression Analysis Status Report—Phase I Risk Assessment*. GA-A13617 General Atomic Company.
172. V. Joksimovich, *et al*. (1978). *HTGR Accident Initiation and Progression Analysis—Phase II Risk Assessment* GA-A15000, General Atomic Company.

173. Crooks, J. L. and Vissing, G. S. (1974). *Diesel Generator Operating Experience in Nuclear Power Plants*, USAEC Report OOE-ES-002.
174. *The Reactor Safety Study—An Assessment of Accident Risks in U.S. Commerical Nuclear Plants*, WASH-1400 Appendix III, October 1975.
175. Scaletta, F. P. *A Reliability Estimate for the Auxiliary Primary Coolant Shut-Off Valve*, General Atomic Company, 1974, unpublished.
176. Cairns, J. J. and Fleming, K. N. (1977). *STADIC: A Computer Code for Combining Probability Distributions*, GA-A14055, General Atomic Co.
177. Melvin, J. G. and Maxwell, R. B. (1974). *Reliability and Maintainability Manual*, AECL-4607, Chalk River Nuclear Laboratories.
178. Hannaman, G. W., (1977). *GCR Reliability Data Bank*, General Atomic Co.
179. Cave, L. and Gow, R. S. (1977). Experience of the reliability of steam and electrically driven circulators for GCR's, *Nuclear Engineering International* **22**.
180. Sullivan, W. H. and Poloski, J. P., (1980). *Data Summaries of Licensee Event Reports of Pumps at U.S. Commercial Nuclear Power Plants, January 1, 1972 through April 30, 1978*, NUREG/CR-1205.
181. Poloski, J. P. and Sullivan, W. H. (1980). *Data Summaries of Licensee Event Reports of Diesel Generators at U.S. Commercial Nuclear Power Plants, January 1, 1976 through December 31, 1978*, NUREG/CR-1362.
182. Hubble, W. H. and Miller, C. F. (1980). *Data Summaries of Licensee Event Reports of Valves at U.S. Commercial Nuclear Power Plants, January 1, 1976 through December 31, 1978, Vol. 1—Main Report, Vol. 2—Appendices A through N, Vol. 3—Appendices O through Y*, NUREG/CR-1363, Vol. 1, 2, 3.
183. Miller, C. F. and Hubble, W. H. (1981). *Data Summaries of Licensee Event Reports of Selected Instrumentation and Control Components at U.S. Commercial Nuclear Power Plants from January 1, 1976 to December 31, 1978*, NUREG/2037.
184. Murphy, J. A. (NRC) letter to Mr. David Carlson, 'Component Failure Rates to be Used for IREP Quantification', September 26, 1980.
185. Monnie, D. I. (1980). *Unavailability of Offsite Power at Nuclear Power Plants*, EGC-EA-5256.
186. *The Time Index*. Published periodically by Newspaper Archive Developments Ltd., Reading, UK.
187. *Britannica Book of the Year*. Published annually by Encyclopaedia Britannica Inc. (internationally).
188. Martz, H. F. and Waller, R. A. (1982). *Bayesian Reliability Analysis*. John Wiley, New York, USA.

References for further reading Appendix 8 see page 251.

Index

Accident(s)
  behaviour, 1
  definition, 1
  domino effect (*see* Consequences)
  explosion at chemical plant, Flixborough, 11–12
  Hartley Colliery, 7 (*see also* Coal mining)
  near accident, 1
  post-accident analysis and assumptions, 44
  sequence, 16–18, 215, 235
  survey (*see* Historical)
Air traffic control (*see* Aircraft)
Aircraft
  air traffic control reliability and maintainability, 27
  development, 3
  equipment, consensus estimating of failure rate, 96–100
Assessed performance
  envelope approach, 41, 177
  statement of, 33, 37
Assessment
  acceptance, 13
  deterministic, 24–25, 202
  general process, 24, 31–32
  high risk plant protection, 22, 37
  human-influence factor, 73
  independent process, 27, 31, 39
  methods of, 21, 32–33, 37
  performance characteristics, 39, 40
  posterior, 37, 83
  prior, 38
  quantified reliability, 12, 202
  related quantities and assumptions, 45–46
  reliability methodology, 22 (*see also* Reliability)
  report preparation (*see* Safety assessment)
  requirement, 24
  strategy and reliability growth, 84–85, 174–179
  validation (*see* Validation)
Availability (*see also* Fractional dead time)
  definition, 9
  losses of, 10
  mean unavailability, 167
  of information, 46
  operational experience and prediction, 167–173
  outage time, 168–169, 228
  pictorial representation, probabilities of events and outage time, 171–173
  plant overall configuration (*see* Plant)
  quantification, human error in testing (*see* Human reliability)
  risk criteria factors, 13

Bayes theorem
  decision-making, 39
  derivation, 230–231
  form of, 114–115
  updating prior (*see* Bayesian techniques)
Bayesian techniques (*see also* Bayes theorem)
  application, 115–120, 203
  comparison with classical results, 153, 166, 203
  conjugate prior model and gamma distribution, 118, 151
  engineering experience, 120, 206
  exponential distribution example, 116
  general approach, 114–120, 203
  growth example of a protective equipment (*see* Growth model)
  modification of prior to posterior, 115–116, 151
  need for prior (*see* Prior)
  posterior estimates (*see* Posterior)

practical considerations, 117–118
reliability estimation, 139
single decision parameter monitoring, 118, 167
updating, 22
updating prior, 115, 145

Catastrophe
critical equilibrium and stability, 6, 40
Hartley Colliery disaster, 7
instability hole, 6, 198
technological development interaction, 3
theory, 6
Catastrophic failure region, 48–49, 60
Chemical plant(s)
automatic shut off, 30
design limitations, 85
economic loss, 11–12, 37
explosion at Flixborough, 12, 38
high integrity shutdown system, 12, 18
protective trip initiators, failure-rate prediction and observation, 108
risk to human life, 11–12
safety goals, 27, 37
single or multi-stream, 37
Coal mining
Hartley Colliery disaster, 7
Markham Colliery winding accident, 8
Commissioning
early-life defects, breakdowns and rectification, 183–184
performance achievement, 37
stage, 24, 90
tests on site, 162, 163
Common-mode failure (*see also* Redundancy, Diversity)
beta factor method, 191, 243, 257
coal pit shafts, 8
common cause failure, redundancy and diversity, 86, 205, 252, 254
defensive strategy, engineering and management, 187–188, 252
design and maintenance task errors, 189
example of limitation on assessment, 191, 203, 204, 206
identification of common-mode errors (*see* Human factors)
in commissioning, 163
in measurement, 20, 249
model to evaluate common cause factor, 262–265
prevention and minimization, 187–191, 252
testing 2-out-of-3 system, 74
Communication
and quantitative reliability criteria, 202
information transfer distortion, 46, 89
of information, 46, 205
preparation of safety assessment report, 80–83
Component(s)
failure rate analysis in equipment, 57
mean failure rates, 229
operator as loop component, 61, 62
Consensus
estimating by, 95–100, 203, 206
failure-rates for nuclear equipment, 96–97
aircraft equipment, 96–100
Consequence(s)
and abnormal event, 2 (*see also* Catastrophe)
domino effect, 16, 27, 37
risk rate formulation, 38
Control (*see also* Shutdown system)
and instrumentation, 20, 27, 62
design process and modification feed-back, 89
information, 177–179 (*see also* Growth)
integral of error, 89
integrated structures, reliability control system analogy, 193–200, 204
manual versus automatic (*see* Human reliability)
overall control of reliability, 90–91 (*see also* Growth model)
phase-advance, 89
reliability growth, 174–179, 182, 206
rod actuating mechanism, failure-rate prediction, 127–131
in service failures, 145, 182
life testing, 131–142
test failures, 143–144, 146
rod blockage fault tree (*see* Fault trees)
rod system, gas-cooled reactor, 123–127
system approach for human operator, 63
system feed-back, 18, 19, 89, 186
Correlation
assessment process, 33
classical and Bayesian reliability estimates, 141, 153, 166
comparison of reliability data (*see* Data)

Correlation—*continued*
failure-rate confidence band, 139
life testing and theoretical assessment of failure rate, 138
plant system operational unavailability and prediction, 167–173
predicted and practical results 107–114, 138, 163 (*see also* Consensus)
predicted and practical results, protective trip initiators, 108–110
engineering equipments and systems, 110–113, 203
sample sizes and analysis techniques, 114
statistical data and material aspects, 113–114
Cost(s)
accidental nuclear releases, 38
availability of information, 36
design-space regions, 27
Flixborough accident, 38
function 26, 27 (*see also* Optimization, Catastrophe)
probability of failure, 27
total life-cycle costs, 27, 228

Data (*see also* Reliability data banks)
banks, 16, 205
calculations for protective system, 149–153, 255–258
collection of, 39, 91
comparison of reliability data from different sources, 268–269
confidence intervals, 138–141
considerations, field experience (*see* Field experience)
decision-making (*see* Decision-making)
Delphi techniques for lack of data, 100, 206
discrete event analysis, 145
example for reliability quantification, 256, 263
for safety assessment, 102
generic, 104, 106, 186, 229
hard, 22, 94, 114, 203
index of electronic equipment operability, 70
κ factors (*see* Failure)
needs for human reliability (*see* Human reliability)
numerical data problems in derivation, 241–243, 255–258
plant management, 101
prior, 2, 38
posterior, 2, 39
reduction equipment (*see* Instrumentation)
reliability of plant, 9
reliability management and control data base, 186
soft, 22, 94, 114, 203
subjective, 22, 94–100, 114, 203, 204 (*see also* Bayesian techniques)
system synthesis and hierarchical level, 54
Decision-making
adaptive analyses, 37
adequacy and available information, 36, 39, 186
and problem solving, 33–37 (*see also* Safety assessment)
basic process of problem solving, 34
data, 37
design major decisions, 37, 176
divergent errors, 37
good decision and a safe outcome, 36
integrated structure requirement, 192
logical expression, 35
material considerations, 35
modification of equipment design, 194, 198, 200
planning future tests 183 (*see also* Growth)
plausibility and logic, 34
relability control strategy (*see* Safety life cycle)
reliability management and control data base, 186, 188, 205, 206
scale of values, 36
uncertainties, 36, 37
utility theory, 36
Decomposition techniques (*see* Systems)
Design
access to components, 10
adaptive analyses (*see* Decision-making)
adequacy and analysis, 8, 79
assessment, 24–25, 79–80
basic reference design, 27
centralization, 27
conceptual, 25, 26, 79, 176
control and modification feed-back, 89, 183, 198
coordinate space, 27, 38
cost-function minima, 27
diversity, 31, 252
estimate, 48–49
function and specification, 24

functional performance, 24, 37, 41, 176
independent safety assessment (*see* Safety assessment)
limitations 85 (*see also* Common-mode failure)
optimization and catastrophe, 7, 26
plant configuration, 37, 252
preparation of item inventory for data collection, 113
primary and secondary (*see* Diversity)
procurement, 27
selection, 26–31
ship (*see* Marine)
solution and adequacy, 22
staff structure, 27
stage control, 10, 204
structural model, 26
subjective data, 94, 203
task errors, 89, 189
technological development, 3
trial and error, 2
Diakoptics (*see* Systems)
Disposability
component and higher system levels, 72
definition, 71
versus maintenance (*see* Maintenance)
Distribution(s) (*see also* Distribution by name)
actuating mechanism life testing results, 136
consensus estimates, 99
item deviations, 51–52, 67
ratio predicted to actual failure-rates, 110–113
Diversity
and reliability limitations, 85–88, 190
coal pit shafts, 8
in measurement, 20 (*see also* Common-mode failure)
in sensing channels, 31
primary and secondary groups, 21
Domino principle
growth programme, 37
obtaining data, 93
sequence and consequence, 16, 37, 202
Duane approach
applied to accelerated life testing, 141–142
model, 181, 194, 202

Failure(s)
and consequences, 2
and safety, 2
catastrophic (*see* Catastrophic failure)
common-mode (*see* Common-mode failure)
human, 16
in service, 145, 255
κ factors, 107, 127
life testing results (*see* Life testing)
mechanical, 18
modes analysis (*see* Equipment)
nuisance and major safety defects, 184
partial, 202, 206
physical, 18, 21
probability, 11
random, 179
rate, environment and stress, 107, 109
revealed, 53, 131, 158, 221
single failure criterion (*see* Strategy)
systematic, 179
to safety (*see* Safety)
types and analysis, 144, 163
unrevealed, 53, 131, 158, 221
Fault tree(s)
common-mode failure predominance, 246
construction, 215
control-rod blockage example, 247
core auxiliary cooling system example, 254, 261
definition, 214
modelling, 54, 202, 206, 215
pressurized water reactor system example, 241
reduction method analysis example, 241–246
symbols, 215
Field experience
aircraft equipment, 96, 98
analysis, exponential and Weibull distributions, 143–145
data considerations, 142–145, 203, 204, 241, 250
in service failures (*see* Failures)
manual diesel generator starts, analysis, 255–258
nuclear equipment, 96, 97
properties of system, 39
Fractional dead time (*see also* Availability)
application to hazard analysis, 56
mean, 53–54

Gamma distribution
asymptotic behaviour of gamma percentiles, 119, 140, 232–233

Gamma distribution—*continued*
number of repairs completed in a total time, 168
prior on failure-rate, 118, 139
Gaussian distribution (*see* Normal distribution)
Growth (*see also* Growth model)
and control, 90–91, 174–179, 193
and distortion of information, 89
control loops, 193–194, 198
during accelerated life testing, 141–142
example of a protective equipment (*see* Growth model)
example of reliability trends, 182
generation time, 174–175
improvement factor, 180, 187
modifications, 146, 177, 179, 183, 190, 197
nuclear reactor and stages, 174
plant system information stages, 164
programme of, 37, 79, 145, 146, 163
properties of system, 39, 202
reliability growth process, 180, 203
stability, 198
time delays, 197
tracking trajectory, 37, 83, 84
Growth model
application of Duane approach (*see* Duane approach)
comparison of required and measured failure-rates, 197
discrete events, 145
divergent errors and decision-making, 37
dynamic behaviour, 194
example of a protective equipment, 199–200
faults, systematic and random, 195, 202, 204
time delays, effects, 197–198
using control techniques for reliability, 90–91

Hazard probability, 56 (*see also* Fractional dead time)
High-integrity protective systems
analysis by synthesis, 54
domino principle, 15, 18, 202
prior estimate, 37
Historical
aspects, 21
chemical plants, 11–12
coal mining, 7–8
events and consequential safety, 16, 201
information, 39, 201, 202, 205
power generation, 8–11
reliability, qualitative and quantitative, 201, 206
ship design (*see* Marine)
survey of catastrophic events, 3–6, 201
Human factor(s)
analysis of channel trip test, 75, 248
and system hardware, 78
approach to error and consequence analysis, 248–249
identification of common-mode errors, 249
in maintenance and testing, 71–73, 206
interfaces, 23, 205, 206
man-machine system, 60, 75, 131, 205
matrix logic array analysis, 75
needs for further research, 250–251
operator action, 60–61
operator response, 61–68
reliability (*see* Human operator)
task error model, 188–189, 206
testing of protective system, 248–250
Human operator
analysis of channel trip test (*see* Human factors)
artificial stimuli, 64
continuous task, 63
control-system approach, 63 (*see also* Transfer function)
data reduction, 62, 75
ergonomics, 61
methods-time study approach, 64–65
optimizing ability, 63
plant emergency shutdown, 61
protective system component (*see* Protective system)
reliability, 69–71, 248
response, visual and audible, 66–68
vigilance testing, 65–68
Human reliability
analytical assessment of, 69, 205, 250
case history plant analysis, manual versus automatic control, 79, 225–228
data needs, 250–251
equipment operability, 70, 131
example analysis of test system, 73–77, 248
interaction with engineering system, 73, 205, 248

matrix approach, 75–77, 226
quantification, human error in testing, 248–250
synthesis, 69–71
task counting method, 78, 249
technique for human error rate prediction (THERP), 71

Industrial safety procedures, 1
Information (*see also* Data)
acquisition and organization, 46, 89, 205
between operator and machine, 62
communication of (*see* Communication)
distortion (*see* Interface, technical)
for safety assessment, 80, 83, 177–179
input to data bank, 105
limited, 2, 38
prediction process, 2
state of nuclear reactor, 62
updating changes, 192
what if questions, 2, 177
Installation
stage, 24
tests on site, 162
Instrumentation
and control, 62
data reduction equipment, 62, 75
design centralization, 27
interface problems and assumptions, 45
Insurance accident payout, 12, 38
Interface(s)
human factors, 23, 205
human influence factors, 73
problem, stage by stage assessment, 25, 80, 190
proof-testing, 73
system boundaries, 44–46
technical, 23, 205

Life testing
accelerated, 137, 180
classical and Bayesian results compared, 141
comparison with theoretical assessment, 138
control rod actuating mechanism, analysis of results, 134
results, 133, 137–138
tests, 131–153
Duane interpretation 141–142 (*see also* Duane approach)
Logic network, 2-out-of-3 configuration 31 (*see also* Majority voting)
Lognormal distribution
beta factor for common-mode failure, 246
consensus estimating, 99
ratio of predicted to observed failure-rate, 110
response time, 58, 67
Loss
financial, 3, 11–12
human life, 3, 4–6, 11–12
insurance, 12
of emergency services, 12

Maintenance
definition of maintainability, 71
process, 24
proof testing principle, 43
task errors, 89, 189, 205
testing and repair, 53 (*see also* Human factors)
versus disposability, 71–73
Majority voting
common cause failure, 74, 84
system, 20, 236
testing 2-out-of-3 system, 73, 74
2-out-of-3 system, 20, 236
Manufacture
communication in, 89
performance achievement, 37, 179
stage, 24
test and fix (*see* Test)
Marine
International Convention for Safety at Sea (SOLAS), 6
Titanic disaster, 6, 201
Measurement(s)
example protective system analysis, 56–60, 84, 234–251
indirect techniques and assumptions, 45, 236
probability and confidence, 146
redundancy and common-mode failure, 85

Normal distribution
consensus estimating, 99
life testing variation (Gaussian), 137
probability function, 50–51
response time, 58, 146
Nuclear power
damage to property (*see* Safety legislation)

Nuclear power—*continued*
generating stations, 8–11
programme, UK, 9–10
safety goals, 26, 27
Nuclear reactor(s)
automatic protective system (*see* Protective system)
consensus estimating of equipment failure-rate, 95–97
control rod actuating mechanism, description, 122–126
failure-rate prediction, 127–131
control rod system, gas cooled type reactor, 122, 126–127
reliability growth trends, 182–183
energy conversion, 1
fast, 10, 84, 174, 176
gas cooled type, 9–10, 121–122, 149, 252
high temperature gas cooled reactor (HTGR), core auxiliary cooling system (CACS), 252–269
liquid shutdown system, 156–157
manual shutdown analysis, 63–67
operator response, 66–68
parameters and measurements, 30–31
pressurized water reactor (PWR), automatic protective system, 234–251
prototype design process, 26
shutdown system, 21, 126
steam generating heavy-water reactor (SGHWR), 153–157
trip time delay, 42, 43, 60, 121–122

Operation(al)
experience and prediction, 162–173 (*see also* Correlation)
feed-back from, 90
marginal, 49
normal and abnormal conditions, 37
performance achievement, 37
reliance on operator action, 60, 248, 258
Optimization
ability of human operator (*see* Human operator)
cliff-edge type discontinuity, 27
cost-function minima, 27, 39, 40
critical equilibrium (*see* Catastrophe)
critical protective system, 6–7 (*see also* Catastrophe)
design centralization, 27

Pareto distribution, service times and queuing, 183
Performance
achievement, 37, 47–49, 202
assessment (*see* Assessed performance)
characteristics, 40–44, 176
design constraints, 24
functional, 24
requirement, 47–49
spectrum (*see* Performance spectrum)
theoretical prediction of, 39
variability, 50–52, 202
Performance spectrum
example analysis, 56–60
method, 48–50, 202, 206
probability model, 48
Plant(s)
analysis of operating systems, 120–121
damage, 1
deterministic and probabilistic approaches, 120
early-life defects and breakdowns, 185–186
fault conditions, 24, 37
financial loss, 1
high risk, 1
human loss, 1
kinetics, 45, 121, 235, 252
manual versus automatic control, case study for steam raising unit, 225–228 (*see also* Human reliability)
operating conditions, 1, 2
overall configuration, 37
performance limits, 18, 19
safety life cycle (*see* Safety life cycle)
tailoring design, 2
Plant process
abnormal conditions, 1
chemical, 11–12
conversion process, 1
emission of poisonous materials, 10–12
Poisson distribution
applied to life testing, 136
number of failures in given time, 168
sampling and confidence, 159
Posterior (*see also* Prior)
deriving posterior information, 147, 203
estimates and tracking, 37
gamma conjugate, 151
monitoring trend of mean, 118
need for, 83

practical considerations, 118, 190
Power generation
conventional, 8
nuclear power, 9, 234
plant running costs, 10
Prior (*see also* Posterior)
example of prior failure-rate prediction, 127, 151
existence of prior system, 39
expert opinion, 100
need for, 37–39
practical considerations, distribution, 117, 203
quantification, basis of, 79
safety assessment and prior estimate, 177
updating (*see* Bayesian techniques)
use of gamma distribution, 118
use of prior prediction, 147, 152, 179
Probability
and confidence in measurement, 146
assessment (*see* Safety assessment)
calculation and uncertainties, 246, 251, 258
criterion, 27
cumulative probability function, 52, 59
hazard analysis, 56
inductive use of, 36
non-accident events, 2
of failure for protective systems, limitations, 85–87
of protective system functioning, 38, 147, 159, 201–202, 251, 254
statement for accident condition, 146
Production, test and fix growth process, 179
Proof testing
channel trip analysis, 73–77
process of, 53
provision for, 43, 238, 239
time between tests, 54, 56, 58, 86
Protective channel(s)
redundancy and common-mode failure, 85, 187, 190
sensing fault conditions, 20, 31, 147, 156, 236
trip proof testing (*see* Proof testing)
2-out-of-3 (*see* Majority voting)
Protective requirement, statement of, 33, 174, 236, 252
Protective system(s)
allocation of targets, 84, 202
analysis by synthesis, 54
automatic shutdown, 18, 19, 146, 235
basic principles for assessment, 88
common-made failure, 85–86, 187–191, 202, 203
consensus estimating of equipment failure rate (*see* Nuclear reactors)
dead time (*see* Fractional dead time)
example analysis, 56–60, 73–77, 149–153, 234–251
fault tree techniques (*see* Fault trees)
gas-cooled reactor protection, 9, 19, 30, 146–149
generalized, 19–21
high integrity shutdown (*see* Chemical plant)
high temperature gas cooled reactor, core auxiliary cooling system, 252–253
human operator as component, 64
instability hole (*see* Catastrophe)
limitations to reliability, 58, 204
optimization (*see* Design, Optimization)
posterior, need for, 834, 190
pressurized water reactor protection, 234–239
prior assessment, 38, 39
proof testing, 43, 53, 73
rare events, 85
redundancy and diversity, probability of failure, 85–87, 190
reliability growth and control, 174, 206 (*see also* Growth)
response time, 37, 40, 60
safety circuits, 20, 147, 236
sensing channel, 30–31, 234, 236
shutdown, 21, 30, 149, 156, 236, 239
steam generating heavy-water reactor protection, 153–157
variability, 25

Rare event(s)
analysis of, 39, 203, 206
common-mode failure, 85–86, 188, 206
data, 95
management decision-making, 188
synthesis, 131
Redundancy (*see also* Common-mode failure)
and reliability limitations, 85–87
coal pit shafts, 8
example, 147, 236
mechanisms, 21
single line components, 8
testing and failure, 74, 75

Reliability
  and availability, 9
  and safety methodology, 4–5
  application, example protective system, 56–60, 84
  control phases, 88, 174–179
  definition, 9
  development, 3, 93, 202
  factors affecting, 12–19
  function of requirement and achievement, 47
  growth, 83, 84, 141–142, 174, 179
  human operator (*see* Human reliability)
  management and control, 22, 88, 89, 188, 204 (*see also* Safety life cycle)
  model, 52–56
  philosophy (*see* Reliability philosophy)
  policy (*see* Reliability policy)
  programme, overall, 184–187
  quantified (*see* Assessment)
  stability, 40, 85, 198, 204
  total technology, 15
Reliability characteristics (*see also* Safety life cycle)
  bath-tub curve, 177
  Bayesian reliability estimation, 139–141, 151, 153, 165
  classical reliability estimation, 138, 141, 150, 153, 165
  comparison of reliability data (*see* Data)
  data calculations for protective system, 149–153, 204, 205, 206
  environmental and stress factors, 107, 127
  estimating by consensus, 95–100
  estimating in appropriate form, 114
  failure probability evaluation, automatic protective system, 234–251
  failure-rate prediction, control rod actuating mechanism (*see* Control)
  mean failure rates for components and instruments, 229
  monitoring and estimating, 94–95, 167, 185, 204
  predicted and practical results (*see* Correlation)
  reliability prediction, core auxiliary cooling system, 252–269
  theoretical prediction of liquid shutdown system, 157–167
Reliability data bank(s)
  definition, 101
  event data store, 103
  flow of data and use, 105, 186, 204
  item inventory, 103
  plant and reliability management, 101, 186, 205
  reliability data store, 103, 205
  wide ranging, SYREL, 16, 102
Reliability philosophy
  catastrophe, 3
  historical 3 (*see also* Historical)
  motivation, 3
  psychological reaction, 3, 12
  technological development, 3
  unreliability leading to disrepute, 20
Reliability policy
  growth strategy, 83–85, 174–179
  historical, 2
  implementing, 85–88, 176, 185
  safety, 2, 3, 60, 206
Repair
  early-life defects (*see* Commissioning)
  rate probability model, 265–266
  rectification and restoration, 179
  testing and maintenance, 53, 58, 72
  time characterisitic, 53, 58, 72
Response
  best estimates, 40
  capability, 40
  deterministic, 40
  envelope approach, 41
  exponential device, 41
  input ramp function, 41
  interaction of loops, stability, 198
  modelling time variations, 146
  operator action, 60–61, 64
  time constant, 41, 42
  time lag, 40
  time of, 37, 40, 42, 146
  to fault conditions, 42, 121, 235, 252, 258
  transfer characteristic, 43–44
  transfer function, 40–41
  workshop timing tests, control rod actuating mechanism, 146
Risk
  acceptable, 12
  accident, 2
  assessment formulation, 38–39
  factors affecting, 12–19, 206
  financial, 2
  optimization of design, 7

probability, 2, 202
rate, 38

Safety
and availability, 8
and reliability (*see* Reliability)
automatic protection (*see* Protective system)
circuits (*see* Protective systems)
codes and guides, 79–80, 222–224 (*see also* Safety legislation)
control, 16, 174–179
control systems, 10, 122
design adequacy (*see* Design)
domino consequence, 16, 37, 122, 202
failure to danger, 20, 57, 158, 163 (*see also* Failure, unrevealed)
failure to safety, 20, 57, 147, 163
implementation, 85
policy formulation, 10, 176
prevention measures, 18
requirements, 79–80, 83

Safety assessment (*see also* Growth, Safety life cycle)
basic principles for assessment (*see* Protective systems)
boundary value, 33, 41, 79, 157, 162, 177
building up prior estimate, 177
categories of recommendation, 82
communication of assessment, 46, 202
contents of report, 81
criteria, 10–11, 27, 32, 174, 185, 204
design submission, 32
deterministic, 24, 146, 202
direct method of proof, 32
implementation, safety and reliability, 85
independent, 27 (*see also* Assessment)
indirect method, 33
interface problems, 44–46, 80
life testing results (*see* Life testing)
maintaining assessment current, 82, 185
methods of, 32–33, 202
probabilistic approach, 10–11, 153, 202
probability of failure, limitations, 85
process, 31–32, 49–50
radioactive fission products, 10
report, 80–83
requirements, 79–80, 83
sample testing estimates, 162
system targets, allocation, 83–85

Safety assessor(s)
and reliability tasks, 186
communication of information, 46, 80, 202
development of prior assessment, 38, 186
independent multi-disciplinary approach, 31, 32
need for organized memory, 177–179
preparation of safety assessment report, 80–83
quantification and discipline, 79, 202, 205

Safety legislation
catastrophe, interaction with, 6, 201
codes and guides, 79–80
damage to property, 2
duty of care, 1
Hartley Colliery disaster, 7–8
Health and Safety at Work Act, UK, 1, 12
Occupational Safety and Health Act, USA, 1
protection and technological development, 6, 201
Safety life cycle (*see also* Growth)
common-mode failure, 187
dynamic behaviour and model, 194–198 (*see also* Growth model)
early-life defects and breakdowns, 185
evaluation of system, 22, 37, 149–153
evolution and change, 192, 193
in-built learning technique, 94, 177–179, 206
integrated structure, 191–192
interfaces, 23, 186
management and control, 22, 176, 186, 206
matrix for design performance parameters, 176
nuisance and major safety defects, 184
optimization of costs, 27
overall reliability programme, 184–187, 204
phases and control, 193, 198, 206
plant safety and reliability monitoring, 167, 176, 204, 206
reliability control phases, 88–90, 174–179, 186
reliability control strategy, 198–200, 206

Sample testing
application to shutdown valves, 159–161
information, 2, 39, 159
Shutdown system(s)
control rod, 122 (*see also* Control)
control rod, reliability trends, 182–183
liquid, 156
mechanisms, 21
valve, reliability prediction, 157–159
Strategy(ies)
adequacy of safety and reliability, 176
allocation of targets, 83–85
backward-looking, 38, 204
defensive, common-mode failures, 187
domino effect, 202 (*see also* Domino principle)
forward thinking and looking, 24, 38, 204
maintainability, 71
man-machine, 60, 71, 78
of growth, 83
process of partitioning, 26
rare events, 39
reliability control, 198–200
single failure criterion, 24
synthesis, 54, 64, 202
tearing-up systems for analysis (*see* Systems)
tracking reliability characteristics, 185, 204
Subjective, engineering judgement, 24, 114, 187, 203 (*see also* Bayesian techniques)
Synthesis (*see* Strategies)
System(s)
decomposition techniques, 55, 250
dynamic characteristics, 39
engineering equipments and systems, predicted and practical results (*see* Correlation)
existence of prior, 39
fault tree techniques, 54 (*see also* Fault trees)
hardware concept, 37, 78
logical flow diagram, 58–59
maintenance and testing, 73
modelling, 54, 69
paper-work concept, 37, 38
reliability limitations, 58–87
synthesis, 54, 58
tearing-up, 55
theoretical prediction of performance, 39–46
topological models, diakoptics, 55
System(s) reliability
example analysis (*see* Protective systems)
National centre of (UK), 16

Test(ing) (*see also* Proof testing, Sample testing)
and fix, 179, 200
and restoration, 53, 58, 205 (*see also* Human factors)
analysis and experiment, 181
Duane approach, 180–183
for vigilance of human operators, 65
life testing control rod actuating mechanism (*see* Life testing)
loop rig tests on shut down valves, 161
operator role, 61
protective system and analysis, 248–250 (*see also* Human factors)
site tests on shutdown valves, 162
workshop tests, control rod actuating mechanism, 146 (*see also* Response)
Time to failure
field data considerations (*see* Field experience)
mean, 52, 144
Tracking
control loops, 198
failure-rate, 193, 200, 204, 206
growth programme trajectory, 37, 204, 206
human operator continuous task, 63
organizing information, 177–179, 192
plant-safety monitoring, 167, 186, 206
ratio of growth, 180, 182
reliability control system analogy, 193–200, 204
reliability parameter changes, 176, 185
single decision parameter, 118
Transfer function
for human operator, 63
generalized transfer characteristic, 43–44
Laplacian operators, 40, 44
response, 40–41, 58, 198
Trip
signals for shutdown, 21, 30–31
under fault conditions, 42, 43

Unavailability (*see* Availability)

Validation, reliability methodology, 22, 185, 203 (*see also* Assessment)
Vapour clouds, accidents, 6, 202–213
Variability
  component additive function, 51
  component product function, 51
  example for probability of failure calculation, 246
  non-systematic, 50
  operator response, 66–68
  region, 48–49, 50, 67
  systematic, 50

Weibull distribution
  analysis of test failures, 144–145
  consensus estimating, 99